SERIES

ENTOMOLOGICA

EDITORS

E. SCHIMITSCHEK & K. A. SPENCER

VOLUME 14

DR. W. JUNK B.V. — PUBLISHERS — THE HAGUE 1977

THE WORLD OESTRIDAE (DIPTERA), MAMMALS AND CONTINENTAL DRIFT

by

N. PAPAVERO

Dr. W. Junk bv Publishers – The Hague 1977

ISBN-13: 978-94-010-1308-6   e-ISBN-13: 978-94-010-1306-2
DOI: 10.1007/978-94-010-1306-2

Cover design Max Velthuijs

CONTENTS

INTRODUCTION

The taxonomic delimitation of the Superfamily Oestroidea is extremely difficult and most authors disagree as to the limits of the group (see HORI, 1967, for a historical survey of Calyptratae classification). ROBACK (1951) included in Oestroidea the families Oestridae (with Hypodermatidae as subfamily), Cuterebridae and Tachinidae. STONE *et al.* (1965) included the Calliphoridae, Sarcophagidae, Tachinidae, Cuterebridae and Oestridae (with Hypodermatidae as a subfamily). ZUMPT (1957) divided the Oestroidea into: (i) Oestridae, with subfamilies Cephenemyiinae, Oestrinae and Hypodermatinae; (ii) Gasterophilidae (with subfamilies Gasterophilinae, Gyrostigminae and Cobboldiinae); and (iii) Calliphoridae, including, among others, the subfamily Cuterebrinae.

I am considering as Oestroidea, for practical purposes, only two groups:

1. The Cavicolae (Fam. Oestridae)–larviparous flies; the 1st stage larvae are deposited by the females directly into the nasal cavities of the mammal hosts; the larvae complete their development in several internal cavities.

2. The Cuticolae (Fams. Hypodermatidae, Rutteniidae and Cuterebridae)–oviparous flies; the egg is laid near the lairs of the host, on an intermediary vector, mainly haematophagous flies (in the unique case of *Dermatobia hominis*), or directly on the host's skin, the egg being then adapted to adhere to the hairs; the 1st stage larva penetrates through the skin of the host, completing its development subcutaneously.

This work will deal only with the Cavicolae (Oestridae s.s.). The Oestridae have a curious distribution: they occur in most areas of the Holarctic and Ethiopian regions, and in a few areas of Australia, being conspicuously absent from the Neotropical and Oriental regions.

The classification of this family is based on the excellent studies of BRAUER in the past century, and of several authors, but especially GRUNIN and ZUMPT, in this century. The current classification is however somewhat conservative, not considering the phylogenetic, zoogeographical, ecological and biological aspects of the group. The recent accumulation of data on Oestridae, mammals, continental drift, palaeoclimatology and palaeoecology now makes possible a tentative analysis and interpretation of the evolutionary phenomena which

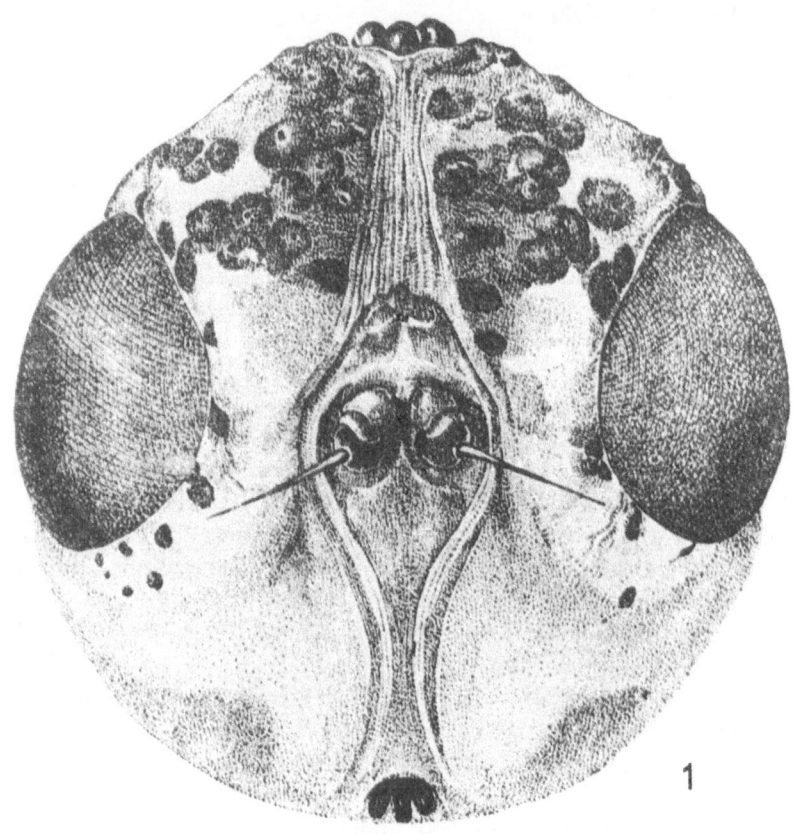

Fig. 1. Head of an oestrid (after GRUNIN, 1966), showing reduced mouthparts.

have operated and still are operating in this group.

The main objective of this work is to search for answers to the following questions:

1. Which is the most probable suprageneric classification of the Oestridae ?

2. How is the present geographical distribution of the family explained ? Or, how is the absence of oestrids in certain continents explained ?

3. Why do oestrids parasitize only a relatively small fraction of the living terrestrial mammals ?

4. Is there any correlation between the phylogeny of the Oestridae and that of their hosts ?

5. Are there any relationships between the zoogeography (past and present) of the Oestridae and their mammal hosts ?

6. What is the evolutionary history of the Oestridae ?

PART A. HISTORY, FOLKLORE, BIOLOGY

I. HISTORY OF KNOWLEDGE ABOUT OESTROIDEA

1. Antiquity and Middle Ages

The oldest document known that probably records parasitism by an oestroid is the Kahun papyrus, discovered by FLINDERS PETRIE in 1889. This papyrus, written during the 12th dynasty (2130-1930 B.C.), dealt with several topics of veterinary medicine. In Chapter IV, lines 17-33 (Fig. 2) there was an interesting reference to the 'nest of a worm' or 'uterus of a worm', interpreted by OEFELE (1901: 523-524) as referring to a case of cutaneous myiasis in a calf, caused by *Hypoderma bovis* (Linnaeus). Several fragments were missing in this papyrus and the lacking passages were completed [in square brackets] by OEFELE, who gave the following translation:

'Betrachtung des Rindes mit dem Neste des einen Wurmes. Wenn ich betrachte das Rind mit Nestern des einen Wurmes [es läuft im Anfange] ohne zu ruhen. [Nicht beruhigt sich] sein Gemüth davon. Nachdem es (das Rind) dann gerannt ist die Nase zur Erde senkend, wird es (das Ei der *H. bovis*) niederfallen auf dasselbe (das Rind). Du musst diagnosticieren, dass verborgene Körper (die Anfänge der Dasselbeulen) davon [vom Niederfallen] sind. Ich operiere dasselbe (das Rind): Eindringen muss ich mit

Fig. 2. Part of the Kahun Papyrus, the first reported reference to *Hypoderma bovis* (after OEFELE, 1901).

meiner Hand in das Innere seiner Beulen. Ein Krug Wasser steht an meiner Seite. Es ist die Hand eines Assistenten, welche reinigt eine Partie seine Hand in diesem Kruge mit Wasser so oft widerholt, als Schleim an der Hand ist, bis du entleerst Blut, das verbacken ist mit Stücken oder mit Milchwasser. Du erkennst die Krankheitsbeseitigung, sobald Milchwasser kommt zugleich mit der Tochter (der Larve der *Hypoderma bovis*). Deine Fingern beim operieren (?) in seinem Neste (?)... Er wird verbunden (?) mit Pflaster (?).'

If OEFELE's interpretation of the text is correct, the ancient Egyptians had a surprising and unique knowledge of the biology of the fly. Other authors do not agree – NEFFGEN (1905), for instance, believed it was a case of conjunctivitis. OEFELE cites, to corroborate his interpretation, a quotation from the same Egyptian text in the Byzantine *Geoponika* of the 10th Century, as referring to *Hypoderma*.

Another very old text relating to oestroids was found among the books of the great library of Assurbanipal, probably dating from the 7th Century B.C. The work called *Har-ra=Hubullu* is a Sumerian-Accadian dictionary, compiled from Sumerian texts, especially liturgical books. It deals with several animals. Tablet 14 endeavours to include all the wild, land-dwelling animals known at the time, listing 179 names of mammals, 75 of reptiles and amphibians, 121 of insects, and 32 of other invertebrates. In lines 283-289 of Tablet 14, referring to the group of 'worms' there is mention of the 'Igigula', a malady of the eyes, which has been interpreted as an ophthalmomyiasis caused by *Oestrus ovis* LINNAEUS, a common occurrence in Mesopotamia (BODENHEIMER, 1960: 116).

Among the ancient Greeks these flies were very well-known, but frequently confused with other flies, especially tabanids (MAC LEAY, 1825; LEACH, 1826; MOULÉ, 1909; JANSEN, 1967). No author has ever associated the adult with the parasitic larva, and the larvae of *Hypoderma* were even assigned to adult tabanids. The Greeks described very carefully the terror caused by these flies to cattle (Fig. 3), which would flee at their approach, in order to escape the 'sting' or 'stab'. *Oistros* was the name given to that pest. Following GIL FERNANDEZ (1959:157):

'La palabra (*oistros, – ou*, m.) se descompone en *ois – tros* y se encuentran sus parientes en *oima – *oisma*, "ímpetu, furia", lat. *ira* (de *eira*), lit. *aistrà*, "pasión violenta", a. nord. *eisa*, "lanzarse vivamente hacia delante"'; and further on, 'el sentido primitivo del término sería, pues, "el que causa furor". Tal es la opinión de BOISACQ E., 1950 [*Dictionnaire étymologique de la langue grecque*, 4e. ed.; Heidelberg], WALDE-POKORNY A., 1930-32 [*Vergleichendes Wörterbuch der idg. Sprachen*, Berlin, Vol. I, p. 107], POKORNY J., 1951 [*Indogermanisches etymologisches Wörterbuch*, Berna, p. 300] y HOFFMAN J. B., 1949 [*Etymologisches Wörterbuch des Griechischen*, München].'

'Una opinión divergente de la que no se hacen eco estes autores es la emitida por FÉLIX HARTMANN, 1927 [*Zeitschrift für vergleichende Sprachenforschung auf dem Gebiete der indogermanischen Sprachen*, Göttingen, 54: 289, n° 1]. Según el, *oistros* estaría en relación

Fig. 3. Reaction of cattle to the approach of *Hypoderma bovis* (after FIGUIER, 1875).

con *oidos*, "hinchazón", *oidma*, "ola que se hicha", al. *Eiter*, "pus", y a. esl. *jadŭ*, "veneno". Desde el punto de vista fonético no hay ninguna dificultad que se oponga a este enjuiciamiento, según el cual el sentido primitivo del término sería "el que produce hinchazón" (-*tros* en una y otra hipótesis tendría valor causativo).'

Probably the oldest reference to the *oistroi* is found in HOMER's Odyssey (XXII, 299-301) – when Ulysses and his companions were planning to kill the pretenders to the hand of Penelope, these 'fled to the other end of the court like a herd of cattle maddened by the gad-fly in early summer, when the days are at their longest' (cf. BUTLER, 1952: 309).

The legend of Io, daughter of the River Inachus, shows how ancient was the knowledge of this parasite. Jupiter having fallen in love with Io, transformed her into a heifer after covering her with a cloud and thus avoiding Juno's jealousy. The goddess, always suspecting, and marvelling at the beauty of the heifer, asked Jupiter about her. Not to be discovered, he had to comply with his wife's wish, and the poor heifer was then placed by Juno under the guard of Argos, the hundred-eyed giant. Io was only freed when Mercury killed Argos, spreading the hundred eyes of the giant over the tail of a peacock. Juno, infuriated by this action, sent a terrible *oistros*, with 'buzzing wings', to persecute the unfortunate heifer, who was so harassed by the incessant attacks of the fly that she crossed the sea and after many wanderings came to the banks of the Nile, where her primitive shape was restored by Jupiter.

This legend was sung by several authors, among them Aeschylus (*Prometheus Bound*; cf. COOKSON, 1952: 45-46):

> 'Again the prick, the stab of gadfly-sting!
> O earth, earth, hide,
> The hollow shape – Argus – that evil thing –
> The hundred-eyed –
> Earth-born-herdsman! I see him yet; he stalks
> With stealthy pace
> And crafty watch not all my poor wit baulks!
> From the deep place
> Of earth that hath his bones he breaketh bound,
> And from the pale
> Of Death, the Underworld, a hell-sent hound
> On the blood-trail,
> Fasting and faint he drives me on before,
> With spectral hand,
> Along the windings of the wasteful shore,
> The salt sea-sand!
> List! List! the pipe! how drowzily it shrills!
> A cricket-cry!
> See! See! the wax-rebbed reeds! Oh, to these ills
> Ye Gods on high,

> Ye blessed Gods, what bourne? O wandering feet
> > When will ye rest?
> O Chronian child, wherein by aught unmeet
> > Have I transgressed
> To be yoke-fellow with Calamity?
> > My mind unstrung,
> A crack-brained lack-wit, frantic mad am I,
> > By gad-fly stung,
> Thy scourge, that tarres me on with buzzing wing!
> > Plunge me in fire,
> Hide me in earth, to deep-sea monster fling,
> > But my desire –
> Kneeling I pray – grudge not to grant, O King!
> > Too long a race
> Stripped for the course have I run to and fro;
> > And still I chase
> The vanishing goal, the end of all my woe;
> > Enough have I mourned!
> Hear'st thou the lowing of the maid cow-horned?'

Answered Prometheus:

> 'How should I hear thee not? Thou art the child
> Of Inachus, dazed with the dizzying fly.
> The heart of Zeus thou hast made hot with love
> And Hera's curse even as a runner stripped
> Pursues thee ever on thine endless round...'

Aeschylus referred to the fly as *oistros* (verses 567 and 880, among others), and also as *myops* (verse 675), thus treating both terms as synonyms.

The terror imposed by the *oistroi* on cattle was so well known to the Greeks, that several words translated this feeling. Thus, the verbs *oistrío, oistréo, oistrélateo, oistrobóleo*, meant 'to become infuriated, subjected to an attack of passion, fury, terror, or desire; to sting, to stab'; the substantives *oistrema* and *oistromanía* meant 'sting, stab that renders furious, or everything that is subjected to a violent passion'; and the adjectives *oistréeis, oiestrélateo, oistrodínetos, oistrodónetos, oistródonos, oistromanés, oistropléx, oistrofóros, oistródes*, etc., meant 'tormented by a sting or stab; thence, furious, ardent, desirous, etc.' (MOULÉ, 1909: 251; BAILLY, 1894: 1362).

Used in these several meanings, those terms are found in Greek authors of the 6th and 5th Centuries B. C., as Sophocles (*Trachiniae*, 655; 1254; *Oedipus Rex*, 1318; *Electra*, 5; cf. JEBB, 1952); Euripides (*Hyppolitus*, 1300; *Iphigenia at Aulis*, 77; *Bacchantes*, 129; 1229; cf. COLERIDGE, 1952), Plato (*Republic*, 577; *Phaedrus*, 251; cf. JOWETT, 1952), and Hippocrates (cf. BAILLY, 1894: 1362).

Confusion between *Hypoderma*, characterized by the loud buzzing flight, and tabanids, more abundant, easier to observe, and the real responsibles for the 'stings' in cattle was widespread among Greek authors, and lasts to our day. In German, 'Bremse' may refer both to oestroids and tabanids; the same occurs with the French 'taon'. RÉAUMUR (1738: 504) had already commented upon this fact:

'Les gens de la campagne sont mieux instruits de la nature & de la cause de ces bosses [the warbles caused by the larvae of *Hypoderma*], qu'ils ne le sont de plusieurs autres faits d'histoire naturelle, qu'ils seroient également à portée d'observer; ils sçavent très-bien que chacune renferme un vers, & même ils sçavent que ce ver vient d'une mouche, que lui-même se transforme en mouche; à la vérité, ils connoissent mal cette mouche, du moins tous les paysans qui m'ont paru les mieux instruits, & qui me l'ont voulu désigner, m'on dit qu'elle étoit comme un taon, & ils donnent le même nom au ver même. Comme les taons sont de toutes les mouches celles qui sont les plus acharnées sur leurs boeufs, & qui en tirent le plus de sang, il étoit assez naturel de penser que ces mêmes mouches quand elles étoient vers, s'étoient nourries sous la peau du boeuf.'

Aristotle (384-332 B.C.) was apparently the first author to cite Cephenemyiinae larvae from the interior of the skull of deer. In his 'History of Animals' (II, xv), he said:

'All deer, however, have living maggots inside their heads: they infest the hollow region under the tongue and near the vertebra to which the head is attached: these maggots are as big as the biggest grubs; they grow in a bunch huddled up together, about twenty of them.' (cf. PECK, 1965).

According to SAINT-HILAIRE (1883: 172, note 9) this sentence could be an interpolation by DITTMEYER, as it is included in the middle of a discussion about the gall of the deer.

Also, several references to the *oistroi* are found in Aristotle (cf. PECK, 1965, 1967, 1970):

1. 'The two-winged [insects] have their sting in front, for example, the fly (*myia*), the horse-fly (*myops*), the gadfly (*oistros*), and the gnat (*empis*).' (*Hist. Anim.*, I, v, 20); 2. 'The trumpet-shells and the purpuras have this proboscis too, and in them it is firm and hard. And just as the *myopes* and the *oistroi* can bore through the hides of quadrupeds, this organ is even more powerful and strong in proportion...' (*Hist. Anim.*, IV, iv); 3. 'From the small flat creatures (platéon zodaríon) that run about on the water of rivers comes the gadfly (*oistroi*); that is why gadflies are most abundant near waters where these creatures are.' (*Hist. Anim.*, V, xix); 4. 'The *myopes* are produced out of timber' (*oi de myopes gignontai ek ton xylon*) (*Hist. Anim.*, V, xix); 5. 'Other [insects] drink blood ('*aimabora*), as the *oistroi* and the *myops*' (*Hist. Anim.*, VIII, xiii); 6. 'Also, there are among the *oistroi* and *myopes* creatures that can pierce through the skin of the human body, and some can actually puncture animal hides as well.' (*Parts of Animals*, II, xvii).

Therefore, under the names of *empis* and *oistros* Aristotle referred much more probably to Chironomidae and Culicidae; as to '*myops*' it was probably a tabanid. Under *oistros* Aristotle also included some crustaceans parasitic on fishes (*Hist. Anim.*, VIII, xx, 10; VIII, xv, 3; VI, xvi, 3; VIII, xvii, 4-5), and a bird that fed on insects, of unknown identity (op. cit.; VIII, v, 5).

Oistroi of oxen were mentioned by Apollonius Rhodius (3rd Century B.C.), in his *Expedition of the Argonauts* (I, 1265; II, 277) and of fish by Oppianus (2nd Century B.C.), in his *The Fishing* (II, 521-532).

And as already noted by OSTEN SACKEN (1893a, 1894, 1895), some of the observations of the early Greeks about the spontaneous generation of 'wasps and hornets' from the flesh and skin of mules, horses and donkeys, could refer to *Hypoderma bovis*. Archelaus (5th Century B.C.), Antigonus (3rd Century B.C.), Nicander (2nd Century B.C. in his *Theriaca*, a treatise on insect venoms and the wounds inflicted by them), Plutarch and Varro (1st Century B.C.), Servius Maurus Honoratus (called Servius Grammaticus; 4th Century A.D.) and Olympiodorus of Thebe (5th Century A.D.), wrote about this subject.

Among the Romans, the *oistros* was called *Asilus*, as in Virgil (Georgics, III, 148; cf. RHOADES, 1952: 71):

'Round wooded Silarus and the ilex-bowers
Of green Alburnus swarms a winged pest –
Its Roman name *Asilus*, by the Greeks
Termed *Oestrus* – fierce it is, and harshly hums,
Driving whole herds in terror through the groves,
Till heaven is madded by their bellowing din,
And Tanager's dry bed and forest banks.
With this same scourge did Juno wreak of old
The terrors of her wrath, a plague devised
Against the heifer sprung from Inachus.
For this too thou, since in the noontide heats
'Tis most persistent, fend thy teeming herds,
And feed them when the sun is newly risen,
Or the first stars are ushering in the night.'

Virgil most certainly referred here to *Hypoderma*. Seneca (1st Century B.C.) also agreed:

'Hunc, quem Graeci oestron vocant, pecora perangentem et totis saltibus dissipantem, asilum nostri vocabant.' (*Epistolae*, LVIII, 1).

The same said Valerius Flaccus (1st Century A.D.):

'Continuo, volucri seu pectora tactus asilo
Emicuit Calabris taurus per confranga saeptis
Obvia quaeque ruens, tali se concitat ardens
In juga senta fugae.'

Knowledge about the *oistroi* or *asili*, already confused among the Greeks, turned out to be more complicated among the Roman writers of the 1st Century A.D. Pliny, the Ancient (23-79 A.D.), for instance, in his *Natural History* (cf. RACKHAM, 1967), repeated Aristotle, i.e., that deer had 'worms' in the head:

'Cervis in capite inesse vermiculi sub lingua inanitate, et circa articulum, qua caput jungitur, numero XX produntur' (Stags are stated to have maggots to the number of twenty in the head beneath the hollow of the tongue and in the neighbourhood of the juncture of the head with the neck) (*Nat. Hist.*, XI, xlix).

Pliny applied the name '*oestrus*' to the drones of bees (op. cit., XI, xvi):

'Quippe nascuntur aliquando in extremis favis apes grandiores, quae caeteras fugant. Oestrus vocatur hoc malum: quonam modo nascens, si ipsae fingunt?' (It is a fact that sometimes larger bees are born in the extremities of the combs which drive away all the rest. This mischievous creature is called *oestrus* – being born in what possible manner if the female bees themselves shape it?).

And in another part of his work (XI, xxxiv), the Roman naturalist considers *asilus* and *tabanus* as synonyms:

'... reliquorum quibusdam aculeus, ut asilo (sive tabanum dici placet), item culici et quibusdam muscis, omnibus autem his in ore et pro lingua' (some of the rest have a sting, for instance the *asilus* (or, if you like, *tabanus*), and also the gnat and some flies, but with all of these it is in the mouth and serves as a tongue).

Oestrus was also applied to drones by Columella (*De Re Rustica*, IX, xiv):

'Eodemque tempore progenerantur in extremis partibus favorum amplioris magnitudinis quam sunt caeterae apes, eosque nonnulli putant esse reges. Verum quidam Graecorum auctores *oistrous* appellent ab eo quod exagitent, neque patiantur examina conquiescere.'

This same text, with minor modifications, was also quoted by Rutilius Taurus Aemilianus Palladius (*De Re Rustica*, VI, x; 4th Century A.D.).

Saint Isidore of Seville (Isidorus Hispalensis), the encyclopaedist and historian (A.D. c. 570-631), in his great work, *Originum sive etymologiarum Libri XX*, written between 622-633, where he tried to preserve the remnants of Greek and Roman culture against the barbarians, also quoted this text:

'Costros [sic] Graeci appellant, qui in extremis favorum partibus maiores creantur apiculae, quas aliqui reges putant: sed Graeci eos oestros appellant et necari iubent, quia requiem concitiunt quiescentis examinis.'

Other authors writing in Latin (Juvenal, P. Papinius Statius, Sextus A.

Propertius, Sextus Pompeus Festus, and Hysichius) mentioned the *oestri* or *asili* in several contexts, as did Greek authors of the 5th and 6th Centuries, such as Thyphiodorus (*The Fall of Troy*, 361 and ff.) and Coluthus (*The Rape of Helen*, 43).

An interesting text was cited by Aelianus (3rd Century A.D.), in his *De Natura Animalium* (IV, 51; VI, 37) – the *oistros* is like a large fly (*myia megisté*), strong and robust, and has a powerful sting (*kentron ischyron*), which comes out from the mouth ('*ertemenon tou stomatos*), and which emits a sound like a buzzing (*bombódes*); the *myops*, in its turn, is similar to the canine fly (*kynomyia*), produces a louder buzzing and has a mouth sting which is less developed. These were probably two different genera of Tabanidae.

An extremely picturesque story about *Oestrus ovis* appeared in the 6th Century, in the book of Alexander Tralianus, *De Arte Medica* (quoted by RÉAUMUR, 1738: 553; also KIRBY & SPENCE, 1822: 157). A young Athenian, Democrate, was subjected to frequent epileptic fits; one day he decided to consult the oracle at Delphi, to learn of some remedy for his ailment. The Pythia, according to Alexander Tralianus, gave her answer in two versions:

'Quos madidis cerebri latebris procreare Capella,
Dicitur humores, vermem de vertice longum'.

Or,

'De grege sume Caprae majoris ruris alumnae
Ex cerebro vermes; ovis dato tergora circum
Multiplici vermi pecoris de fronte revulso.'

MOUFFET (1634: 284) and TOPSELL (1658: 1107, 1122) offer another version of the answer of the Pythia:

'Take a tame goat that hath the greatest head,
Or else a wilde goat in the field that's bred,
And in his forehead a great worm you'l finde,
This cures all diseases of that kinde.'

Poor Democrate was entirely baffled by the extremely obscure words of the Pythia. However, an old man of 98 years, very versed in the language of the gods, came to his rescue. Having praised the medical science of Apollo, he disclosed to Democrate the hidden meaning of the remedy against epilepsy – the 'worms' referred to by the Pythia were in fact engendered in the head of goats and sheep, near the base of the brain; these 'worms' were expelled by the goats through sneezing. Democrate should obtain them before they touched the ground.

Fig. 4. Reaction of sheep to an attack by *Oestrus ovis* (after FIGUIER, 1875).

for that purpose extending pieces of cloth in front of the goats, where the worms would tumble. Two of these worms should be wrapped up into a piece of skin of a black sheep, and the little sac thus formed worn around the neck. It is not known whether the cure operated or not.

The 'oestrum' is cited again in Chapter 6 of the book '*De Universo*', written by RABANUS MAURUS, born in 776, abbott and teacher at Fulda and later (847) Bishop in Mainz (BODENHEIMER, 1928: 119-120).

In the 10th Century, during the reign of Constantine VII Porphyrogenitus, a compilation of ancient texts relating to agriculture was undertaken, based on previous compilations, especially by Vindonius Anatolius of Beirut, Didymus of Alexandria, and Cassianus Bassus. That work, entitled *Geoponika, ai perí georgías eklogai* (ca. 950) has been attributed either to Cassianus Bassus (cf. BECKH, 1895) or to the emperor Constantine VII himself. There again reference is made to the *myopes* that cause terror to cattle, the means of avoiding them (sprinkling the grazing areas with infusions of bay fragments), as well as to the use of plaster for curing the warbles, seemingly a translation and adaptation of the Egyptian text cited at the beginning of this chapter (OEFELE, 1901: 524; BECKH, 1895: 473, 483).

In the 11th Century, Raschi (Rabbi Schlomo Jizchaki; 1030-1105) gave a commentary on the Talmud, and referred to the 'Grabeleisch' (*Hypoderma bovis*), which was a worm placed between the skin and the flesh of the oxen (LEWYSOHN, 1858: 317, 318; MOULÉ, 1912: 581-582; BODENHEIMER, 1928).

GEORGIUS PACHYMERUS (1242-1310), a Byzantine author, refuted the ideas of Antigonus, Pliny, Plutarch, Nicander, Aelianus, Archelaus, Varro and others, on spontaneous generation, stating that wasps did not originate from the skin or flesh of horses, but actually *from the brain.* Is this a possible allusion to the species of *Rhinoestrus* parasitic on horses?

During the Middle Ages several authors wrote about oestrids, designated as *ester, odestrum, oester, oestrum,* and *orestes* (Medieval Latin, from *oistros* or *oestrus*), as *arzillus, aselus, asilo, asilius* (from the Latin *asilus*), or even as *bastaban* (from the Latin *tabanus*), *musca modica* (?) and *scabro* (ROLLAND, 1911).

A reference which may be the first one about *Cephalopina titillator,* parasitic on camels, is found in the writings of an Arab author of the 14th Century, Kamal ad-Din Muhammad ibn Musa ad-Damiri (1344-1405) (BODENHEIMER, 1928: 144). He was born in one of the two towns called Damira, near Damietta, and spent his life in Egypt. He became Professor of Tradition at the Shafi'ite school of law in the Rukniyya at Cairo, and also at the mosque el-Azhhar. In his 'Life of Animals' (*Hayat al-Hayawan*) he lists in alphabetical order 931 animals mentioned in the Koran, a compendium of the traditions and the poetical and proverbial literature of the Arabs.

Under 'Nagaf' he says: 'A worm, which lives in the nose of the camel and the sheep. Abu 'Ubaida said about it: 'it is also the white worm, which lives in the dates, and this worm is called an-nagaf'. His book was translated into English by JAYAKAR (1906-1908).

2. From the 16th Century to the Systema Naturae (1758)

After the discovery of the Americas, the European settlers became acquainted with other oestrideous larvae parasitic on man and animals. Compiling the knowledge of natural history of the Aztecs, in the 16th Century, Fray BERNARDINO DE SAHAGÚN, in his *Historia General de las Cosas de Nueva España*, recorded the existence of 'gusanos que se crían en los brazos o miembros de los conejos y ratones'; those worms can be seen 'metidos dentro de la carne y miran hacia fuera'. The larvae (*Cuterebra*) were called *nacaocuilin*, derived from *nacátl* = flesh, and *ocuilin* = worm (VOGELSANG & MARTÍN DEL CAMPO, 1947: 50).

Missionaries and travellers in Latin America also mentioned larvae parasitic on man, known among the Indians by several names (see GUIMARÃES & PAPAVERO, 1966). Those larvae were to be described by LINNAEUS Jr., in 1781, as *Oestrus hominis*.

In Europe, several authors wrote about these flies. JOHANN THOMAS FREIGIUS, in his 'Questiones physicae' (1579), treats *asylus, tabanus* and *oestrum* as synonyms.

MOUFFET (1634: 935-936) declared that 'the fly called *oestrum* is of a yellowish colour, who when it enters the ears of an ox causeth him to run mad; he carries before him a very hard, stiff and well-compacted sting, with which he strikes through the ox his hide. They follow oxen and horses and young cattle by scent of their sweat, because they cannot reach them with their sight, being very weaksighted. They are generated of the worms that come out of the wood putrefied (or, according to another authority, from horse-leeches).'

In the 17th Century the scientific phase of research began in Europe. The great Italian naturalist, FRANCESCO REDI, published drawings of the larval stages of the parasite of sheep (*Oestrus ovis*), as well as drawings of a larva of Cephenemyiinae (REDI, 1668). VALLISNIERI (1713, 1733) obtained for the first time, by rearing larvae, an adult of *Oestrus ovis*.

The great researcher of the biology of insects, RÉAUMUR, studied the larvae parasitic on sheep and deer, and *Hypoderma bovis* (1734, 1738, 1740). It is curious to note the explanation given at that time by hunters

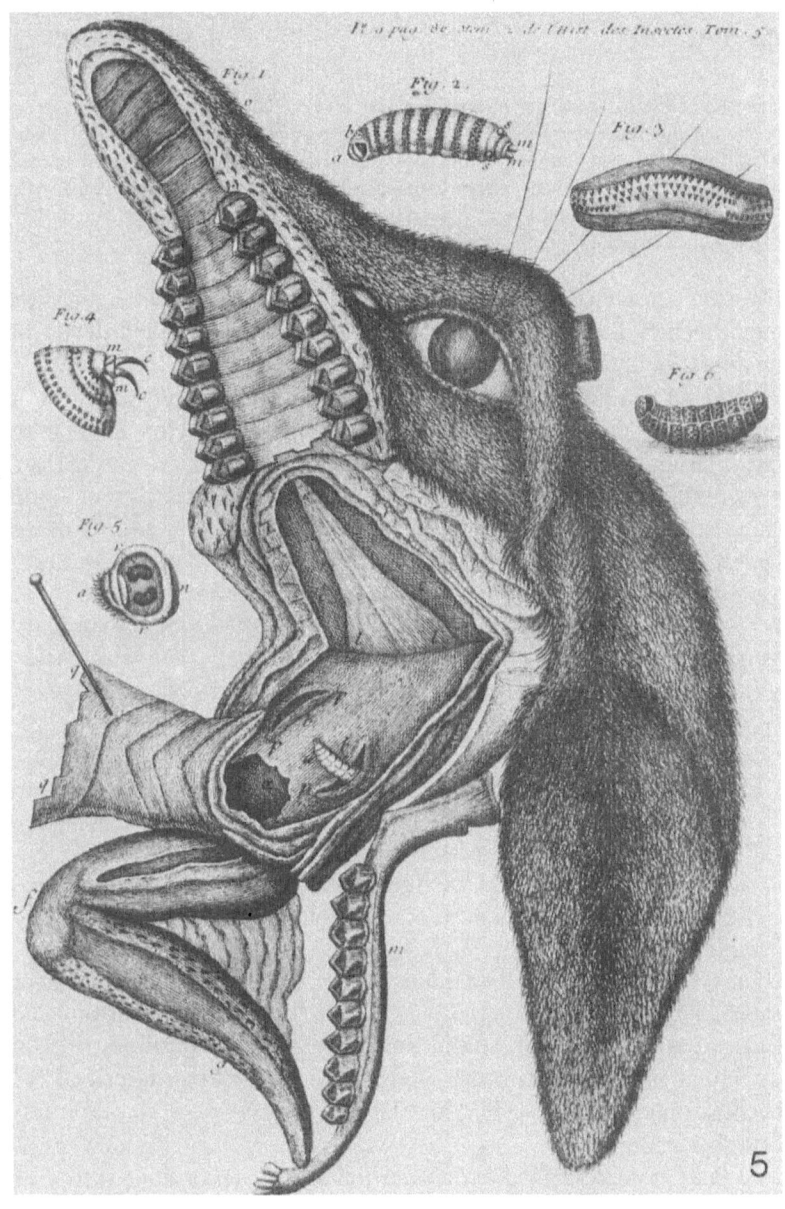

Fig. 5. Head of deer showing location of Cephenemyiinae larvae (after Réaumur, 1738).

to the presence of oestrideous larvae in the skull of cervids (RÉAU-MUR, 1738: 527):

'Aussi les chasseurs sçavent non-seulement que les cerfs sont très-sujets à avoir de ces tumeurs, & que l'intérieur de chaque est occupé par un ver, auxquel ils donnent le nom de *taon*; ils ont de plus fait un système physique sur l'usage dont ces vers sont au cerf. Ils ont imaginé qu'ils servent à faire tomber son bois; qu'il vient un temps où les vers s'acheminent sous la peau, pour se rendre à la racine du bois; que tous la rongent de concert; & que le bois dont l'appui à été coupé, est obligé de tomber.'

RÉAUMUR showed the impossibility of such a phenomenon, and magnificently illustrated the situation of the larvae in the interior of the deer's head (Fig.5).

LINNAEUS, on his trip to Lapland, studied the cutaneous parasites of the reindeer, known to the Laps by the name of 'curbma' (LINNAEUS, 1737). He published that paper several times (1738, 1741, 1747, 1749, 1757, 1761, 1762, 1772), on certain occasions giving different Latin names to the parasite (1741; in BRYK, 1924). In the 10th edition of his *Systema Naturae* (1758) he fixed definitely the concept of *Oestrus*, creating that genus for those flies with reduced mouthparts, whose larvae were parasitic on the skin, head or digestive tract of some mammals; his genus *Oestrus* included the species *bovis, tarandi, nasalis, haemorrhoidalis* and *ovis*, parasitic on the ox, reindeer, horse and sheep.

3. From the Systema Naturae to the revision of Bau (1906)

The publication of the *Systema Naturae* led to a fever of scientific exploration and classification of living beings. Cataloguing of faunas became fashionable among naturalists; descriptions of new species and new biologies appeared.

DE GEER (1776) described some larvae parasitic on the digestive tracts of horses. PALLAS (1776, 1777), following an adventurous and fruitful journey through Asiatic Russia, described several oestrid larvae found in mammals, some of the latter also being described for the first time. MODEER (1786) and FISCHER (1787) discussed other oestroid species.

CLARK (1797) described a few additional parasites of the intestine of horses. In his monograph on *Oestrus* (1815) he proposed a subdivision of the genus into: (i) Gastricolae (Chylivorae); (ii) Cuticolae (Purivorae); and (iii) Cavicolae (Lymphivorae). He was the first author to divide the genus *Oestrus*, erecting the genus *Cuterebra* for some North American cuticole species.

LEACH (1818) proposed a new 'Arrangement of the Oestrideous insects', where he created the category 'Oestridea', therefore being considered as the author of the family Oestridae (and co-ordinate categories). He proposed equally a division of *Oestrus*, erecting the genus *Gasterophilus* for those species with larvae parasitic on the digestive tract of horses.

The great French entomologist, P. A. LATREILLE, also recognized a suprageneric category for this group of flies, the 'Tribu des Oestrides', placing it in his 'Famille des Athéricères'. LATREILLE noticed how artificial was the grouping of the species united under *Oestrus*, and suggested a more natural system. On that occasion (1818) he erected the genera *Cephenemyia*, *Oedemagena*, *Hypoderma* and *Cephalomyia*; he restricted the genus *Oestrus* to the gastricole species (with the same concept as LEACH's *Gasterophilus*), and recognized *Cuterebra* CLARK as valid.

CURTIS (1826) fixed *ovis* LINNAEUS as the type-species of *Oestrus* LINNAEUS and *Oestrus equi* CLARK as the type-species of *Gasterophilus* LEACH, thus rendering *Cephalemyia* LATREILLE (monobasic for *ovis*) a synonym of *Oestrus*.

With the proposed new genera, CLARK's 1815 system became as follows: 1. Gastricolae: *Gasterophilus* Leach; 2. Cuticolae: *Cuterebra* Clark, *Oedemagena* Latreille, *Hypoderma* Latreille; 3. Cavicolae: *Oestrus* Linnaeus (=*Cephalemyia* Latreille), *Cephenemyia* Latreille.

Travelling through South Africa, DELEGORGUE observed for the first time oestrideous parasites of African antelopes; he registered his obsevations (1847) thus:

'Les animaux sauvages de cette partie de l'Afrique avaient cela de particulier, que toutes les espèces étaient habités par des quantités considérables d'oestres. Les *Catoblepas gnu* et *gorgon* en laissaient tomber à tout moment par les narines. L'*Acronotus lunata*, dont je sciais fréquemment les cornes avec une partie du crâne, présentait sous leur naissance une cavité qui en était toujours remplie.'

New genera were added by BRAUER (*Oestromyia*, 1860) and SCHINER (*Pharyngomyia*, 1861).

BRAUER had started his studies on Oestridae as early as 1858; in 1863 he published a monograph on the group, where, in addition to a compilation of all the knowledge accumulated to his time, he added his own, extremely valuable, observations. He extended, however, the concept of the family too far, embracing entirely unrelated groups, resulting in considerable confusion in the literature.

BRAUER's 'Oestriden' were divided into 7 groups: (i) *Ctenostylum*; (ii) *Therobia* and *Aulacocephala*; (iii) *Gasterophilus*; (iv) *Hypoderma* (with *Oedemagena* as subgenus), and *Oestromyia*; (v) *Cephalomyia* (emend.) and *Oestrus*; (vi) *Cephenomyia* (emend.) and *Pharyngomyia*; and (vii) *Rogenhofera*, *Cuterebra* and *Dermatobia*. The inclusion of some entirely unrelated genera, based almost exclusively on the reduction of the mouth-parts, made BRAUER'S 'Oestriden' an artificial group. *Ctenostylum*, for instance, is a Pyrgotid (STEYSKAL, 1967). In his 1863 monograph BRAUER also described a great number of new species and placed others in synonymy.

Following the publication of the monograph, several new genera were proposed: *Endocephala* Lioy, 1865 (a synonym of *Cephenemyia*); *Pharyngobolus* Brauer, 1866; *Microcephalus* Schnabl, 1877; *Gyrostigma* Brauer, 1886; *Cobboldia* and *Dermatoestrus* Brauer, 1892; *Spathicera* Corti, 1895, and *Portschinskia* Semenov, 1902.

In 1887 BRAUER segregated *Rogenhofera*, *Cuterebra* and *Dermatobia* into a separate family, Cuterebridae. The Oestridae were restricted to the Old World. In 1889 BRAUER & BERGENSTAMM presented a new key for the classification of the Oestridae, dividing them into Cuterebridae and Oestridae Typicae. The latter were organized into:

1. Gastricolae: *Gasterophilus*, *Gyrostigma* and *Cobboldia*;
2. Cavicolae, with 3 sections:
2.1. Genuinae: *Oestrus*, *Rhinoestrus*, *Pharyngobolus*, *Cephalomyia*, *Pharryngomyia* and *Cephenomyia*;
2.2. Dubiosae: *Therobia*, *Aulacocephala*, *Microcephalus*, *Oestroderma*;
2.3. Cuticolae: *Hyoderma*, *Oedemagena* and *Oestromyia*.

In 1895 AUSTEN erected the genus *Bogeria*, near *Cuterebra*. In 1906, in the section of Oestridae for the *Genera Insectorum*, BAU adopted the same organization proposed by BRAUER & BERGENSTAMM, eliminating however some of the 'dubiosae' genera, his arrangement resulting as follows:

I. Abteilung–Cuterebridae: *Rogenhofera*, *Cuterebra*, *Bogeria* and *Dermatobia*;
II. Abteilung–Oestrinae Typicae: (i) Gastricolae: *Cobboldia*, *Spathicera*, *Gyrostigma* and *Gasterophilus*; (ii) Cavicolae: *Cephalomyia*, *Oestrus*, *Rhinoestrus*, *Pharyngobolus* and *Cephenomyia*; (iii) Cuticolae: *Oestromyia*, *Dermatoestrus*, *Strobiloestrus* and *Oedemagena*; (iv) Dubiosae: *Microcephalus* and *Oestroderma*.

BAU's scheme was followed, with minor modifications. until very recently.

During this period great innovations in the higher classification of the Calyptratae had been introduced by GIRSCHNER. He divided the Calyptratae into two 'families'–Anthomyidae and Tachinidae. The Anthomyidae (1893) comprised the groups Coenosiinae, Muscinae and Gasterophilinae (the latter then proposed as a new subfamily). The Tachinidae (1896) included the groups Oestrinae, Hypoderminae (proposed as new subfamily), Calliphorinae (including *Cephenemyia* and *Pharyngomyia*), and others which are not considered here. GIRSCHNER's system was adopted by BEZZI & STEIN (1906) in the catalogue of palae-arctic Diptera, but there the Gasterophilinae were moved to the Tachinidae.

4. From 1908 to the classification of Townsend (1935-38)

During this period important modifications were introduced – the phase of colonial exploration of Africa was followed by the phase of scientific exploration, with the resulting description of many new taxa. Consequently, new changes had to be incorporated into the system of classification of the Oestroidea. New genera were proposed: *Gedoelstia* Rodhain & Bequaert (1913), *Kirkia* Gedoelst (1914), *Cephalopsis* Townsend (1912), *Loewioestrus* Townsend (1918), *Atrypoderma* Townsend (1919) and *Pallasiomyia* Rubtsov (1939).

An extremely important discovery was made by FROGGATT (1913) – the existence of oestrid larvae in the kangaroo (*Macropus*), described as *Oestrus macropi*. For the reception of this species TOWNSEND (1916) erected the genus *Tracheomyia*.

ENDERLEIN (1911) established the subfamily Cobboldiinae, in the Tachinidae, together with the Gasterophilinae.

BAU (1929) subdivided *Cuterebra* into several subgenera: *Procuterebra*, *Orthocuterebra*, *Metacuterebra* and *Paracuterebra*.

AUSTEN (1930) erected a new subfamily, Rutteniinae, placing it in the Tachinidae, together with the Oestrinae; AUSTEN placed the Cobboldiinae and Cuterebrinae in the Muscidae.

The greatest revolution in the classification of the Calyptratae was proposed by TOWNSEND (1935, 1936, 1938). In 1935 TOWNSEND divided his Superfamily Muscoidea into various families, among them the Cuterebridae, Gasterophilidae and Hypodermatidae; he considered as Cuterebridae the tribes Chaulioestrini, Cephenemyiini, Portschinskiini, Cuterebrini, Pseudogametini, Neocuterebrini (=Rutteniinae of AUSTEN), Pharyngomyiini, Cobboldiini, Oestrodermatini and Dermatobiini, erecting therefore 8 new tribes (nowadays distributed in 7 families!). The Hypodermatidae were divided into 3 tribes, 2 of them new: Hypodermatini, Strobiloestrini and Oestromyiini. In 1936 he divided the Superfamily Oestroidea into Gymnosomatidae, Oestridae, Prosenidae, Rutiliidae, Tachinidae, Dexiidae and Exoristidae. The Oestridae (1936: 17) were split into 9 tribes, 7 of which are now considered as Tachinidae, and are thus reduced to only the Rutteniini and the Oestrini.

5. From 1947 to the present.

The greatest specialist of this period is K. YA. GRUNIN, who has been contributing to our knowledge of the Oestroidea since 1947 in greatly detailed papers with excellent illustrations. In addition to descriptions of new genera and species, he produced several papers of broader aim, on the importance of the 1st larval stage in phylogeny (1950), on the structure of the anterior spiracle of the larvae (1951), and a very important paper on the phenomenon of aggregation (1959). In 1953 his first book was published, on the larvae of Oestridae parasitic upon domestic animals. His magnificent revision of the palaearctic Oestridae appeared in 1957 and was translated into German in 1966, in the series 'Fliegen der palaearktischen Region'.

GRUNIN divided the Oestridae into 2 subfamilies: (i) Cephenomyiinae, with *Cephenomyia* (emend.) and *Pharyngomyia*; and (ii) Oestrinae, with *Oestrus, Rhinoestrus* and *Cephalopina*.

In 1962 he published the revision of the Hypodermatidae, fixing the limits of the family, organizing the classification, and finally eliminating the problem of the 'Oestridae Dubiosae'. In that work (translate into German in 1964) he divided the Hypodermatidae into 4 tribes: (i) Oestromyiini, with *Oestromyia*; (ii) Portschinskiini, with *Portschinskia*; (iii) Oestrodermatini, with *Oestroderma*; and (iv) Hypodermatini, with *Hypoderma, Oedemagena, Pavlovskiata, Pallasiomyia, Przhevalskiana* and *Crivellia*.

The inclusion of valuable data on the biology, morphology, anatomy, distribution, control, etc., of the Oestroidea, makes GRUNIN's works fundamental for the study of the group.

In 1953 PARAMONOV described for the first time the adult of *Tracheomyia macropi*; this species was redescribed (adult and larvae) by GRUNIN, in 1961.

In 1962 BENNETT & SABROSKY revised the Nearctic species of *Cephenemyia*.

ZUMPT (1957) offered a scheme of classification for the Oestroidea, although not going into greater detail: 1. Oestridae, with the subfamilies Cephenemyiinae, Oestrinae and Hypodermatinae; 2. Gasterophilidae, with the subfamilies Gasterophilinae, Gyrostigminae and Cobbolddiinae; 3. Calliphoridae, including among others the subfamily Cuterebrinae.

This same scheme was adopted by him in a book on the Old World myiasis-producing flies (1965), which includes all the information available on the Oestroidea.

II. POPULAR KNOWLEDGE ABOUT THE OESTROIDEA

1. Utilization as food

Among the eskimos of the Ahearmiut tribe (Hudson Bay) larvae of *Oedemagena tarandi* (Linnaeus), subcutaneous parasites of the caribou, are greatly appreciated as food; the hunters bring them to their wives and children as appetizers. GABUS (1956: 280), who lived among the Ahearmiut, reports:

'With a busy air, Ayutnar carefully extracts round flat larvae from between the skin and flesh of the caribou; these larvae are called 'kummak'. He offers me some:
 - No!
 - They are very good! They taste like milk!
 - No!
 - You eat raw caribou; the 'kummak' also eat raw caribou... Their flesh is the same! I don't understand why you don't want to eat "kummak"!'

During the Vega Expedition to the Arctic Circle, the great explorer ERLAND NORDENSKIÖLD reported a similar habit among the tribe Chukchi, from the Bering Strait. The Chukchi take the subcutaneous parasites of the reindeer (*Oedemagena tarandi*) as food; the larvae are known as 'gorm'; the Chukchi also do not dislike adult insects. The larvae are sculptured by the Chukchi in ivory (Figs. 6-7) obtained from marine mammals (BERGIER, 1941: 129, 226).

2. Popular names

ROLLAND (1881, and especially 1911), GRUNIN (1957), GANSSER (1951) and GUIMARÃES & PAPAVERO (1966) compiled a number of popular names applied to the Oestroidea, which are listed below, with names obtained from additional sources. Abbreviations employed are - H (*Hypoderma bovis* or *lineatum*); O (*Oestrus ovis*); D (*Dermatobia hominis*).

(i) Names applied to adults

1. arrebenta-buey (H; Spanish)
2. arzillo (H; Italian dialect)
3. asillo (H; Italian dialect)
4. asiolo (H; Italian dialect)

Figs 6. A Chukchi, eater of *Oedemagena* larvae.

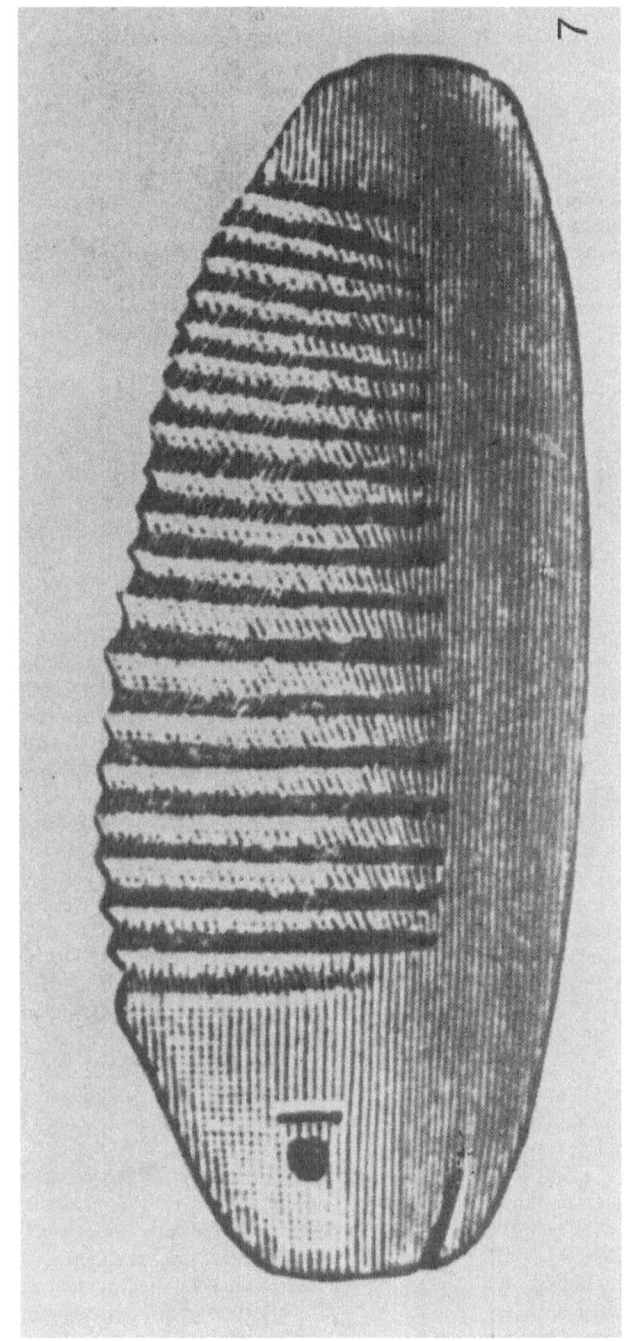

Fig. 7. Larva of *Oedemagena tarandi*, sculptured in ivory by the Chukchi (after BERGIER, 1941).

5. assillo (H; Italian dialect)
6. babouino (H; France: Bouches-du-Rhône)
7. barbarin (O; France: Berry)
8. barbèsin (O; France: Ariège)
9. barbin (O; Provençal)
10. beaw (H; English dialect)
11. belogolovkaya mukha (*Rhinoestrus purpureus*; Russian)
12. belogolovik (*Rhinoestrus purpureus*; Russian)
13. bèrbial (O; Languedoc)
14. berlin (?O; France: Centre)
15. besiggin (?; Italian dialect)
16. Biesfliege (H; German)
17. Bisebrummel (H; German)
18. Blindfliege (H; ancient German)
19. bot-fly (Oestridae in general; English)
20. bou (H; Dutch, dialect of Vriesland)
21. bouin (H; France: Poitou)
22. bouine (H; France: Berry)
23. Braas (H; German dialect)
24. bree (H; English)
25. breeze (H; English)
26. breeze-fly (H; English)
27. Bremo (H; ancient German)
28. Brems (H; Dutch)
29. Bremsche (H; ancient German)
30. Bremse (H; German)
31. Breosa (H; Anglo-Saxon)
32. Brese (H; ancient English)
33. brèsin (O; France: Maine-et-Loire)
34. brimsa (H; Anglo-Saxon)
35. briosa (H; Anglo-Saxon)
36. Brisbummel (H; German dialect)
37. brize (H; English dialect)
38. Bröms (H; Danish, Swedish)
39. bumbore (H; English)
40. burè (H; France: Poitou)
41. burin (H; France: Poitou)
42. burrel-fly (H; English)
43. cantarina (H; France: Rhône, Loire)
44. cleg (H; English)
45. cloidhe (H; Scottish Gaelic)
46. cloidheag (H; Scottish Gaelic)
47. clyr (H; Welsh)
48. clyrin (H; Welsh)
49. coss (H; Italian dialect)
50. coss da bouv (H; Italian dialect)
51. creabaire (H; Scottish Gaelic)
52. creadhal (H; Scottish Gaelic)

53. creithleag (H; Scottish Gaelic)
54. crëyr (H; Welsh)
55. crouba (H; France: Jura)
56. curbma (*Oedemagena tarandi*; Lapland)
57. Dasselfliege (Hypodermatidae; German)
58. Dasselmücke (H; German)
59. Engerlingsfliege (or – plage) (H; German)
60. sialë (H; France: Loire)
61. espara (H; Basque)
62. gad-bree (H; English)
63. gad-fly (H; English)
64. ghiza (H; Rumanian)
65. gleithire (H; Scottish Gaelic)
66. goundouleau (H; France: Aveyron)
67. gourdon (H; France: Nièvre)
68. gourgouli (O; France: Gard)
69. grey-fly (O; English)
70. gudbree-fly (O; English)
71. Hautbremse (H; German)
72. heel-fly (H; English)
73. Herle (H; ancient German)
74. Horle (H; ancient German)
75. Hornessel (H; ancient German)
76. Hornesser (H; ancient German)
77. Hornhsche (H; ancient German)
78. Hornisse (H; ancient German)
79. Hornusz (H; ancient German)
80. horzel (H; Dutch)
81. horzelvlieg (H; Dutch)
82. hudbremse (H; Danish, Norwegian)
83. Hummil (H; ancient German)
84. Huombil (H; ancient German)
85. isalho (H; Languedoc)
86. isalo (H; Provençal)
87. kozlyak (*Crivellia silenus*; Russian)
88. kruchak (O; Russian; GRUNIN, 1953: 23)
89. kudbremse (H; Danish, Norwegian)
90. kumyr (*Cephalopina titillator*; Russian)
91. laguerais (?; France: Semur, Côte d'Or)
92. marangouin (D; French Guiana)
93. mberuaró (D; Tupi-Guarani tribes; from *mberu* = fly; *ñaró* = angry; BERTONI, 1926; also meruaró).
94. mirunta (D; Peru; applied more to the disease; BARRAILLIER, 1892)
95. moche chagoignotte (H; France: Côte d'Or, Saint-Didier)
96. mosca del tarolo (H; Italian dialect)

96. mosca del tarlo (H; Italin dialect)
97. moscardo (H; Portuguese)
98. moscardón del buey (H; Spanish)
99. moucha tantarina (H; France: Rhône, Loire)
100. mouche bordine (H; France: Berry)
101. mouche bouine (H; France: Berry)
102. mouche du talon (H; Canadian French; from 'heel-fly')
103. mouche gouine (H; France: Charente-Inférieure)
104. mouche tantalique (H; France: Rhône, Loire)
105. mouche-ver (H; France: Maine-et-Loire)
106. mouscartt (H; France: Landes, Gironde)
107. mousche bouvine (H; ancient French)
108. mouche bouyne (H; ancient French)
109. mousco barbèsimo (O; Provençal)
110. mousco bouino (H; France: Bouches-du-Rhône)
111. mousco bóy'na (H; Provençal, Languedoc)
112. mousco dé bioou (H; Provençal, Languedoc)
113. mosque-bère (H; France: Béarn)
114. mousque-caysque (H; France: Oleron, Salies)
115. mouche-hisse (H; France: Orthez)
116. mou-cha da bouv (H; Romanche)
117. nonne (?; France: Côte d'Or)
118. Ochsenbremse (H; German)
119. Oksenbremse (H; Danish, Norwegian)
120. old maid (H; English)
121. ox-fly (H; English)
122. pilu (or pilyu) (*Oedemagena tarandi*, USSR)
123. pishchevodnik (*H. lineatum*; Russian)
124. Premo (H; ancient German)
125. Primissa (H; ancient German)
126. rèla-bu (H; France: Haute-Savoie)
127. Rinderbreme (H; Swiss)
128. runderhorzel (H; Flemish)
129. runderhorzelplaag (H; Dutch, Flemish)
130. Schafbremse (O; German)
131. Schafenger (O; German)
132. schapenbremse (O; Dutch)
133. schapenhorzel (O; Dutch)
134. severnyi nosoglotochnik (*Cephenemyia trompe*; Russian)
135. severnyi podkozhnik (*Oedemagena tarandi*; Russian)
136. souffle (H; Switzerland: Canton de Murat)
137. stout (H; English)
138. strechie (H; Rumanian)
139. střeček (H; Czech)
140. surra (H; France: Davos)
141. syanu (*Cephenemyia trompe*; USSR)
142. tafano (H; Italian dialect)
143. tagueneau (H; French dialect)
144. takouk (H; Arabic)
145. tantarina (H; France: Rhône, Loire)
146. taon (H; French)
147. taon des tanneurs (H; Canadian French)
148. tarlo (H; Italian dialect)
149. tarolo (H; Italian dialect)
150. torcel (D; Central America; GRUBE, 1860; SANTAMARIA, 1942)
151. tórsalo (D; same as above)
152. trumpet-fly (O; English)
153. tupe (D; Ecuador)
154. Ura (D; Brazil, Argentina)
155. varon (H; French Switzerland)
156. varron (H; French)
157. Währenübel (H; Swiss)
158. wame (H; English)
159. warble-fly (Hypodermatidae; English)

(ii) Names applied to the larvae

1. Aengerling (H; German)
2. barros (H; Spanish)
3. beef-worm (D; Guatemala; BEQUAERT, 1938)
4. bekuru (D; Kaingang Indians of Brazil; VAL FLORIANA, 1920)
5. berne (H in Portugal; D in Brazil)
6. berro (H; Portugal)
7. Biesmandl (H; German)
8. Bieswurm (H; German)
9. bijspauw (H; Flemish)
10. bijsworm H; Dutch)
11. bikuru (D; Kaingang Indians of Brazil; VAL FLORIANA, 1920)

12. Bisendewurm (H; ancient German)
13. Bisewurm (H; ancient German)
14. borro (D; Bolivia; BOLLE, 1958)
15. boulèno (H; France: Gard)
16. bot (English)
17. Buckelwurm (H; German)
18. cattle-grub (H; English)
19. colmoyote (D. Mexico and Guatemala; from the Náhuatl *ocuilin* = worm, and *moyotl* = fly; (o)cuil(in)-moyotl = worm of the mosquito; SANTAMARIA, 1959)
20. coss (H; Romanche)
21. dlam (H; Arabic)
22. Engerling (H; German)
23. garbou (O; France: Bigorre)
24. gatta (H; Italian dialect)
25. gattina (H; Italian dialect)
26. gorm (*Oedemagena tarandi*; Chukchi dialect)
27. grub (English)
28. gusano de monte (D; Central America, Ecuador, Colombia, and Venezuela)
29. gusano de mosquito (D; Venezuela)
30. gusano de zancudo (D; Venezuela)
31. gusano macaco (D; Venezuela)
32. gusano peludo (D; Bolivia)
33. herboz (H; France: Fribourg)
34. huid-larve (H; Dutch)
35. ikitugn (D; Kaingang Indians of Brazil; VAL FLORIANA, 1920)
36. kitudn (D; Kaingang Indians of Brazil; VAL FLORIANA, 1920)
37. kozii podkozhnik (*H. silenus*; Russian)
38. kturn (D; Kaingang Indians of Brazil; VAL FLORIANA, 1920)
39. kummak (*Oedemagena tarandi*; Ahearmiut dialect)
40. liarches (H; France: Lyon)
41. Made (H; German)
42. maggot (English)
43. môlon (H; France: Quarouble, Nord)
44. morèle (H; France: Deux-Savoies)
45. moyocuil (D; Mexico; from the Náhuatl *moyotl* = mosquito, and *ocuilin* = worm; moy(otl)ocuil(in) = mosquito-worm)
46. mulot (H; France: Dagny-Lambercy, Lambercy, Aisne)
47. muskietenworm (D; Surinam)
48. nuche (or nunche) (D; Colombia, Venezuela)
49. obyknovennyi podkozhnik (H; Russian)
50. ouarbô (H; Wallon dialect)
51. ouarpé (H; Wallon dialect)
52. ouèrbâ (H; Wallon dialect)
53. ovechii ovod (O; Russian)
54. ovod (Russian)
55. ovod korotysh (*Rhinoestrus latifrons*; Russian)
56. ovod maloshii (*Rhinoestrus usbekistanicus*; Russian)
57. Piswurm (H; German dialect)
58. proof-maggot (H; English dialect)
59. sërène (H; France: Berry)
60. suglacuru (or suylacuru) (D; Mayna Indians, South America; LA CONDAMINE, 1745)
61. taon (H; French)
62. teter (H; Flemish)
63. ton (H; Ancient French, cf. d'Yanville, 1788: *Traité de Vénerie*, p. 412)
64. ura (D; Tupi-Guarani tribes; Brazil, Argentina; cited already in 1629 by MONTOYA)
65. vaèy'naviâ (H; France: Jura)
66. varan (H; France: Annecy)
67. varanbon (H; France: Annecy)
68. varnavoué (H; France: Haute Savoie)
69. varon (H; French)
70. varoun (H; France: Boulogne-sur-Mer)
71. varron (H; France)
72. ver bouvier (H; French)
73. ver du boeuf (H; French)
74. véranbí (H; France: Belfort)
75. verblé (H; France: Pas-de-Calais, Saint-Pol)
76. vere (H; Swedish)
77. verm (H; Italian dialect)
78. voere (H; Danish, Norwegian)
79. voerebolla (H; Danish, Norwegian)
80. Waernegel (H; Anglo-Saxon)
81. Werren (H; (Swiss))
82. wolve (H; English)
83. wormil (H; English dialect)
84. wormul (H; English dialect; cf. *Trans. Linn. Soc.* 1797: 297)
85. wornail (H; Dialect of Dorset)
86. wornil (H; Dorset)
87. wornul (H; English dialect; *Trans. Linn. Soc.* 1797: 297)

(iii) Names applied to the tumors caused by the larva of *Hypoderma*

1. adultamientos (Spanish)
2. Beulen (Swiss, German)
3. bosse (French)
4. botta (Italian dialect)
5. boussouflure (French)
6. bügna (Italian dialect)
7. Dassel (German)
8. Dasselbeule (German)
9. gosp (English)
10. grub (English)
11. horzelbulten (Dutch)
12. knobbels (Flemish)
13. medranças (Portuguese)
14. territt (France: Landes)
15. teter (Flemish)
16. vairon (Italian dialect)
17. vara (Danish, Norwegian)
18. varbulde (Ancient Swedish)
19. varbylden (Danish, Norwegian)
20. verblé (France: Pas-de-Calais, Saint Pol)
21. vere (Danish)
22. vermon (Italian dialect)
23. währnen (Swiss)
24. warble (English)
25. warble-grub (English)
26. Weenbulle (German)
27. Wehrlen (Swiss)
28. woru (Danish, Norwegian)

(iv) Words relating to the terror caused by *Hypoderma* to cattle

1. asiar (Italian dialect)
2. bisâhe (French dialect)
3. bisen (German)
4. biser (French dialect)
5. breskenn (Breton)
6. breskign (Breton)
7. bresquennein (Vannetais)
8. bresquigner (Vannetais)
9. bziner (Wallon; by extension, 'une bzinoire' = a girl that walks at night, of bad habits)
10. djilà (Italian dialect)
11. jilar (Italian dialect)
12. mouche (la) (French dialect; by extension, 'mouche' = somebody in a hurry)
13. peudre (French dialect)
14. peudreen (French dialect)
15. peuren (Vannetais)
16. ouider (Breton)
17. siölar (Italian dialect)

3. Folklore

The following stories and popular sayings were extracted from ROLLAND (1881, 1911):

1. 'Terre d'oestres' – sandy, dry land.
2. 'Une mouche bovine' – a spy.
3. 'Il remue la tête comme s'il avait le *bouin*' (France: Vienne) – a restless individual.
4. 'Le *tagueneau* est plus fort que le taureau' (proverb of Ille-et-V.).
5. 'Quand la *nonne* s'élève haut dans les airs pour jouer c'est signe de beau temps pour le lendemain' (France: Côte d'Or, Semur).
6. 'Quand les *laguerais* piquent les chevaux c'est signe de pluie' (France: Côte d'Or).
7. When a shepherd wants to chase away a cow that invaded a neighbour's pasture, he only has to shout: '*Moucha tantarina*, pica la bovina' (France: Loire) for the cow to run away immediately.

8. As to origin of the Vannetais words *peudreen* and *peudre* (from *poudre* = powder), and of the expression '*la mouche*', ROLLAND (1911: 178) gives the following story:

'On appelle *la mouche* la terreur panique, l'épouvante subite qui parfois, dans nos foires, s'empare, sans cause apparente, de la masse entière du bétail et la rend furieuse. Le cri sinistre: *La mouche! La mouche!* retentit aussitôt partout; les hommes s'agitent aussi et frappent de leurs bâtons sur les cornes des animaux pour les contenir. S'ils n'y parviennent, boeufs et vaches se précipitent au hasard, renversant tout ce qui se trouve sur leur passage. Il en résulte une confusion inexprimable, et souvent les accidents les plus graves en sont la suite. Dans beaucoup des lieux, les paysans attribuent *la mouche* à la poudre [whence *peudre, peudreen*] de foie de loup, que les voleurs répandent afin de profiter du désordre pour faire leur coup' (France: Berry).

In the region of Bretagne (France) a different explanation for the same occurrence is given:

'On croit que ces étranges agitations sont produites par une énorme *mouche* noire, grosse comme un courbeau, qui traverse tout à coup les airs avec un bourdonnement épouvantable.'

9. As to the origin of the term *peuren* (sudden terror of horses), ROLLAND (*op. cit.*) reports the following story, extracted from the poem 'Foér Voriadec (La Foire aux chevaux)', VANNES, no date, pp. 22-24:

'La chose (est) attribuée à un coup de feu mystérieux se faisant entendre à midi juste. Pour mettre fin au désordre, il n'y a qu'un moyen... mettre en croix deux manches de fouet. Dès que ce signe est fait, les animaux se radoucissent et reprennent leur calme.'

III. BIOLOGY OF THE OESTROIDEA —
THE PHENOMENON OF AGGREGATION

Both Cuticolae and Cavicolae depend, during the larval stage, entirely upon their hosts, and are dispersed by the mammals as they move around. Consequently, when dropping to the ground to pupate, the mature larvae of one population will be scattered over a more or less large area. Upon emergence, adults will appear in different areas, in most cases very distant from each other.

Now several problems arise: the adults have highly reduced mouth-parts (Fig. 1) and do not feed, only licking up a few drops of water (CATTS & GARCIA, 1963); their energy, stored during the larval and pupal stages, is to be used in flying, mating, and, in the case of females, in developing the eggs or the larvae and in the search for a suitable host. The situation is further complicated by the fact that oestroids have a very short adult life and their populations are relatively small. The small number of individuals of a single population in a large area would render the meeting of the sexes difficult.

To ensure reproduction against those several drawbacks – limited number of individuals, scattered distribution, short life span, limited amount of non-renewable energy – a mechanism was developed, the aggregation phenomenon.

As soon as they emerge from the puparia, males and females fly towards the highest place in their range, generally to high mountain peaks with trees. There males wait for the females, remaining at the same site to the end of their lives. The females come to this spot, are fertilized, fly immediately downwards and search for a suitable host.

BRAUER (1863) was the first to observe the aggregation habit among males of *Cephenemyia*, correlating this behaviour with copulation; the fact was later verified and corroborated by MÖSCHLER (1935) and ULLRICH (1939). Aggregation was verified among all groups of Oestroidea (GRUNIN, 1959; CATTS, 1964), except for the Rutteniidae.

Early observers believed that only males aggregated in certain sites. These authors were led to this conclusion because either only males were collected (BOIE, 1850; STECK, 1932; TOWNSEND, 1935; THOMANN, 1947; CHAPMAN, 1954), or because females were captured in very low percentages (2-10%). In spite of this, most authors were unanimous in considering aggregation as a reproductive activity; only ULLRICH (1939) denied that possibility, notwithstanding his observation of four mating pairs of *Cephenemyia ulrichi* Brauer at a site where males of this species were gathering. As noted by GRUNIN (1959), the apparent rarity of females in aggregation sites is only due to their flying away soon after fertilization, whereas the males remain there to the end of their lives. MÖSCHLER (1935), LINDNER (1954) and GANSSER (1956, 1957) suggested making use of this phenomenon for the control of cattle pests with insecticides.

Aggregations take place especially during the hottest hours of the morning (GRUNIN, 1959; CATTS, 1964; GUIMARÃES, 1966). The males gathered at the highest place in the locality are generally very mobile, with the exception of *Rhinoestrus purpureus, Cephalopina titillator* and *Oedemagena tarandi*. If the weather is good, males fly for long periods from one place to the other, swarming in the air or frequently alighting on rocks. The activity of the males depends not only on the weather but also on the total number of individuals present in the area. For example, when there are two or more individuals of *Cephenemyia ulrichi* they keep flying; a single male remains quiet, in the same place (MÖSCHLER, 1935). It seems that the agitation of the male is necessary to ensure copulation – in experimental conditions both *Hypoderma bovis* and *H. lineatum* mated successfully when forced to fly (WEINTRAUB, 1956, 1961). The males are powerful fliers and in sunny weather their activity does not cease even under strong winds. With weak or moderate winds their animation is maximal (BRAUER, 1863; MÖSCHLER, 1935; ULLRICH, 1939). CATTS (1964: 580) found maximum male activity at temperatures between 20 and 30° C.

Males assume a characteristic 'waiting attitude' (GRUNIN, 1959: fig. 2; CATTS, 1964: 580, fig. 1) and try to chase any other insect or object of about their own size passing in front of them.

The number of individuals present at a site varies greatly – THOMANN (1947) observed an exceptional case of thousands of individuals of *Cephenemyia stimulator* in one place; GRUNIN (1959) captured 11 males of *Oestromyia marmotae* and apparently exhausted the local population; the maximum number of males of *Hypoderma bovis* found at a single site is 40 (GANSSER, 1956, 1957).

However, not all oestroids aggregate in high places. Some congregate in plains, near the lairs of the hosts or on low bushes (CAPELLE, 1970). Some, like *Oestrus ovis*, form small aggregations of males and females in crevices of walls of stone or timber houses (GRUNIN, 1959). *Hypoderma bovis* and *H. lineatum* favour pastures, plains, and mountain slopes (BOIE, 1850; TIEF, 1888; STECK, 1932; KÜHL, 1949; GANSSER, 1956, 1957; GRUNIN, 1959) and prefer four types of environment: (i) trails along pastures, over sunlit stones; (ii) sandy or clayish soils, near rivers and creeks; (iii) well illuminated crevices protected from the wind, during the early part of the day; and (iv) cattle-trodden areas, with abundant cattle dung; as pointed out by GANSSER (1957), *Hypoderma* is not always found in places apparently suitable for aggregation. Finally, *Oestromyia leporina* prefers abrupt mountain slopes with sparse rocks inhabited by its host, *Ochotona alpina* Pallas (GRUNIN, 1949).

Regardless of its several variables, the aggregation habit is a constant among the Oestroidea. On the other hand, this mechanism brings two disadvantages (CATTS, 1964): (i) not all the sites in one area may be occupied by both sexes; (ii) individuals concentrated at one site are more vulnerable to predators.

These two disadvantages are counterbalanced by two other mechanisms: intraspecific chasing (CATTS, 1964) and mimetic resemblance.

1. Intraspecific chasing

The first males to arrive at a certain site of aggregation chase other males of the same species, which are then forced to look for unoccupied sites. By this simple mechanism the population becomes more dispersed, simultaneously ensuring occupation of several sites and avoiding heavier predation. This process was discovered in *Cephenemyia jellisoni* and *C. apicata* and very aptly described by CATTS (1964: 580-581):

'After landing, males adopted a characteristic "waiting attitude". From such perches males flew in rapid pursuit of any passing object of about their own size. This response was useful in making observations; tossing stones near waiting flies caused them to chase the stone and reveal their presence.'

'Certain branches appeared to be selected as landing sites. Males chased stones on 20-35 consecutive tosses and after each flight returned to one of several landing sites. In pursuit, the fly followed the stone until the latter dropped into distant vegetation. Then the fly would return to a previously occupied landing site. Most chasing was intraspecific, with flies in sustained pursuit of each other for 2-3 minutes. Generally, from below, the flight pattern appeared to be in the shape of a narrow figure-of-eight curved to fit the margin of the aggregation tree. Waiting males perched close to each other (within 6 in.), but when one took flight, all waiting males at the site entered into the chase and flew rapidly around the perimeter of the tree. At the end of the chase, flies landed in rapid succession until all had resumed the waiting attitude on the tips of branches.'

'Chasing appeared to limit the total number of male flies at an aggregation site. The maximal capacity for any site may be related to the dimensions of the aggregation trees. At one site, with diameter and height of 25 and 10 ft respectively, a maximum of five males was recorded repeatedly. At a smaller tree, with diameter and height of 5 and 6 ft respectively, not more than two males occurred simultaneously. After males were captured and removed from a site, the void generally was filled within 20 minutes by new flies.'

'Marked males tended to remain at a specific site throughout one day's activity and

33

usually were seen at the same site on later days. The longest period between marking and final resighting of any male was six days (average 4.4 days). The total marked and released was 37.22% of these being resighted on later dates.'

'As bot fly populations increased, there was a corresponding increase of intraspecific chasing at aggregation sites. During such periods a few solitary male flies circled or landed near the observer standing or moving on the hilltop.'

'When both species were active, intrageneric chasing occurred. *C. jellisoni* seemed to dominate aggregation sites at summit locations. When both species (*C. jellisoni* and *C. apicata*) were at the same site, their activities were stratified; *C. jellisoni* flew about the upper branches and *C. apicata* flew only about the base of the site. Where surrounding vegetation limited the fly activity to upper branches only, *C. jellisoni* was seen repeatedly to chase away males of *C. apicata*. In flight, males of the darker species, *C. jellisoni*, appeared bigger and had a deeper sounding flight tone than males of *C. apicata*. Actually, there was no consistent measurable difference in the gross size (wing span and body dimension) of a series of ten males of each species.'

2. Mimicry

This is another mechanism which probably comes into play to prevent heavy predation upon aggregated individuals. Adults of several oestroids resemble certain wasps and bees. PORTSCHINSKY (1906-1915, 1913) called attention to the fact that adults of *Oestrus ovis* look like small stones or bird droppings, and reported similar observations for *Rhinoestrus purpureus*. Unfortunately, there are very few data on this topic but a note should be given to interest specialists on this subject.

From what is known about the aggregation phenomenon, two conclusions can be drawn: 1. This mechanism has probably appeared several times independently in the course of evolution of the several groups of Oestroidea, as indicated by its several modalities. It is also known to occur among other groups of insects, such as Hymenoptera and Lepidoptera (HUDSON, 1906; RICHARDS, 1927; CHAPMAN, 1954; DODGE & SEAGO, 1954; MELL, 1956), and several families of Diptera (for example, in the Gasterophilidae, BRAUER, 1863; WALTON, 1930; ULLRICH, 1939; GRUNIN, 1959). Even among males and females of the apterous cerambycid beetle, *Dorcadion hispanicum* Muls., this phenomenon was observed (POULTON, 1904). 2. Sight must play a fundamental role in the adult life of oestroids, as demonstrated by their efficiency in locating aggregation sites. This fact must be kept in mind, as it will be used in the discussion of the evolution of this group in later sections.

Species reported in the literature as forming aggregations are the following:

1. Cuterebridae
    1. *Dermatobia hominis* (Linnaeus Jr.) – GUIMARÃES, 1966

2. *Pseudogametes semiater* (Wiedemann) – LUTZ, 1917

3. *Cuterebra polita* Coquillett (= *americana* Fabr.) – CAPELLE, 1970

2. Hypodermatidae

1. *Oedemagena tarandi* (Linnaeus) – HADWEN, 1926

2. *Hypoderma bovis* (Linnaeus) and *H. lineatum* (Villers) – BOIE, 1850; TIEF, 1888; STECK, 1932; KÜHL, 1949; GANSSER, 1956, 1957; GRUNIN, 1959; CATTS, GARCIA & POORBAUGH, 1965

3. *Oestromyia leporina* (Pallas) – GRUNIN, 1949

4. *Oestromyia marmotae* Gedoelst – GRUNIN, 1959

3. Oestridae

1. *Pharyngomyia picta* (Meigen) – BRAUER, 1863; ČEPELÁK, BUČEK & MANDELÍK, 1972

2. *Cephenemyia stimulator* (Clark) – ZELLER, 1842; BRAUER, 1863; ULLRICH, 1939; THOMANN, 1947

3. *Cephenemyia ulrichi* Brauer – MÖSCHLER, 1935; ULLRICH, 1939

4. *Cephenemyia trompe* (Modeer) – HADWEN, 1926; POLYAKOV, 1965

5. *Cephenemyia auribarbis* (Meigen) – ČEPELÁK, BUČEK & MANDELÍK, 1972

6. *Cephenemyia apicata* Bennett & Sabrosky – CATTS, 1964

7. *Cephenemyia jellisoni* Townsend – CHAPMAN, 1954; CATTS, 1964

8. *Cephalopina titillator* (Clark) – GRUNIN, 1959

9. *Rhinoestrus purpureus* (Brauer) – BRAUER, 1863; GRUNIN, 1959

10. *Oestrus ovis* Linnaeus – GRUNIN, 1959.

IV. REFERENCES

ARTURE (–), (1753): Observations sur l'espèce de ver nommé Macaque. *Mém. Acad. R. Sci. Paris* 1752: 72-73.

AUSTEN, E. E., (1895): On the specimens of the genus *Cutiterebra* and its allies (Family Oestridae) in the collections of the British Museum, with description of a new genus and three new species. *Ann. Mâg. nat. Hist.* (6)15: 377-396, 1 pl.

AUSTEN, E. E., (1930): On a new dipterous parasite (Family Calliphoridae, subfamily Calliphorinae) of the Indian Elephant, with notes on other dipterous parasites of elephants. *Proc. zool. Soc. Lond.* 1930: 677-688, 2 figs.

BAILLY, M. A., (1894): Dictionnaire Grec-Français, xxxii + 2227 pp. Hachette, Paris.

BARRAILLIER, E., (1892): Viaje a Andamarca y Pangoa. Fechado en Jauja a 22 de noviembre de 1892. *Boln Soc. geogr. Lima* 2:?

BARTHÉLEMY-SAINT HILAIRE, J. (ed.), (1883): Histoire des animaux d'Aristote, 1: cclxxiv + 314 pp. Hachette, Paris.

BAU, A., (1906): Diptera, Fam. Muscaridae, Subfam. Oestrinae. *Genera Insect.* 43: 1-31, pl.

BAU, A., (1929): Versuch einer Teilung der Gattung *Cuterebra* (Diptera, olim Oestridae) in vier Untergattungen. *Zentbl. Bakt. ParasitKde.* 77(25-26): 542-544, 1 fig.

BECKH, H. (ed.), (1895): Geoponica sive Cassiani Bassi scholastici De Re rustica Eclogae, xxxvii + (1) + 641 pp. B. G. Teubner, Lipsiae.

BENNETT, C. F. & C. W. SABROSKY, (1962): The Nearctic species of the genus *Cephenemyia* (Diptera, Oestridae). *Can. J. Zool.* 40: 431-448, 7 figs., 1 pl., 1 table.

BEQUAERT, J. C., (1938): Notes on the Arthropoda of medical importance in Guatemala. *Publs Carnegie Instn* 499: 223-228.

BERGIER, E., (1941): Peuples entomophages et insects comestibles. Études sur les moeurs de l'homme et de l'insecte, 231 pp., illus. Imprimérie Rullière Frères, Avignon.

BERTONI, M. S., (1926): La civilización guarani, 3 (Conocimientos. La higiene guarani y su importancia cientifica y practica. La medicina guarani. Conocimientos cientificos). Asunción.

BEZZI, M. & P. STEIN, (1907): Schizophora, Eumyidae. Schizometopa. In: T. BECKER, M. BEZZI, K. KERTÉSZ, & P. STEIN, Katalog der paläarktischen Dipteren, 3: 189-828.

BLANCHARD, R., (1896): Contribution à l'étude des diptères parasites (3e. série). *Annls Soc. ent. Fr.* (7)6: 641-677, pl. 17, figs. 10-13.

BODENHEIMER, F. S., (1928): Materialen zur Geschichte der Entomologie bis Linné, 1: x + 498 pp., 155 figs. W. Junk, Berlin.

BODENHEIMER, F. S., (1960): Animal and man in Bible Lands, 232 pp. Collection de Travaux de l'Académie Internationale d'Histoire des Sciences (n° 10). E. J. Brill, Leiden.

BOIE, F., (1850): Entomologische Beiträge. *Stettin. ent. Ztg* 11(1): 29-32, 212-216, 359-360.

BOLLE, W. R., (1958): Neue Wege zur Bekämpfung der *Dermatobia hominis* mit Neguvon und Assuntol. *Vet.-med. Nachr.* 4: 193-206. (Also published in 1960: Nuevos caminos en la lucha contra la *Dermatobia hominis* con Neguvon y Asuntol. *Revta vet. venez.* 51 (9): 255-257).

BRAUER, F. M., (1858): Neue Beiträge zur Kenntnis der europäischen Oestriden. *Verh. zool.-bot. Ges. Wien* 8: 449-470, pl. 10, figs. 1-8, pl. 11, figs. 1-7.

BRAUER, F. M., (1860): Neue Beiträge zur Kenntnis der europäischen Oestriden. *Ibid.* 10: 641-658.

BRAUER, F. M., (1863): Monographie der Oestriden, 292 pp., illus. Wien.

BRAUER, F. M., (1866): *Pharyngobolus africanus*, m. Ein Oestride aus dem Rachen des afrikanischen Elephanten. Nachtrag zur Monographie der Oestriden. *Verh. zool.-bot. Ges. Wien* 16: 879-883, pl. 19, figs. 1-2.

BRAUER, F. M., (1883): Zweiflügler des kaiserlichen Museums zu Wien. III. *Denkschr. Akad. Wiss., Wien (Math.-nat. Kl.)* 47(1): 1-100.

BRAUER, F. M., (1886): Nachträge zur Monographie der Oestriden. I. Ueber die von Frau A. ZUGMAYER und Hrn. F. WOLF entdeckte Lebensweise des *Oestrus purpureus*. *Wien. ent. Ztg* 5: 289-304.

BRAUER, F. M., (1887): Nachträge zur Monographie der Oestriden. II. Zur Charakteristik und Verwandtschaft der Oestridengruppen in Larven und volkommenen Zustande. *Ibid.* 6: 4-16.

BRAUER, F. M., (1892): Ueber die aus Afrika bekannt gewordenen Oestriden und insbesondere über zwei neue von Dr. HOLUB aus Südafrika mitgebrachten Larven aus dieser Gruppe. *Ber. Akad. Wiss., Wien (Math.-nat. Kl.)* 101(1): 4-16, pl. 1.

BRAUER, F. M. & J. VON BERGENSTAMM, (1889): Die Zweiflügler des kaiserlichen Museums zu Wien. IV. Vorarbeiten zu einer Monographie der Muscaria Schizometopa (exclusive Anthomyidae). Pars. I. *Denkschr. Akad. Wiss., Wien (Math.-nat. Kl.)* 56(1): 69-180, 11 pls.

BUTLER, S., (1952): The Iliad of Homer and the Odyssey. In: HUTCHINS, R. M. (ed.), Great Books of the Western World, 4: vii + 322 pp. Enc. Britannica Inc., Chicago.

CAPELLE, K. J., (1970): Studies on the life history and development of *Cuterebra polita* (Diptera: Cuterebridae) in four species of rodents. *J. med. Ent.* 7: 320-327.

CATTS, E. P., Jr., (1964): Field behavior of adult *Cephenemyia* (Diptera: Oestridae). *Can. Ent.* 96: 579-585, 7 figs.

CATTS, E. P., Jr., & R. GARCIA, (1963): Drinking by adult *Cephenemyia* (Diptera: Oestridae). *Ann. ent. Soc. Am.* 56(5): 660-663.

CATTS, E. P., Jr., R. GARCIA, & J. H. POORBAUGH, (1965): Aggregation sites of males of the common cattle grub, *Hypoderma lineatum* (de Villers) (Diptera Oestridae). *J. med. Ent.* 1: 357-358.

ČEPELÁK, J., G. BUČEK, & D. MANDELÍK, (1972): [Further observations on the biology of warble flies in the area of Kováčovo Hills (Diptera, Oestridae)]. *Polnohospodárstvo* 18(9): 812-823, 6 figs., 5 tables. (In Czech).

CHAPMAN, A., (1954): Studies on summit frequenting insects in western Montana. *Ecology* 35: 41-49.

CLARK, B., (1797): Observations on the genus *Oestrus*. *Trans. Linn. Soc. Lond.* 3: 289-329, pl. 23.

CLARK, B., (1815): An essay on the bots of horses and other animals, 72 pp., 2 col. pls. London.

CLARK, B., (1827): On the insect called *oistros* by the ancients, and of the true species intended by them under this appellation: In reply to the observations of W. S. MACLEAY, Esq., and the French naturalists. *Trans. Linn. Soc. Lond.* 15(2): 402-411.

COLERIDGE, E. P., (1952): The plays of Euripides, pp. 199-447. In: HUTCHINS, R. M., (ed.), Great Books of the Western World, 5 (*Aeschylus, Sophocles, Euripides, Aristophanes*), ix + 649 pp. Enc. Britannica Inc., Chicago.

COOKSON, G. M., (1952): The plays of Aeschylus, pp. 1-9 *Ibid.*

CORTI, E., (1895): Aggiunte alla fauna ditterologica della Provincia di Pavia. *Boll. Soc. ent. ital.* 26 (1894): 389-395.

CURTIS, J., (1826): British entomology: Being illustrations and descriptions of the genera of insects found in Great Britain and Ireland, 3: pls. 99-146. London, '1823-1840'.

DE GEER, C., (1776): Mémoires pour servir à l'histoire des insectes, 6: 523 pp., 30 pls. Stockholm.

DELEGORGUE, A., (1847): Voyage dans l'Afrique australe, 1838-1844, 1: 16 + 580 pp.; 2: 624 pp., 6 portraits, 2 maps. Paris.

DODGE, H. R. & J. M. SEAGO, (1954): Sarcophagidae and other Diptera taken by trap and net on Georgia mountain summits in 1952. Ecology 35(1): 50-59, 4 figs., 5 tables.

ENDERLEIN, G., (1911): Neue Gattungen und Arten aussereuropäischen Fliegen. Stettin. ent. Ztg 72: 135-209.

FIGUIER, L., (1875): Les Insectes, 585 pp., 619 figs. Hachette, Paris.

FISCHER, J. L., (1787): Observationes de oestro ovino atque bovino factae, 69 pp., 4 col. pls. Lipsiae.

FROGGATT, W. W., (1913): The kangaroo bot fly, Oestrus macropi, sp. n. Agric. Gaz. N.S.W. 24(7): 565-568, 1 pl.

GABUS, J., (1956): Os esquimós. Nas margens da Baía de Hudson, 319 pp., illus. Livraria Tavares Martins, Porto.

GANSSER, A., (1951): Dasselfliegen. Biologie, Schäden und Bekämpfung von Oestriden, mit besonderer Berücksichtigung schweizerischer Verhältnisse, 128 pp., 62 figs. Basel.

GANSSER, A., (1956): Warble flies and other Oestridae, biology and control, 63 pp. 48 figs. The Hide and Allied Trades Improvement Society, Surrey.

GANSSER, A., (1957): Zur Biologie der Dasselfliege und der Bekämpfung der Dasselplage durch Abfangen der Dasselfliegen. Schweizer Arch. Tierheilk. 99(1): 17-27.

GEDOELST, L., (1914): Note sur un genre nouveau d'oestride. Bull. Soc. Path. exot. 7(3): 210-212.

GIL FERNANDEZ, L., (1959): Nombres de insectos en griego antiguo, xii + 263 pp. Consejo Superior de Investigaciones Cientificas, Instituto 'Antonio de Nebrija', Manuales y Anejos de 'Emerita', vol. 18. Madrid.

GIRSCHNER, E., (1893): Beitrag zur Systematik der Musciden. Berl. ent. Z. 38: 297-312, 3 figs.

GIRSCHNER, E., (1896): Ein neues Musciden-System auf Grund der Thoracalbeborstung und der Segmentierung des Hinterleibes. IIIte Z. Ent. 1: 12-16, 30-32, 61-64, 105-112.

GRUBE, E., (1860): Beschreibung einer Oestridenlarve aus der Haut des Menschen. Arch. Naturgesch. 26: 9-16, pl. 1, figs. 4-5.

GRUNIN, K. YA., (1948): Ovod (Oestrus caucasicus, sp. n.), parazitiruyushchnii na dagestanskom ture (Capra cylindricornis Blyth.). (A botfly (O. caucasicus) parasitic on Capra cylindricornis). Dokl. Akad. Nauk. SSSR (n.s.) 61(6): 1125-1127, fig.

GRUNIN, K. YA., (1949): (On a physiological peculiarity of hypodermal bot flies of the genus Oestromyia Br.). Ibid. 67: 193-196.

GRUNIN, K. YA., (1950): Lichinki 1. stadii sem. Oestridae i Hypodermatidae i ikh znachenie dlya uscanovleniya filogenii. (First stage larva of Oestridae and Hypodermatidae and their influence on phylogeny). Parazit. Sb. 12: 225-271, figs., pls.

GRUNIN, K. YA., (1951): Stroenie perednikh dykhalets lichinok ovodov. (Sructure of the anterior spiracle of larval bot flies). Ent. Obozr. 31(3-4): 463-466, figs.

GRUNIN, K. YA., (1953): Lichinki ovodov domashnik zhivotnikh SSSR. (Larvae of Oestridae from domestic animals in the USSR). In: Akad. Nauk SSSR, Oprideliteli po Faune SSSR, 51: 1-124, 139 figs. Moscow-Leningrad.

GRUNIN, K. YA., (1957): Nosoglotochnie ovoda (Oestridae). In: Zool. Inst., Akad. Nauk SSSR, Fauna SSSR. Nasekomye Dvukrylye 19(3): 1-145, 230 figs., 5 tables. (=Novaya Serya, n° 68). Leningrad.

GRUNIN, K. YA., (1959): (Aggregation of bot fly males on the highest points in the locality and its cause). *Zool. Zh.* 38: 1683-1688, 3 figs.

GRUNIN, K. YA., (1961): O lichinke nosoglotochnogo ovoda kenguru (*Tracheomyia macropi* Frog.) (Diptera, Oestridae) iz Avstralii. (On the larva of the kangaroo nasal bot fly, *Tracheomyia macropi* Frog., from Australia). *Ent. Obozr.* 40(4): 929-933, 10 figs.

GRUNIN, K. YA., (1962): Podkozhnye ovoda (Hypodermatidae). In: *Inst. Zool., Akad. Nauk SSSR. Fauna SSSR. Nasekomye Dvukrylye* 19(4): 237 pp., 358 figs., 6 tables. (=Novaya Serya, n° 82). Leningrad.

GRUNIN, K. YA., (1964): Hypodermatidae (Fam. 64b). In: LINDNER, E., (ed.), *Die Fliegen der paläarktischen Region,* 8: 1-40, figs. 1-52, 6 tables (=Lfg. 252), 41-72, figs. 57-172 (=Lfg. 254), 73-96, figs. 173-264 (=Lfg. 255), 97-128, figs. 265-344 (=Lfg. 257), 129-154, figs. 345-358 (=Lfg. 261). Stuttgart.

GRUNIN, K. YA., (1966): Oestridae (Fam. 64a), in: LINDNER, E., (ed)., *Die Fliegen der paläarktischen Region,* 8: 1-32, 90 figs., 4 tables (=Lfg. 264), 33-64, figs. 91-166 (=Lfg. 265), 65-96, figs. 167-240 (=Lfg. 266). Stuttgart.

GUIMARÃES, J. H., (1966): Nota sôbre os hábitos dos machos de *Dermatobia hominis* (Linnaeus Jr.) (Diptera, Cuterebridae). *Papéis Avulsos Zool. S. Paulo* 18(25): 277-279.

GUIMARÃES, J. H. & N. PAPAVERO, (1966): A tentative annotated bibliography of *Dermatobia hominis* (Linnaeus Jr., 1781) (Diptera, Cuterebridae). *Arqos Zool. S. Paulo* 14(4): 223-294. 2 pls.

HADWEN, S., (1926): Notes on the life history of *Oedemagena tarandi* L. and *Cephenomyia trompe* Modeer. *J. Parasit.* 13: 56-65. (Also 1932: Zametki ob istorii ovodov severnogo olenya *Oedemagena tarandi* L. i *Cephenomyia trompe* Modeer (Referat LYUBIMOVA, M. i S. AL'FA). *Sb. Olenev., tundr. Vet. Zootekhn.,* Moscow, pp. 296-298).

HORI, K., (1967): Comparative anatomy of the internal organs of the calyptrate muscoid flies. V. Consideration on the phylogeny of the Calyptratae. *Sci. Rep. Kanazawa Univ.* 12(2): 215-254, 1 fig.

HUDSON, G. V., (1906): Notes on insect-swarms on mountain-tops in New Zealand. *Trans. N.Z. Inst.* 38 (1905): 334-336.

JANSEN, J., Jr., (1967): On the identity of the Greek parasitic insect 'oistros'. *Ent. Ber., Amst.* 27(2): 30-36.

JAYAKAR, A. S. G., (1906-1908): Ad Damiri's Hayat al Hayawan (A zoological lexicon). 2 vols. London.

JEBB, R. C., (1952): The plays of Sophocles, pp. 95-195. In: HUTCHINS, R. M., (ed), Great Books of the Western World, 5 (Aeschylus, Sophocles, Euripides, Aristophanes), ix + 649 pp. Enc. Britannica Inc., Chicago.

JOWETT, B., (1952): The dialogues of Plato, *Ibid.* 7: vii + 814 pp.

KIRBY, W. & W. SPENCE, (1822): An introduction to entomology: Or elements of the natural history of insects, 1: xxiii + (3) + 518 pp., 3 pls. 4th. ed.). London.

KÜHL, R., (1949): Beiträge zur Biologie und Bekämpfung der Rinderdasselfliegen *Hypoderma bovis* und *Hypoderma lineatum. Anz. Schädlingsk.* 22(5): 74-78.

LA CONDAMINE, C. M. DE, (1745): Rélation abrégée d'un vovyage fait dans l'intérieur de l'Amérique Méridionale, depuis de la côte de la Mer du Sud, jusqu'aux côtes du Brésil & de la Guiane, en descendent la Rivière des Amazones. Veuve Pissot, Paris.

LATREILLE, P. A., (1818): Articles 'Oestre', 'Oestridés', and 'Oestrides', pp. 264-274. In: Société de Naturalistes et d'Agriculteurs, Nouveau dictionnaire d'histoire naturelle appliquée aux arts, à l'agriculture, à l'économie rurale et domestique, à la médecine, etc., 23: 1-612, pls. Deterville Libraire, Paris.

LEACH, W. B., (1817): On the genera and species of eproboscideous insects, 20 pp., 3 pls. Edinburgh. (Also in *Mem. Wernerian nat. Hist. Soc.* 2: 547-566, 1818).

LEACH, W. B., (1818): On the arrangement of oestrideous insects. *Mem. Wernerian nat. Hist. Soc.* 2(2): 567-568, (1817).

LEWYSOHN, L., (1858): Die Zoologie des Talmuds. Eine umfassende Darstellung der rabbinischen Zoologie, unter steter Vergleichung der Forschungen älterer und neuerer Schriftsteller. Frankfurt a.M.

LINDNER, E., (1954): Aussichstürme zur Vernichtung der Rachenbremsen des Wildes. *Aus d. Heimat, Bremerhaven* 62: 5-6, 128-132.

LINNAEUS, C., (1737): Flora Lapponica, exhibens plantae per Lapponian crescentes, secundum systema sexualis, collectas in itinere impensis Soc. Reg. Litter. et Scient. Sveciae A. 1732 instituto. Additis synonymis & locis natalibus omnium, descriptionibus & figuris rariorum, viribus medicatis & oeconomicis plurimarum, xxxvii + 372 pp., 12 pls. Amstelodami.

LINNAEUS, C., (1739): On Renarnas Brömskulor i Lapland. *K. svenska VetenskAkad. Handl.* 1: 121-132. (2nd ed., 1741, vol. 1: 119-130; republished in BRYK, 1924, part 2).

LINNAEUS, C., (1741): *Oestrus rangiferinus.* Descriptus a Carolo Linnaeo. (K. Vetensk. Soc.) *Acta Soc. R. Sci. Uppsal.* 1741: 102-115. (Also in BRYK, 1924, part 2a).

LINNAEUS, C., (1746): Fauna Svecica, sistens animalia Sveciae regni: Quadrupedia, Aves, Amphibia, Pisces, Insecta, Vermes, distributa per classes & ordines, genera & species. Cum differentiis specierum, synonymis, auctorum, nominibus incolarum, locis habitationum, descriptionibus insectorum, xxvii + 411 pp., 2 pls. Stockholm.

LINNAEUS, C., (1747): Wästgötta-Resa, på Riksens hägloflige Ständers befallning förrättad år 1746. Med anmärkningar uti Oeconomien, Naturkunnegheten, Antiquiteten, Invånarnes Seder och Lefnadssätt, med tillhörige figurer, xi + 284 + (20) pp., 5 pls. Stockholm.

LINNAEUS, C., (1749): Von den Bremsebeulen in den Häuten des Rennthier in Lappland. *K. Svenska VetenskAkad. Handl.* 1: 145-157. (2nd ed., 1768, vol. 1: 145-157).

LINNAEUS, C., (1755): Tumeurs qui se forment en Lapponie dans la peau des rennes et le remède. NOUVELLISTE OECONOMIQUE ET LITTÉRAIRE 8: 10-19.

LINNAEUS, C., (1757): De Horzel der Rendieren. Uitgezogte Verhandel, uit de nieuwste werken van de Societeiten der Wetenschappen in Europa 1: 641-660.

LINNAEUS, C., (1758): Systema naturae per regna tria naturae, secundum classes, ordines, genera, species, cum characteribus, differentiis, synonymis, locis, Ed. 10: 824 pp. Holmiae.

LINNAEUS, C., (1761): Fauna Suecica, sistens animalia Sueciae regni: Mammalia, Aves, Amphibia, Pisces, Insecta, Vermes, distributa per classes, et ordines, genera et species. Cum differentiis specierum, synonymis, auctorum, nominibus incolarum, locis natalium, descriptionibus insectorum. Editio altera, aucta, 578 pp. Stockholmiae.

LINNAEUS, C., (1762): De tumoribus in pelle rangiferorum, Curbma Lappis, ab oestro provenientibus. (K. svenska Vetensk. Akad.) *Analecta Transalpina* 1: 24-31.

LINNAEUS, C., (1767): Systema naturae per regna tria naturae, ed. 12, 1(2): 533-1327. Holmiae.

LINNAEUS, C., (1772): Mouche du renne. (K. svenska Vetensk. Akad.) *Coll. acad., Partie Étrang.* 11: 84-85.

LINNAEUS, C., In: BRYK, F., (ed.), Letter from LINNAEUS to PHILIP MILLER, Upsaliae June 20, 1737, giving an account of the *Oestrum* (sic) *Lapponum*, pp. 31-34, in his Linné's gesammelte Schriften entomologischen Inhaltes. Linné als praktischer Entomologe, 104 pp. Stockholm.

LINNAEUS, C., (1781): Mittheilungen über *Oestrus Hominis.* (In Pallas') *Neue nord. Beitr. Phys., Geogr., Erd. u. Völk.* 1: 157-158.

LIOY, P., (1864): I ditteri distribuiti secondo un nuovo metodo di classificazione naturale. *Atti Ist. veneto Sci.* (3) 9: 499-518, 569-604, 719-771, 879-910, 989-1027, 1087-1126, 1311-1352; 10: 59-84.

LONDRES, (?), (1854): On insect larvae under the human skin. *Mon. J. med. Sci.* 13: 371.

LUTZ, A., (1917): Contribuições ao conhecimento dos oestrideos brazileiros. *Mems Inst.*

Oswaldo Cruz 9: 94-112, pls. 27-29.

MACLEAY, W. S., (1825): On the insect called oistros by the ancient Greeks, and asilus by the Romans. *Trans. Linn. Soc. Lond.* 14(2): 353-359.

MELL, R., (1956): Balz bei Schmetterlingen. *Ent. Z.* 66: 241-248, 259-262.

MODEER, A., (1786): Styng-Flug-Slägtet. *K. svenska VetenskAkad. Handl.* 7: 125-158, 180-185.

MÖSCHLER, A., (1935): Beobachtungen über die Lebensweise und die Schädlichkeit der Elchrachenbremse, *Cephenomyia ulrichi* Brauer, auf der kurischen Nehrung. *Z. ParasitKde.* 7: 572-578.

MONTOYA, A. R. DE, (1629): Tesoro de la lengva gvarani, 14 + 408 pp. Iuan Sanchez, Madrid.

MOUFFET, T., (1634): Insectorum sive minimorum animalium theatrum, 326 pp., London.

MOULÉ, L., (1909): La parasitologie dans la littérature antique. I. L'oistros des grecs. *Archs Parasit.* 13: 251-264.

MOULÉ, L., (1912): Idem. III. Parasites de la peau et des tissus sous-jacents. *Ibid.* 15: 543-555.

NEFFGEN, H., (1905): Die Veterinär-Papyrus von Kahun. Ein Beitrag zur Geschichte der Tierheilkunde der alten Aegypter. Berlin.

NEIVA, A. & B. PENNA, (1916): Viajem scientifica pelo norte da Bahia, sudoeste de Pernambuco, sul do Piauhí e de norte a sul de Goiaz. *Mems Inst. Oswaldo Cruz* 8: 72-224, 1 map, pls. 1-28.

OEFELE, F. VON, (1901): Studien über altägyptischen Parasitologie. Erster Teil: Äussere Parasiten. *Archs Parasit.* 4: 481-530, 22 figs.

OSTEN SACKEN, C. R., (1893a): On the so-called Bugonia of the ancient and its relation to *Eristalis tenax*, a two-winged insect. *Boll. Soc. ent. Ital.* 25: 186-217.

OSTEN SACKEN, C. R., (1893b): Corrigendum concerning 'Bugonia'. *Entomologist's mon. Mag.* 29: 287.

OSTEN SACKEN, C. R., (1894a): The so-called Bugonia of the ancients and its relation to a bee-like fly, *Eristalis tenax. Rep. Smithson. Instn* 1893: 487-500.

OSTEN SACKEN, C. R., (1894b): On the oxen-born bees of the ancients (*Bugonia*) and their relation to a bee-like fly, *Eristalis tenax*, xiv + 80 pp. J. Heerning, Heidelberg.

OSTEN SACKEN, C. R., (1895): Additional notes in explanation of the Bugonia of the ancients, 23 pp., 3 figs.

PALLAS, P. S., (1776): *Spicilegia zoologica*, 11: 1-86. Berolini.

PALLAS, P. S., (1777): Idem, 12: 1-71. Berolini.

PARAMONOV, S. J., (1958): Notes on Australian Diptera. IX. The description of the adult *Tracheomyia macropi* Frog., first endemic Australian oestrid. *Ann. Mag. nat. Hist.* (12)6: 195-199, 3 figs.

PECK, A. L. (transl.), (1965): Aristotle: Historia Animalium, 1 (Books I-III): c + 239 pp. W. Heinemann Ltd., London & Harvard Univ. Press, Cambr., Mass.

PECK, A. L. (transl.), (1967): Parts of Animals, 434 pp.

PECK, A. L., (transl.), (1970): Aristotle: Historia Animalium, 2 (Books IV-VI): vi + 414 pp.

POLYAKOV, V. A., (1965): O meste vstrechi samtsov i samek nosoglotochnogo ovoda severnogo olenya. (On the meeting place of the male and female of the reindeer botfly). *Trudy vses. nauchno-issled. Inst. vet. Sanit. Ektoparazit.* 26: 212-213.

PORTSCHINSKY, J., (1906-1915): Russkii ovod (*Rhinoestrus purpureus* Br.) parazit loshadi, vypryskivayushchii lichinok v glaza lyudei. [*Rh. purpureus*, a parasite of the horse, injecting its larva in the eyes of men]. *Trudy Byuro Ent.* 6(1): 1-44, 1906; (2): 1-41, 1908; (3): 1-47, 1915.

PORTSCHINSKY, J., (1913): Ovechi' ovod' (*Oestrus ovis* L.), ego zhizn', svoistva, sopoby bor'by i otnosheni'e eko k' chelov'ku. [*O. ovis*, its biology and relation with man]. *Ibid* 10(3): 1-63.

POULTON, E. B., (1904): A possible explanation of insect swarms on mountain tops. *Trans. R. ent. Soc. Lond.* 1904 (Proc.): xxii-xxiv.

RACKHAM, H. (transl.), (1967): Pliny, Natural History, 3 (Libri VIII-XI): ix + 611 pp. W. Heinemann, Ltd., London & Harvard Univ. Press, Cambr., Mass.

RÉAUMUR, R. A. F., (1734): Mémoires pour servir à l'histoire des insectes, 1: 4 + 654 pp., 50 pls. Paris.

RÉAUMUR, R. A. F., (1738): Idem, 4: xxxvi + 636 pp., 44 pls.

RÉAUMUR, R. A. F., (1740): Idem, 5: xliv + 728 pp., 38 pls.

REDI, F., (1668): Esperienze intorno alla generazione degl'insetti fatte da Fr. Redi e da lui scritte in una lettera all'illustrissimo Sgr. Carlo Datti, 228 pp., 29 pls. Firenze.

RHOADES, J., (1952): The poems of Virgil, In: R. M. HUTCHINS, (ed.), Great Books of the Western World, 13: vii + 379 pp. Enc. Britannica Inc., Chicago.

RICHARDS, O. W., (1927): Sexual selection and allied problems in the insects. *Biol. Rev.* 2(4): 298-360.

ROBACK, S. S., (1951): A classification of the muscoid calyptrate Diptera. *Ann. ent. Soc. Am.* 44: 327-361, 2 figs., 7 pls.

RODHAIN, J. & J. BEQUAERT, (1913): *Gedoelstia cristata* nov. gen. nov. sp. d'oestride parasite de *Bubalis Lichtensteini* au Katanga. *Rev. zool. afr.* 2(2): 171-186, 4 figs.

ROLLAND, E., (1881): Faune populaire de la France, 3 (Les reptiles, les poissons, les mollusques, les crustacés et les insectes. Noms vulgaires, dictons, proverbes, légendes, contes et superstitions), xv + 365 pp. Maisonneuve, Paris.

ROLLAND, E., (1911): Idem, 13 (Les insectes, Première Partie): 217 pp. Paris.

RUBTSOV, I. A., (1939): Novyi vid kozhnogo ovoda s kosuli i vzaimeotnosheniya blizhaishik vidov s khozuevami. *Nauchno-metodich. Zap. Glavn. Upr. po Zapovedinkam*, Moscow 5: 118-122.

SANTAMARIA, F. J., (1942): Diccionario general de americanismos, vol. 3. Pedro Robredo, México, D.F.

SCHINER, I. R., (1861): Vorläufiger Commentar zum dipterologischen Teile der 'Fauna Austriaca'. III. *Wien. ent. Mschr.* 5(5): 137-144.

SCHNABL, J., (1877): *Microcephalus*, nov. gen. Oestridarum. *Dt. ent. Z.* 21: 49-55.

SEMENOV, A., (1902): Generica quaedam nomina mutanda vel emendanda. *Russk. ent. Obozr.* 2: 353.

STECK, T., (1932): Ein eigenartiges Vorkommen der Dasselfliege (*Hypoderma bovis* L.). *Mitt. schweiz. ent. Ges.* 25(6): 206-207.

STEYSKAL, G. C., (1967): Family Pyrgotidae, In: Depto Zool., Secr. Agric., A Catalogue of the Diptera of the Americas south of the United States 56: 1-8. S. Paulo.

STONE, A., C. W. SABROSKY, W. W. WIRTH, R. H. FOOTE & J. R. COULSON, (1965): A Catalog of the Diptera of America north of Mexico. U.S. Dept. Agric. *Agric. Handb.* 276: iv + 1696 pp. Washington, D.C.

THOMANN, H., (1947): Ueber ein Massenschwärm von *Cephenomyia stimulator* Clark (Dipt.). *Mitt. schweiz. ent. Ges.* 20(4): 304-305.

TIEF, W., (1888): Beitrag zur Kenntniss der Dipterenfauna Kärnthens. 19. Jahresber. k. k. Staatsgymnasiums Villach.

TOPSELL, E., (1658): The history of four-footed beasts and serpents, 1: xiv + 818 pp.; 2: xii + pp. 819-1130, illus. London.

TOWNSEND, C. H. T., (1912): A readjustment of muscoid names. *Proc. ent. Soc. Wash.* 14: 48-53.

TOWNSEND, C. H. T., (1916): New genera and species of Australian Muscoidea. *Can. Ent.* 48(5): 151-160.

Townsend, C. H. T., (1918): New muscoid genera, species, and synonymy. *Insecutor inscit. menstr.* 6: 151-182.

Townsend, C. H. T., (1919): New genera and species of muscoid flies. *Proc. U.S. natn Mus.* 56: 541-592.

Townsend, C. H. T., (1935): Manual of Myiology, 2: 289 pp. Itaquaquecetuba.

Townsend, C. H. T., (1936): Idem, 3: 249 pp. São Paulo.

Townsend, C. H. T., (1938): Idem, 7: 242 pp. (Addenda et Corrigenda, pp. 243-246). São Paulo.

Ullrich, H., (1939): Zur Biologie der Rachenbremsen unseres einheimischen Wildes, Genus *Cephenomyia* [sic] Latreille und Genus *Pharyngomyia* Schiner. Verh. VII. int. Kongr. Ent., Weimar, 3: 2149-2171, pls.

Vallisnieri, A., (1713): Esperienze ed osservazioni intorno all'origine, sviluppi e costumi di varii insetti, 232 pp., 12 pls. Padova.

Vallisnieri, A., (1733): Opere fisico-mediche continenti un gran numero di trattati, osservazioni, ragionamenti e dissertazioni sopra la fisica, la medicina e la storia naturale, 3 vols. Venezia.

Vogelsang, E. G. & R. Martín del Campo, (1947): Parasitología de los Nahoas. *Revta Med. vet. Parasit., Caracas* 6(1-4): 47-52, 2 figs.

Walton, C. L., (1930): The occurrence of males of the horse bot fly. *NWest. Nat.* 5: 224-226.

Weintraub, J., (1956): The mating, oviposition and other activities of warble-fly adults (*Hypoderma* spp.) under laboratory conditions. Abstr. 10th int. Congr. Ent., Montreal, Sect. 12 (med. and vet. Ent.).

Weintraub, J., (1961): Inducing mating and oviposition of the warble flies *Hypoderma bovis* (L.) and *H. lineatum* (De Vill.) (Diptera: Oestridae) in captivity. *Can. Ent.* 93(2): 149-156, 10 figs.

Zeller, P. C., (1842): Dipterologische Beyträge. Zweyte Abtheilung. *Isis* (Oken's) 1842: 807-847, 1 pl.

Zumpt, F., (1957): Some remarks on the classification of the Oestridae s. lat. (Diptera). *J. ent. Soc. Sth Afr.* 20(1): 154-161, pl.

Zumpt, F., (1965): Myiasis in man and animals in the Old World, xv + 267 pp., illus. London.

PART B. SYSTEMATICS AND PHYLOGENY OF THE OESTRIDAE

The classification of the Oestridae presented here is entirely based on the literature, representing only a reinterpretation of the systems proposed by GRUNIN (1957, 1966) and ZUMPT (1965). It is therefore tentative, and will certainly be improved in the future. It tries to give a better picture of the phylogenetic trends within the family, taking into consideration the biological, geographical and ecological aspects of the different taxa. Also the accompanying illustrations are extracted from GRUNIN and ZUMPT.

Hitherto only two subfamilies were accepted: Cephenemyiinae and Oestrinae. I am proposing two more, for the genera *Pharyngobolus* and *Tracheomyia*, respectively, as both adults and larvae differ significantly from those of the *Oestrus*- and *Cephenemyia*-groups. The four subfamilies of Oestridae may be recognized by the following keys.

I. KEY TO SUBFAMILIES

(i) Adults

1. Cell R_5 open; postscutellum not developed; generally very pilose flies; median carina of antennal fovea absent or weakly developed, bare or pilose; vein M_1 with a distinct stump (vestige of M_2); Holarctic . . . . CEPHENEMYIINAE Townsend
   Cell R_5 closed at wing margin or closed and petiolate; generally bare flies; other combinations of characters 2
2. R_5 closed at wing margin, without a petiole (Fig. 9); M_1 with stump vein (vestige of M_2); antennal fovea with carina of uniform height, uninterrupted, dividing the fovea into two depressions

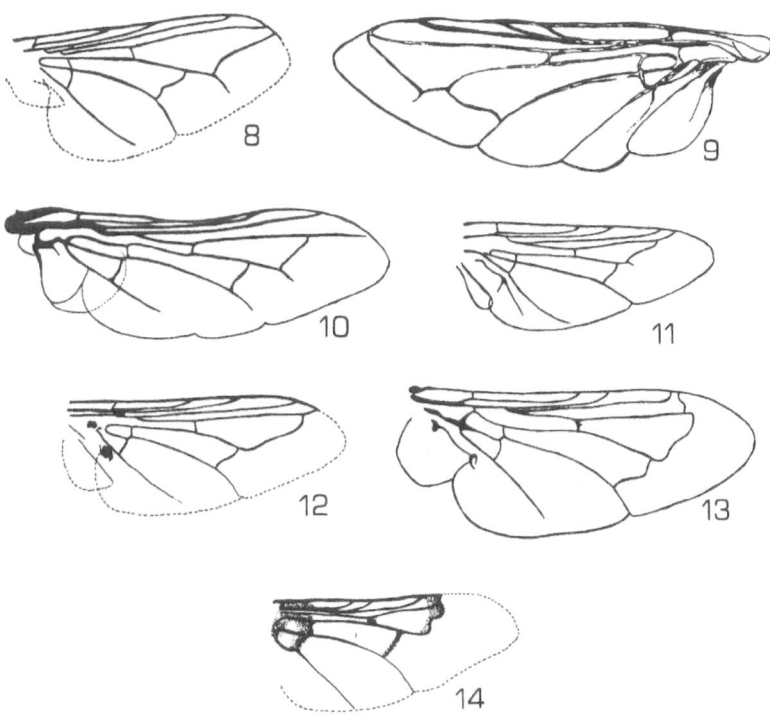

Figs. 8-14. Oestridae wings; Fig. 8. *Procephenemyia trompe*; Fig. 9. *Pharyngobolus africanus*; Fig. 10. *Tracheomyia macropi*; Fig. 11. *Kirkioestrus minutus*; Fig. 12. *Rhinoestrus purpureus*; Fig. 13. *Gedoelstia sp.*; Fig. 14. *Cephalopina titillator*.

where the antennae are located; Ethiopian (Congo forest belt) ..
.... PHARYNGOBOLINAE, subfam. n.
R_5 closed and petiolate, or, if without a petiole, the closed cell ends
far away from wing apex (Figs. 10-14) 3
3. Ocelli present; M_1 with or without a stump vein; median carina of
antennal fovea largely interrupted between the third antennal
segments, or very low medially, or complete and uninterrupted;
Africa and Eurasia .... OESTRINAE Leach
Ocelli absent; M_1 with a stump vein; median carina of antennal
fovea interrupted medially; Australia ....
.... TRACHEOMYIINAE, subfam. n.

(ii) Third stage larvae

1. Dorsal and ventral surfaces of the body with an identical arrangement of spines; parasites of Cervidae; Holarctic ....
.... CEPHENEMYIINAE Townsend
The arrangement of spines on the dorsal and ventral surfaces is different; if similar (*Cephalopina*) then from the 2nd thoracic segment to the 7th abdominal segment the processes have a row of
large conical warts behind the anterior zones of spines; or
(*Pharyngobolus*) segments 3-10, dorsally with 2 anterior rows, one
of large, the other of small and irregularly placed, spines, with
an additional row of large spines on the posterior margin 2

2. Ventral surface of the body with acute spines; parasites of Eutheria 3
Ventral surface of the body with wide stripes of subquadrangular
scales with sinuous margins; parasites of the kangaroo (*Macropus*);
Australia .... TRACHEOMYIINAE, subfam. n.
3. Segments 3-10 dorsally with 2 rows of spines anteriorly, one with
large, the other with small spines, partly irregularly placed; posterior half of segments with only one row of large spines; ventral
surface with a similar arrangement, but the anterior half of the
segments with a third row of irregularly distributed spines; parasites of the African elephant; Africa (Congo forest belt) ....
.... PHARYNGOBOLINAE, subfam. n.
Arrangement of the spines variable, but never as above; parasites
of several groups of Artiodactyla (except Cervidae) and Equidae
(*Equus*); Africa and Eurasia .... OESTRINAE Leach

48

II. SUBFAMILY CEPHENEMYIINAE TOWNSEND

TOWNSEND (1935 : 110, 216) erected the tribe Cephenemyiini for the genus *Cephenemyia* Latreille. GRUNIN (1956, 1957, 1966) gave it subfamily rank and discussed in great detail the morphology, biology, and economic aspects of the group. I am here proposing a subdivision of the Cephenemyiinae into tribes, which can be recognized by the following keys.

1. Key to tribes

(i) Adults

1. Body with long, dense hairs, making the flies similar to *Bombus*; M_2 long; 6th abdominal tergite almost as long as the 5th; Holarctic .... CEPHENEMYIINI Townsend
   Body with moderately dense, short hairs, the integument everywhere visible; thorax and abdomen with silvery-white pollinosity and black patterns: M_2 short and abruptly interrupted; 6th abdominal tergite half the length of the 5th; Palaearctic .... PHARYNGOMYIINI Townsend

(ii) First stage larvae (Figs. 15-16)

1. Larvae spindle-shaped, i.e., the anterior segments of the body are not laterally expanded; spines of the last transverse rows of each zone of the abdominal segments very long, forming a fringe on the posterior margin of the zone (Fig. 15) .... CEPHENEMYIINI Townsend
   Larvae clavate, i.e., the thoracic segments 2-3, as well as the 1st abdominal segment, together with the former, are greatly expanded laterally: spines of the last rows of each zone not elongate and not forming fringes (Fig. 16) .... PHARYNGOMYIINI Townsend

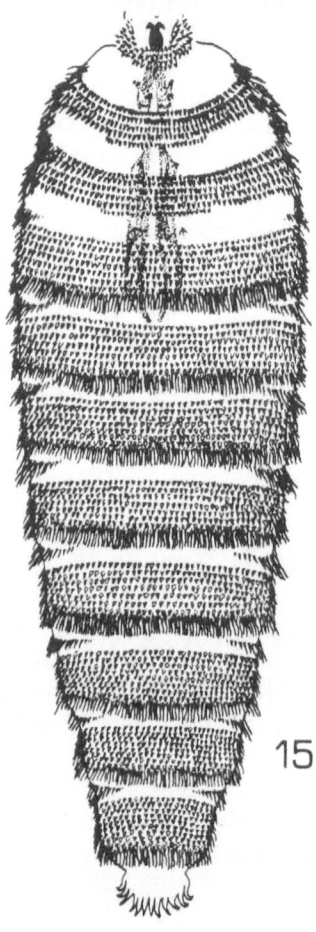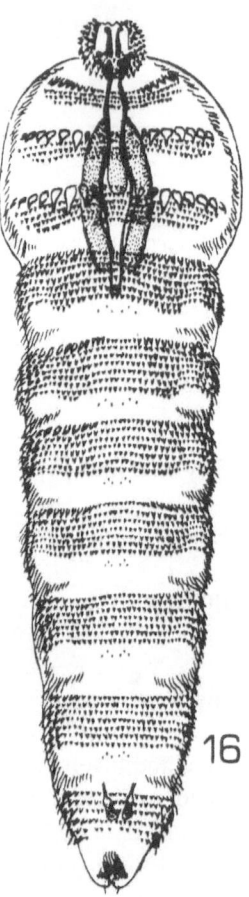

Figs. 15-16. First-stage larvae; Fig. 15. *Procephenemyia trompe* (Cephenemyiini); Fig. 16. *Pharyngomyia dzerenae* (Pharyngomyiini).

(iii) Third stage larvae (Figs. 17-20)

1. Lower lip of the internal margin of both posterior spiracles generally distant from each other as much as the superior lips (Fig. 18);
   rarely more distant between themselves than the superior lips,
   and in that case at most twice the distance between the superior
   lips; black granules on top of the posterior spiracles largely fused
   in the mature larva, forming a crown . . . .
   . . . . CEPHENEMYIINI Townsend
   Lower lips separated from each other by at least three times the dis-

tance existing between the superior lips (Fig. 19); black granules on top of the posterior spiracles not forming a crown in the mature larva . . . . PHARYNGOMYIINI Townsend

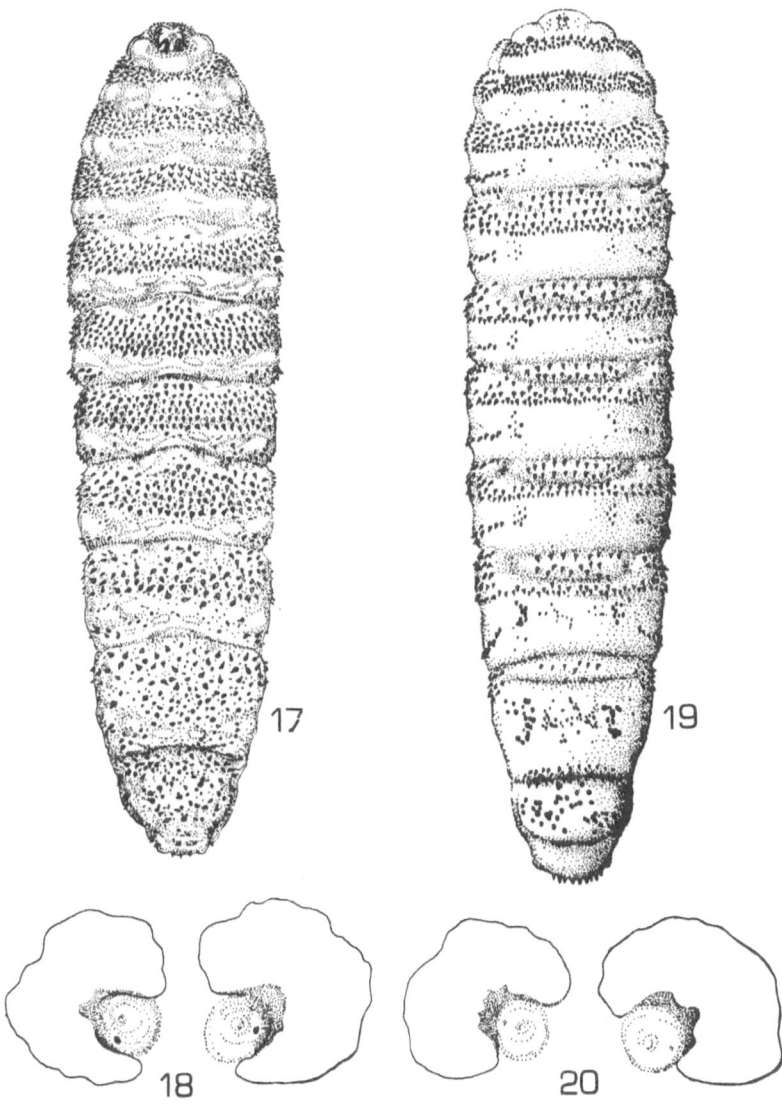

Figs. 17-20. Third-stage larvae and posterior spiracle; Figs. 17-18. *Procephenemyia trompe;* Figs. 19-20. *Pharyngomyia dzerenae.*

2. Tribe Cephenemyiini Townsend

I am proposing here a division of *'Cephenemyia'* into 2 subtribes and 3 genera, which may be recognized as follows:

(i) Adults

1. Median carina of antennal fovea absent or weakly developed, never with hairs; gena and scutellum with yellow pilosity (Subtribe CEPHENEMYIINA Townsend) 2

    Median carina of antennal fovea well developed, with long and fine hairs; pilosity of gena and scutellum reddish; Palaearctic (Subtribe ACROCOMYIINA, new) . . . . *Acrocomyia*, gen. n.

2. Median carina of antennal fovea absent; Palaearctic . . . . *Procephenemyia*, gen. n.

    Median carina of antennal fovea present, although little developed, and without hairs; Holarctic (predominantly Nearctic) . . . . *Cephenemyia* Latreille

(ii) Third stage larvae

1. Base of the fleshy cones of the pseudocephalon contiguous (Fig. 21); surface of occiput significantly wider than long (Fig. 22); spines of dorsal surface considerably larger than those of the ventral surface; anterior spiracle with a narrow opening, not enlarged at base, apically covered by a fragile, weakly pigmented involucre; external diameter of the spiracle from 2 to 3 times as wide as the diameter of the apical sensory organ of the fleshy cones of the pseudocephalon; superior and inferior lips of the internal margin of the posterior spiracle similar in shape and separated from each other by almost the same distance (Fig. 23); mature larva with numerous black spines and dots; the largest part of the superior surface of the spines black; parasites of Odocoileinae (Cervidae); (Subtribe CEPHENEMYIINA Townsend) 2

    Base of the fleshy cones of the pseudocephalon clearly separated; surface of occiput approximately quadrate; dorsal and ventral spines of the body not very different in size; opening of the anterior spiracle wide, the spiracle enlarged basally and covered by a strongly pigmented apical involucre, whose external diameter is from 4 to 5 times as wide as that of the sensory organs of the fleshy cones of the pseudocephalon; superior lips of the internal margin of the posterior spiracle large; the inferior lips small,

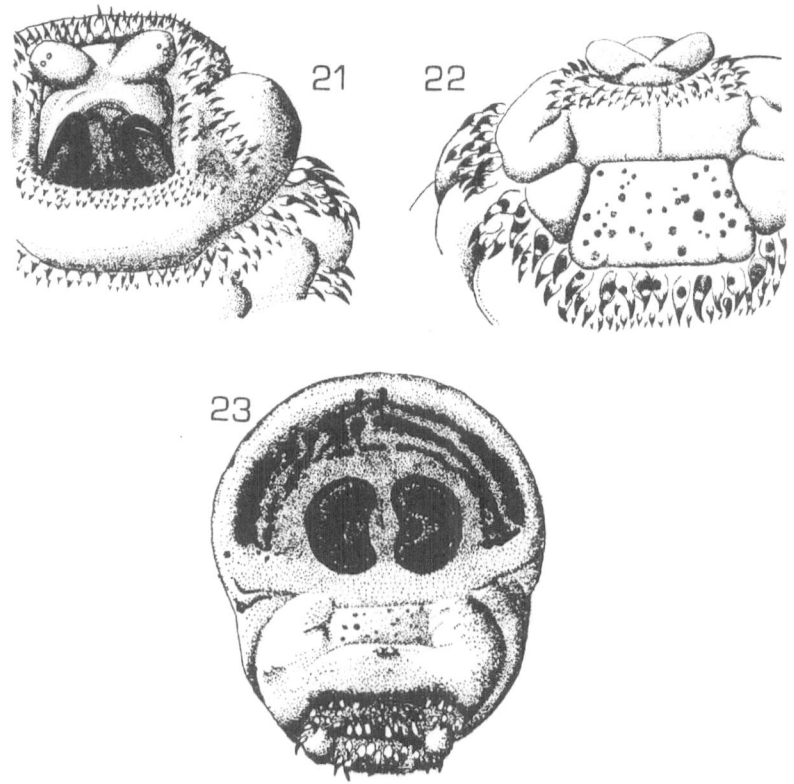

Figs. 21-23. *Cephenemyia ulrichi*; Fig. 21. pseudocephalon, ventral view; Fig. 22. same, dorsal view; Fig. 23. posterior spiracle.

acute, separated from each other by a distance almost twice as long as that between the superior lips; mature larva with a small number of brown dots on the more distal segments, and spines with only the apex dark brown; parasites of Cervinae (Cervidae); Palaearctic; (Subtribe ACROCOMYIINA, new) . . . .
. . . . *Acrocomyia*, gen. n.

2. Median part of the row of spines of the anterior margin of the dorsal surface of the abdominal segment with spines always present; Palaearctic . . . . *Procephenemyia*, gen. n.
Median part of the row of spines of the anterior margin of the dorsal surface of the 7th abdominal segment without spines; Holarctic (predominantly Nearctic) . . . . *Cephenemyia* Latreille

2.1. Subtile *ACROCOMYIINA,* **new**

The Acrocomyiina (type-genus, *Acrocomyia,* gen. n.) differ from Cephenemyiina by the well developed carina of the antennal fovea, with long and fine pilosity, by the different colour of the pilosity of the gena and scutellum, and by several larval characters, as shown above. In addition, the Acrocomyiina are parasitic on Cervidae of the subfamily Cervinae, while the Cephenemyiina are restricted to the Cervidae subfamily Odocoileinae (only secondarily parasitizing Cervinae, in the Nearctic Region).

Genus *Acrocomyia,* **new**

Characters: the same as the tribe. Type-species, *Oestrus auribarbis* Meigen.

Acrocomyia auribarbis (Meigen), **n. comb.**

'Larvae in the cervid skull'(?) of Redi, 1636: 165 (doubtful; could also be *Pharyngomyia picta*; cf. BRAUER, 1863: 193).

? 'Oestre du Cerf' Réaumur, 1734: 85, pl. 9, figs. 2-6.

? *Oestrus elaphi* Schrank, 1781: 418 (*nomen nudum*); 1803: 61; VILLERS, 1789: 349. (BRAUER, 1863: 194 : 'kann auf *Ceph. picta, rufibarbis, stimulator, Hypoderma Diana* and *Acteon* bezogen werden').

Oestrus auribarbis Meigen, 1824: 171, pl. 38, fig. 17.

Oestrus rufibarbis Meigen, 1824: 172 (a variation of *auribarbis*).

Cephenemyia (or *Cephenomyia*) *auribarbis*; MACQUART, 1835: 50; JOLY, 1846: 283; RONDANI, 1857: 23; AUSTEN, 1898: 9; BLOOMFIELD, 1898; GRIMSHAW, 1895, 1900; CAMERON, 1932a: 185 (morphol., biol.); 1932b, 1937; PATTON, 1937: 120 (♂ and ♀ genitalia, 1st stage larva); COLLART, 1952; ZUMPT, 1956: 127; GRUNIN, 1957: 86-89, figs. 127-134 (1966: 53, figs. 135-139) (descr. ♂♀, larvae, biol.); DRÓZDZ, 1961.

Oestrus trompe Modeer of Ratzeburg, 1844: 10, fig. 13 (misident.), and of KELLNER, 1847: 366, and 1853 (misident.).

Cephalomyia cervi Joly, 1846: 291.

Cephenomyia rufibarbis; BRAUER, 1858: 394, 401, 410, pl. 10, fig. 1, pl. 11, fig. 3; 1863: 193, pl. 3; 1892: 79; SCHINER, 1862: 395; GRIMSHAW, 1895: 155.

DISTRIBUTION: Western Europe (Great Britain, Germany, Austria), wherever occurs *Cervus elaphus* Linnaeus.

HOSTS: The main host is *Cervus elaphus* Linnaeus (Cervidae, Cervinae);

sometimes also *Dama dama* (Linnaeus) (Cervinae).

BIOLOGY: Up to 30 larvae are commonly found in the throat of *Cervus*. The work of CAMERON (1932), carried out in Scotland, indicates that the 1st stage larvae are found in the throat of *Cervus* from July to February; the 2nd stage larvae from February to March; the 3rd stage from May to June. GRUNIN (1957, 1966) says that the 3rd stage occurs in *Cervus* from February to April, the larvae beginning to be expelled in March. The pupal period lasts from 3 to 4 weeks. The flight period extends from May to June (inclusive) (BRAUER, 1863). The female fertility is from 500-600 young larvae.

2.2. Subtribe CEPHENEMYIINA Townsend
The Cephenemyiina, as characterized in the above keys, are primarily parasites of Odocoileinae.

Genus *Cephenemyia* Latreille

Cephenemyia Latreille, 1818: 271. Type-species, *Oestrus trompe* Modeer (mon.).
Cephenomyia, emend.
Endocephala Lioy, 1865: 81 (nom. nov. without justification). Type-species, *Oestrus trompe* Modeer (aut.).
Cephalemya Curran, 1934: 414 (error).

REFS.—ALDRICH, 1915 (North Amer. spp.); ANDERSON & OLKOWSKI, 1968; AUDOUIN, 1823; AUSTEN, 1898; BANZHAFF, 1928 (biol.); BARRETT & WORLEY, 1966 (biol.); BAU, 1906a, 1928 (taxon.); BENNETT, 1955; BENNETT & SABROSKY, 1962 (rev. Nearctic spp.); BERGMANN, 1899, 1916, 1917 a-b, 1919, 1920, 1932 (biol.); BEZZI & STEIN, 1907 (cat. Palaearctic spp.); BOLDYREV & USPENSKII, 1936; BOYES, 1964 (chromosomes); BROWNING & LAUPPE, 1964 (biol.); CATTS, 1964 (aggregation, ethology); CATTS & GARCIA, 1963 (biol.); ČEPELÁK, BUČEK & MANDELÍK, 1942 (aggregation); CHILCOTT, 1965 (cat. Nearctic spp.); DRÓZDZ, 1961b; DYK, 1962; EICHHORN et al., 1970; ESPMARCK, 1968 (reindeer defense against attacks); FITCH, 1928; FRINGS & NUSSBAUM, 1909; GLÄSER & STRÖSE, 1963; GLAZENER & KNOWLTON, 1967; GRUNIN, 1947, 1953, 1957, 1966; GRYUNER, 1929b; HADWEN, 1922, 1940; HENNIG, 1952 (larvae); HOFFMAN, 1909; HONESS & FROST, 1942; ISAICHIKOV, 1932; JAMES, 1948 (biol., Myiases); KELLNER, 1846-47, 1853; KNAB, 1913; KNORRE, 1957; LANGMUIR, 1938 (speed of flight); LÜHE, 1906; LYUBIMOV & AL'F, 1932; MISHIN, 1954, 1957; NATVIG, 1918, 1933; NIKOLAEV, 1959a-b;

Pavlovskii (M. M.), 1909; Polyakov, 1965 (aggregation); Samuel, Trainer & Glazener, 1971; Schipp, 1927; Schmidt, 1915, 1918; Sdobnikov, 1933, 1935; Séguy, 1928; Senger, 1963; Smith, 1972; Startsev, 1937; Stewart (N.H.), 1920, 1930; Terent'ev, 1930, 1933a-b; Thompson, 1889; Townsend, 1927; Troschel, 1881, 1882 (biol.); Ullrich, 1936 (biol.), 1939a-b; Voblikova, 1960, 1961, 1962, 1963; Yoshikawa, 1927; Zumpt, 1965 (Old World species).

The Palaearctic species were revised by Grunin (1957, 1966); the Nearctic species were revised by Bennett & Sabrosky (1962). The keys below were adapted from those authors.

Key to species (based on adults)

1. Nearctic species 2
   Palaearctic species 6
2. Mesonotum predominantly yellow-haired, without a transverse black stripe, only a narrow stripe of black hairs above the base of wings; dorsum of abdomen appearing as if entirely covered by black hairs (Eastern USA) . . . . *phobifera* (Clark)
   Mesonotum with a broad black transverse stripe, sometimes narrowly interrupted at the median line by yellow hairs; abdomen with numerous red, yellow, or white hairs 3
3. Abdomen predominantly yellow- or reddish-yellow-haired; tergites 2-3 with black hairs laterally, in a narrow stripe, or hidden by the wing; transverse stripe of mesonotum in both sexes with straight anterior margin, the black hairs abruptly ending at the transverse suture of the mesonotum 4
   Abdomen with 2 or 3 colours, tergite 5 and part of 4 with whitish hairs, interrupted by a median stripe of black hairs, in contrast with the numerous reddish hairs of the proximal tergites; mesonotum normally with a few or many black hairs in front of the transverse suture; in males, the anterior margin of the black transverse stripe irregular 5
4. Front relatively wide, in the male subequal to or slightly wider than one eye, in the female 1.5-1.7 times the width of an eye; abdominal hairs longer than in *jellisoni*, in the males long, light-yellow to orange-yellow; Circumpolar . . . . *trompe* (Modeer)
   Front narrower, in the male obviously narrower than the width of an eye, in the female 1.2 times the width of an eye; abdominal hairs shorter than in *trompe*, in the male orange or reddish-yellow, but rarely obscuring the black ground colour of the body (Western USA) . . . . *jellisoni* Townsend
5. The entire dorsum of the abdomen with light colours, hairs of tergi-

tes 2-3 and proximal half of 4 light-orange, moderately long and
very dense (Southwestern USA) . . . . *pratti* Hunter
Proximal 2/3 of abdomen dark, to the naked eye appearing black, in
contrast with the whitish hairs of the distal tergites; tergites 2-3
in reality with dark hairs, orange, or mixed orange and black, the
black ones more abundant on the median 1/3 of tergites 2-3 and
proximal parts of 2-4, resulting in an appearance of alternate
dark-orange and black stripes (Northwestern USA) . . . .
. . . . *apicata* Bennett & Sabrosky
6. Wing membrane clearly infuscated at the sides of the transverse
vein r-m; all femora with black hairs; length up to 18 mm
(Eurasia: temperate zone) . . . . *ulrichi* Brauer
Spot at sides of r-m very weak, little evident; femora with long yellow hairs on the basal portion; body length up to 16 mm (Circumpolar) . . . . *trompe* (Modeer)

Key to Nearctic species (based on 1st stage larvae)

1. Anal pecten with 8-10 spines . . . . . *pratti* Hunter and *apicata* Bennett
& Sabrosky
Anal pecten with more than 10 spines 2
2. Ventral segments 5-9 with rows of spines, anal pecten with 12 spi-
nes . . . . *phobifera* (Clark)
Ventral segments 5-9 with 7 or more rows of spines, anal pecten with
14 or 16 spines 3
3. Ventral segments 5-9 with 7-8 rows of spines, anal pecten with 14
spines . . . . *trompe* (Modeer)
Ventral segments 5-7 with 12-14 rows of spines, anal pecten with
16 spines . . . . *jellisoni* Townsend

Key to Nearctic species (based on second stage larvae)

1. Upper anal row with 8-10 spines . . . . *pratti* Hunter and *apicata* Bennett & Sabrosky
Upper anal row with 12 or more spines 2
2. Ventral spines pointed, not toothed . . . . *trompe* (Modeer)
Ventral spines not pointed, toothed 3
3. Upper anal row with 12 spines, ventral segments 5-9 with 5-7 rows
of spines, spines as broad as long . . . . *phobifera* (Clark)
Upper anal row with 16 spines, ventral segments 5-9 with 11-12
rows of spines, spines longer than broad . . . . . . *jellisoni* Townsend

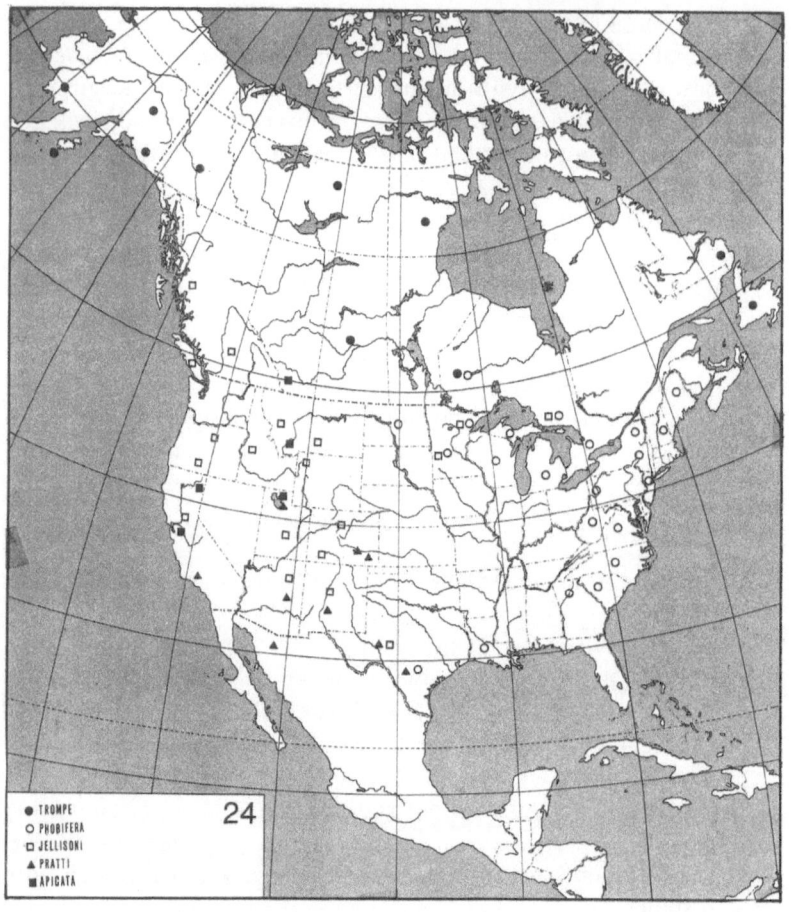

Fig. 24. Geographical distribution of *Cephenemyia* in North America (adapted from BENNETT & SABROSKY).

Key to Nearctic species (based on third stage larvae)

1. Upper anal row with 8-10 spines, anterior spiracle with single aperture 2
   Upper anal row with 12 or more spines, anterior spiracle with several apertures 3
2. Dorsal posterior rows of spines of segment 11 with 18 (12-22) spines .... *pratti* Hunter
   Dorsal posterior rows of spines of segment 11 with 30 (24-40) spines .... *apicata* Bennett & Sabrosky

3. Dorsal posterior rows of spines of segment 11 with 40 (28-47) spines; upper anal row with 14 spines . . . . *trompe* (Modeer)

   Dorsal posterior rows of spines of segment 11 with less than 40 spines on the average; upper anal row with 12 or 16 spines 4

4. Upper anal row with 12 spines; total spines of anal patch 32 (22-40) . . . . *phobifera* (Clark)

   Upper anal row with 16 spines; total spines of anal patch 42 (32-56) . . . . *jellisoni* Townsend

Key to Palaearctic species (based on 3rd stage larvae)

1. Spines of the anterior rows of the segments irregularly placed; median spines of dorsal surface of segment 7 of abdomen absent on anterior margin . . . . *ulrichi* Brauer

   Spines of the anterior rows of the segments regularly placed, forming small transverse stripes; median spines of dorsal surface of segment 7 of abdomen generally absent . . . . *trompe* (Modeer)

Cephenemyia apicata Bennett & Sabrosky

Cephenemyia apicata Bennett & Sabrosky, 1962: 483.

DISTRIBUTION: Canada (Alberta, British Columbia), USA (Montana, California).

HOST: 'Mule deer', *Odocoileus hemionus* (Rafinesque) (Cervidae, Odocoileinae).

Cephenemyia jellisoni Townsend

Cephenemyia jellisoni Townsend, 1941: 161.

REFS. – CAPELLE & SENGER, 1959; COWAN, 1943 (biol., morphol.. in *Odocoileus hemionius columbianus*); HAIR, HOWELL, ROGERS & FLETCHER, 1969 (biol., in *Odocoileus virginianus*); HERMAN, 1945-46; BENNETT & SABROSKY, 1962.

DISTRIBUTION: Canada (British Columbia, Ontario), USA (Arizona, California, Colorado, Idaho, Minnesota, Montana, Oregon, Utah, Wyoming).

HOSTS: 'Mule deer', *Odocoileus hemionus* (Rafinesque); 'white-tailed deer', *Odocoileus virginianus* (Zimmermann); 'moose', *Alces alces* L.; 'elk', *Cervus canadensis* Erxleben (Cervidae: Odocoileinae and Cervinae).

Cephenemyia phobifera (Clark)

Oestrus phobifer Clark, 1815: 69; WIEDEMANN, 1830: 255; MACQUART, 1843: 182 (1843: 25).

Cephenomyia phobifer; Brauer, 1863: 213, 291, pl. 5, fig. 11; ALDRICH, 1905: 415.

Cephenomyia abdominalis Aldrich, 1915: 149.

Cephenemyia phobifera; BENNETT & SABROSKY, 1962: 441; BENNETT, 1962 (biol., in *Odocoileus virginianus*); BLICKLE, 1956 (biol.); GOLINI, SMITH & DAVIS, 1968; HUTSON, 1931; SMITH & BENNETT, 1966.

DISTRIBUTION: Canada (Ontario), USA (Georgia, Maine, Michigan, Minnesota, New Hampshire, New Jersey, New York, North Carolina, North Dakota, South Carolina, Texas, Virginia, Wisconsin).

HOSTS: 'White-tailed deer', *Odocoileus virginianus* (Zimmermann); 'moose', *Alces alces* L. (Cervidae: Odocoileinae).

Cephenemyia pratti Hunter

Cephenemyia pratti Hunter, 1915: 170.

REFS. – KNOWLTON & ROWE, 1936; JELLISON, 1935; BENNETT & SABROSKY, 1962: 443.

DISTRIBUTION: USA (Colorado, Utah, Texas, Arizona, California, New Mexico).

HOSTS: 'Mule deer', *Odocoileus hemionus* (Rafinesque); 'white-tailed deer', *Odocoileus virginianus* (Zimmermann) (Cervidae: Odocoileinae).

Cephenemyia trompe (Modeer)

Oestrus trompe Modeer, 1786: 125; FISCHER, 1787: 78; FABRICIUS, 1794: 231, 1805: 229; COQUEBERT, 1803: 100, pl. 23, fig. 1; PANZER, 1809: 107; LATREILLE, 1809: 342; 1816: 271; 1825: 200; FALLÉN, 1818: 10; MEIGEN, 1824: 170; ZETTERSTEDT, 1821: 39; 1838: 622; 1844: 972; LAESTAD, 1831: 447; WALKER, 1848: 684; BONSDORFF, 1866: 30.

Cephenemyia (or *Cephenomyia*) *trompe*; LATREILLE, 1818; MACQUART, 1835: 50, pl. 13, fig. 20; BLANCHARD, 1840: 608; WESTWOOD, 1843: 179; JOLY, 1846: 292; SCHOLTZ, 1848: 3; BRAUER, 1858: 392, pl. 10, fig. 3; 1860: 650; 1863: 203; 1875: 77; 1887: 76; RONDANI, 1857: 23; SCHINER, 1862: 395; PAVLOVSKII, 1909: 288; YOSHIKAWA, 1927; PALMER, 1929; BERGMANN, 1932; HADWEN, 1926, 1932; TERENT'EV & TERENT'EV, 1933; BOLDYREV & USPENSKII, 1936; SDOBNIKOV, 1937; STARTSEV, 1937;

BREEV, 1950, 1956; GRUNIN, 1950: 233; 1953: 57; 1957: 69, figs. 91-105
[1966: 44, figs. 4, 8, 20, 30, 35, 52, 63, 85, 99, 100-105, 107-113];
MISHIN, 1954; REHBINDER, 1970; GOMOYUNOVA & KALVISH, 1971.

Oestrus tarandi Clark, 1843: 125.

DISTRIBUTION: Circumpolar (Northern Hemisphere).

HOSTS: *Rangifer tarandus, R. arcticus, R. caribou*; secondarily in *Odocoileus hemionus*.

BIOLOGY: According to ZUMPT (1965: 149) 'larviposition takes place while the female hovers for a moment in front of the animal's nose, then darts close and ejects a number of larvae into the nostrils. The reindeer suffer greatly from the fly attacks. When the insect is hovering in front of their noses they assume a terror-stricken look, their eyes staring, their mouth open, and their bodies in a tensely strained attitude. When a reindeer is in this rigid state, the slightest touch on any part of the animal will cause muscular contractions, which shake the whole body, just like an electric shock. When the insect deposits its larvae such a shock follows. It is succeeded by a total relaxation, the deer evidently realizing that it is not likely to be struck twice by the same insect. The animal appears nauseated and walks a few steps with its head elevated, sneezing and showing signs of nasal irritation. The first instar larvae may be found in the nasal cavities; the 2nd instar occurs in the nasal cavities as well as in the pharynx; 3rd are regularly found in the pharyngeal pockets, usually in numbers from 30-60. The mature larvae leave the host via the nostrils and cause a sneezing irritation. The dislodged larvae pupate rapidly, as a rule in 5-6 hours. The pupal period covers 16-31 days, but under adverse conditions, periods of up to 56 days have been observed.'

Cephenemyia ulrichi Brauer

Cephenemyia ulrichi Brauer, 1862: 5; BRAUER, 1863: 199; PORTSCHINSKY, 1874: 195; FREY, 1914; BAU, 1920: 541; MÖSCHLER, 1928, 1935 (biol.); SZIDAT & HEINEMANN, 1937; ULLRICH, 1937 a-b, 1939 (biol.); ZUMPT, 1956: 129; KNORRE, 1957: 569 (biol.); GRUNIN, 1957: 80, figs. 19, 31, 53, 62, 118-137 [1966: 49, figs. 123-134]; ZUMPT, 1965: 152.

Cephenomyia ulrichi kaplanovi Grunin, 1947: 224.

REF. – ZUMPT, 1950: 235.

DISTRIBUTION: Eastern and Central Europe, Far East (not known from Scandinavia; the distribution of this species does not seem to coincide with that of its host).

HOST: *C. ulrichi* is a specific parasite of the elk (*Alces alces* L.).

BIOLOGY: In Central Europe adults are on the wing from the end of

May to mid September; the pupal period lasts 30-34 days and sometimes takes only 21 days. The elk is normally intensively infested, up to
240 larvae of 2nd and 3rd stage having been found in a single animal; on
average, 80-100 larvae are found (in a population of elks in the former
East Prussia, ULLRICH, 1939). In adult elks infestation rarely leads to
death, although the general health condition of the animals is very bad;
in young individuals, if heavily infested, mortality is high.

Cephenemyiina 'incertae sedis'

1. *Cephenemyia macrotis* Brauer, 1863: 211, 279.
LOCALITY: 'North America' (from larvae found in *'Cervus macrotis* Say',
    North Western Territories); cf. OSTEN SACKEN, 1878: 144. BENNETT &
    SABROSKY (1962: 445) comment: 'The name *Cervus macrotis* would
    have referred to some western mule deer. At present it is carried in
    the synonymy of the western *Odocoileus hemionus hemionus* (Rafines-
    que)'; '*C. macrotis* could well have been *jellisoni, apicata,* or *pratti,*
    probably the first.'
2. *Cephenemyia mexicana* Brauer, 1894: 547 (*nomen nudum*).
LOCALITY: 'Mexico. Larva extracted from *Cervus mexicanus* Gmelin.'
3. *Cephenemyia multispinosa* Ullrich, 1935: 43.
REFS. – GRUNIN, 1957: 90, figs. 132-133 [1966: 55, figs. 140-141]; ULL-
    RICH, 1939 a-b; ZUMPT, 1965: 152 (as probable syn. of *auribarbis*).
HOST: *Dama dama.*

Genus *Procephenemyia,* **new**

Procephenemyia differs from the two other genera of Cephenemyiini
by the total absence of facial carina; other characters as in the keys
above. Type-species, *Oestrus stimulator* Clark.

Procephenemyia stimulatrix (Clark), **n. comb.**

Oestrus stimulator Clark, 1815: 69, pl. 5, figs. 28-29; CLARK, 1857: 5542,
    5630; MEIGEN, 1824: 170; COOKE, 1862: 8023.
Oestrus microcephalus Clark, 1815: 37.
Cephenemyia (or *Cephenomyia) stimulator;* MACQUART, 1835: 51; ZELLNER,
    1841: 181; 1842: 839, pl. 7, fig. 72; JOLY, 1846: 292; BRAUER, 1858: 393,
    pl. 10, fig. 2; 1860: 647; 1863: 206; 1875: 77; 1887: 76; SCHINER, 1862:
    395; PORTSCHINSKY, 1874: 195; ALTUM.(?), 1881; TROSCHEL, 1882: 119;

BECHER, 1882: pl. 4, fig. 27; WACHTL, 1886: 305; SIEBECK, 1902: 428; MÖSCHLER, 1935: 572; SHPRINGHOLZ-SCHMIDT, 1937; ULLRICH, 1939: 2149; THOMANN, 1947a-b; BOUVIER, BURGISSER & SCHNEIDER, 1952: 265 (biol.); ZUMPT, 1956: 128; DYK & DYKOVÁ, 1962: 193; GRUNIN, 1957: 77, figs. 106-109 [1966: 48, figs. 84, 114-117]; ZUMPT, 1965; DUDZIN-SKY, 1964, 1970a-b (biol.).

Cephenemyia trompe (Modeer) of CLARK, 1843: 125, and SAXESEN, 1850 (misident.).

Cephenemyia capreoli Henning, 1855: 305 (larva).

Oestrus biangulatus Cooke, 1857: 5438.

Cephenemyia stimulatrix; RONDANI, 1857: 28.

DISTRIBUTION: Palaearctic Region, wherever the host occurs.

HOST: *Capreolus capreolus* (Cervidae, Odocoileinae: Capreolini).

BIOLOGY: ZUMPT (1965: 150) summarized the biology of this species: '*stimulator* is evidently strictly host-specific to the Roe Deer (*Capreolus capreolus*). The life history almost coincides with that of *C. trompe*. According to ULLRICH (1936) first instar larvae were found in Germany from September to April, second in April and May, and third from May to August. The flies were on the wing from June until the beginning or end of September. The life-span of the males is on average 5 days, but sometimes extends up to 8 days, whereas the females live about 16 days. The development of the larvae from the eggs to the infection stage in the abdomen requires about 14 days. Over 500 larvae may be found in the female abdomen. The adults are abroad mainly in the morning, from 9 a.m. until 1 p.m., with a peak between 11 and 12 a.m., and the males have the habit of swarming around elevated points like rocks, hills, or artificial constructions like survey beacons.'

'The first instar larvae are found in the nasal cavities, mainly between the ethmoturbinalia. The fully-grown first instar larvae and the young second stage larvae then migrate upward to the choanae, the pharynx and sometimes even to the larynx, which together with the pharynx represent the favourite place for the third instar larvae. The mature larvae then migrate back towards the nostrils.'

'The roe deer are similarly disturbed by the larvipositing flies as are the reindeer by *C. trompe*. The pathological reactions caused by the larvae depend upon their number, their localization and the resistance of the host. The larvae feed on the mucous secretions and also on blood, causing a discharge of blood-stained mucus through the nostrils. A blackening of the larynx by great numbers of larvae may lead to the host's death by suffocation. Heavily infested deer suffer from attacks of coughing and loud snortings. Dislodged larvae may be passed to the lungs with severe consequences. The clinical picture is complicated and deteriorated by a simultaneous infection with lung-worms.'

3. Tribe Pharyngomyiini Townsend

The tribe Pharyngomyiini was erected by TOWNSEND (1935: 110, 217), monotypic for *Pharyngomyia* Schiner.

Genus *Pharyngomyia* Schiner

Pharyngomyia Schiner, 1861: 140. Type-species, *Oestrus pictus* Meigen (orig. des.).

In his revision of the Palaearctic Oestridae (1957, 1966) GRUNIN included two species in *Pharyngomyia: picta*, whose larvae parasitize the nasopharingean cavities of practically all genera of Cervidae in Europe, and *dzerenae*, known only from the 3rd larval stage, parasitic on the Mongolian gazelle, *Procapra gutturosa*. The presence of oestrid larvae in *Procapra* had already been reported by PALLAS (1777), who considered them *Oestrus ovis*. BRAUER (1863) thought that the species had been misidentified, and GRUNIN (1950) described the 3rd larval stage as *Pharyngomyia dzerenae*, although commenting that it was probably a new genus, as the differences between this species and *picta* are considerable. However, only the discovery of adults will confirm this fact.

The 3rd stage larvae of these two species may be identified by the following key:

1. Antennal lobes largely separated from each other at base (Fig. 25); last segment of the body with several almost regular rows of large spines anteriorly; mature larva with only the tips of the spines black, on the dorsal surface; parasites of Cervidae . *picta* (Meigen)

   Antennal lobes approximate at base (Fig. 27); last segment of the body with only a few denticles anteriorly; mature larva with all the spines of the body entirely black or dark brown; parasites of *Procapra* (Bovidae, Antilopinae) . . . . *dzerenae* Grunin

Pharyngomyia picta (Meigen)

Oestrus pictus Meigen, 1824: 172; CURTIS, 1828: pl. 106, fig. 1; JOLY, 1846: 286; KELLNER, 1847: 366; 1853: 89.
Oestrus cervi Clark, 1847: 1.

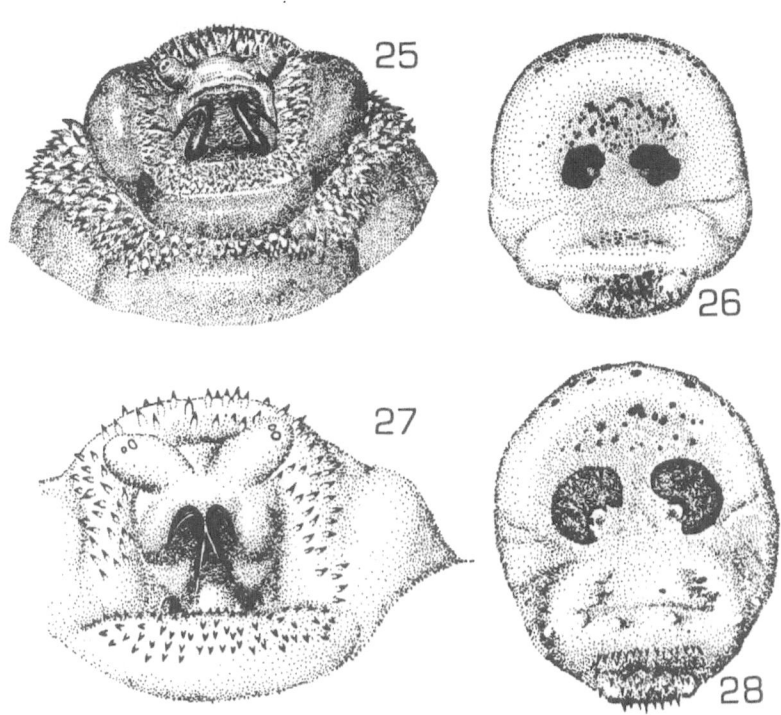

Figs. 25-28. Comparison between antennal lobes and posterior spiracles of *Pharyngomyia picta* (Figs. 25-26) and *'Pharyngomyia' dzerenae* (Figs. 27-28).

Cephalomyia cervi; Reichenbach, 1855.

Cephalomyia picta; RONDANI, 1857: 23; BRAUER, 1858: 395, 401, pl. 11, figs. 4, 4a-b.

Pharyngomyia picta; SCHINER, 1861: 393; BRAUER, 1863: 178-182; AUSTEN, 1898: 10; ULLRICH, 1939: 2149; NIKOL'SKII, 1927; SHAMANSKII, 1931; ISAICHIKOV, 1932; GRUNIN, 1950: 237; 1957: 93-100, figs. 134-155 [1966: 58-61, figs. 142-163]; DRÓZDZ, 1961; ZUMPT, 1965: 143.

DISTRIBUTION: Eurasia, Far East (coinciding with the distribution of *Cervus elaphus* and *Cervus nippon*).

HOSTS: The main host is *Cervus elaphus* L.; other hosts reported are *Cervus nippon, Dama dama, Alces alces* and *Capreolus capreolus*.

BIOLOGY: 'In Central Europe the adults are on the wing from June to August, and BRAUER (1863) said that they are more active while larvipositing than those of *Cephenemyia auribarbis*. The first instar larvae are found in the nasal cavities up to February and March, but the fully grown ones migrate down to the pharynx. The second and third instars are attached, like the *Cephenemyia* larvae, to the wall of the pharynx and larynx, and are found there from March to May or June. Pupation normally takes 30-40 days, but periods of only 3 weeks have also been noted. The pathogenic effects have not been especially studied, but certainly coincide with those of *Cephenemyia* species.' (ZUMPT, 1965: 146).

'*Pharyngomyia*' *dzerenae* Grunin

'*Oestrus* in *Antilope gutturosa*' Pallas, 1777: 50, 57; BRAUER, 1863: 161 (translation of Pallas); SHUMAKOVICH, 1934: 58 (distr.).
Pharyngomyia dzerenae Grunin, 1950: 861 (3rd stage larva).
REFS. – GRUNIN, 1957: 100-102, figs. 156-168 [1966: 61-63, figs. 164-166]; ZUMPT, 1965: 146.
DISTRIBUTION: Mountains of the Altai region.
HOST: 'Dzerena', *Procapra gutturosa* (Pallas) (Bovidae, Antilopinae, Antilopini).

III. SUBFAMILY TRACHEOMYIINAE, **NEW**

Medium-sized flies (length, 9 mm), characterized by the total absence of ocelli (Fig. 29), M_1 with stump vein, cell R_5 closed and long petiolate (Fig. 10), median carina of antennal fovea medianly interrupted; 3rd stage larva unique in the family, presenting ventrally wide stripes of subquadrangular scales with sinuous border (Figs. 30-32); parasites of the kangaroo (*Macropus*).
DISTRIBUTION: Australia. Type-genus, *Tracheomyia* Townsend.

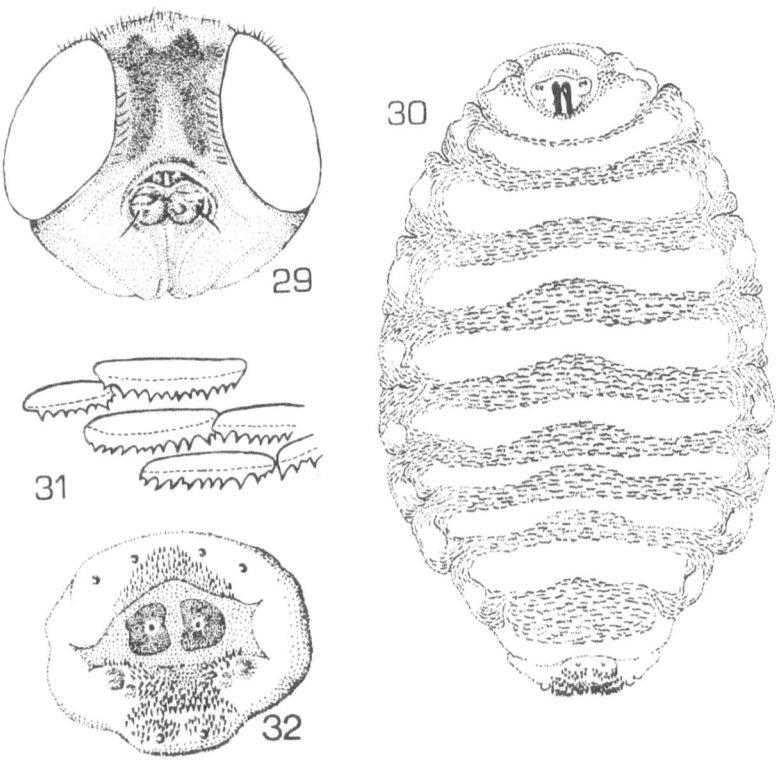

Figs. 29-32. *Tracheompyia macropi*; Fig. 29. Head; Fig. 30. Larva III; Fig. 31. Ventral spines of larva; Fig. 32. Posterior spiracle.

Genus *Tracheomyia* Townsend

Tracheomyia Townsend, 1916: 160. Type-species, *Oestrus macropi* Froggatt (orig. des.).

Tracheomyia macropi (Froggatt)

Oestrus macropi Froggatt, 1913: 565-568.
Tracheomyia macropi; TOWNSEND, 1916: 160; RODHAIN & BEQUAERT, 1916; PARAMONOV, 1953: 198 (descr. of adult, ♀); GRUNIN, 1961: 929 (larva); KOCH, 1961 (larva, biol.); MCCARTHY, 1961 (larva, biol.); ZUMPT, 1965; MYKYTOWICZ, 1964 (larva, biol.).

The larvae of *Tracheomyia macropi* were originally found in 'a large kangaroo', which was erroneously labelled by RODHAIN & BEQUAERT (1916) as *Macropus canguru*. The larvae were found in the trachaea, apparently feeding on mucous secretions. FROGGATT (1913) recorded finding as a rule 1 or 2 larvae per host, and that only occasionally 3-6 larvae were found.

TOWNSEND (1916) erected for this species the genus *Tracheomyia*. The adult was described only in 1953, by PARAMONOV (only the female). KOCH (1961) and MCCARTHY (1961) reported larvae from the kangaroo *Macropus (Megaleia) rufus* (Desmarest), from Western Australia and Queensland, respectively.

MYKYTOWYCZ (1964) examined 384 trachaeae of the kangaroo *Macropus (Megaleia) rufus*, for a period of 11 months, in the localities of Hay and Wilcannia, in New South Wales, and Cunnamulla, in Queensland. 57% of the animals from any one area were found infested. In general, there were no differences between the infestation of both sexes, but males always presented more larvae. The highest number of larvae found in a single animal was 94, in a male weighing 11 kg (tibial length: 390 mm). Young animals were more susceptible than older ones. Larvae were found throughout the year. The analysis of the frequency of occurrence of the different larval stages suggested the existence of two periods of larviposition – the first in May-July, the second in November. Researches undertaken on *Macropus canguru* (Miller), which was examined simultaneously, showed that this species is not parasitized.

From 568 larvae examined, 313 (56%) were found in the superior part of the trachaea, near the pharynx; 238 (41%) in the lower trachaea; and 17 (3%) in the bronchii or bronchioli. The 3rd stage larvae were more frequently found around the pharynx. On some occasions MYKYTOWYCZ found 3rd stage larvae in the stomach content of kangaroos; however, as he commented, 'the animals were not examined immediately after kill; therefore, it is possible that post-mortem dispersal of the larvae could have taken place.'

Macropus robustus Gould, an inhabitant of mountains and hills sparsely covered by vegetation, was also found parasitized, in Woodstock, Western Australia.

There are significant differences between the larvae described by GRUNIN (1961) and MYKYTOWYCZ (1964), possibly indicating that *Tracheomyia* could be represented in Australia by at least two species.

IV. SUBFAMILY PHARYNGOBOLINAE, **NEW**

Relatively large flies (length, 13-15 mm), characterized as follows: ocelli present; M_1 with stump vein; R_5 closed on wing margin (Fig. 9); median carina of antennal fovea uninterrupted, of regular height; larvae (see above) parasitic on the African elephant.

DISTRIBUTION: Africa (Congo, Uganda, Northern Rhodesia, Zambezi valley). Type-genus, *Pharyngobolus* Brauer.

Genus *Pharyngobolus* Brauer

Pharyngobolus Brauer, 1866: 879. Type-species, *africanus* Brauer (orig. des.).

Pharyngobolus africanus Brauer

Pharyngobolus africanus Brauer, 1866: 883, pl. 19, figs. 1, 1a-c.

REFS. – RODHAIN & BEQUAERT, 1919: 388-395, figs. 2a-b; RODHAIN, 1927: 201-203, fig. 6, pl. 1, figs. 1-2; AUSTEN, 1930: 687; ZUMPT, 1958: 9; 1962: 399; 1965: 153, figs. 202-204; ROTH, 1964.

DISTRIBUTION: Africa (Congo, Uganda, Northern Rhodesia, Zambezi valley).

HOST: African elephant, *Loxodonta africana* Blumenbach.

BIOLOGY: Larvae are found in the walls of the pharyngean region; ROTH (1964) found them in the fibrous walls of the main bronchi; RODHAIN & BEQUEAERT (1919) believed that mature larvae were expelled together with faeces. ZUMPT (1965: 153) records information about the way the larvae are expelled: 'Mr. VAN BRUGGEN of the Natal Museum came across a small booklet issued by the Zoological Garden in Basle, Switzerland, which recorded that in 1953 some African elephants, imported from the Congo six months previously, had sneezed out several large maggots. Two of these larvae had been sent to the Museum of Natural History in Basle. They pupated and one female fly hatched.'

The pupal period of the Basle larva lasted from February 22 to March 10, 1953. RODHAIN & BEQUAERT (1919) said that practically all elephants examined in the Congo showed larvae in the pharynx.

V. SUBFAMILY OESTRINAE LEACH

LEACH (1817) was the first to erect a suprageneric category for the *Oestrus*-group. LATREILLE (1818) also proposed a tribe 'Oestrides'.

These are flies of 9-15 mm of length, with the following characters: ocelli present, M_1 with or without a stump vein, cell R_5 closed and petiolate, or closed without petiole, in that case the cell ending far from the wing margin, antennal fovea with variable median carina, body generally with several sculptures (pustulae, pits, projections). Larvae variable, parasitic on several groups of Artiodactyla (except Cervidae) and Equidae (genus *Equus*).

DISTRIBUTION: Ethiopian and Palaearctic; introduced with domestic mammals in several parts of the world. Type-genus, *Oestrus* Linnaeus.

1. Key to tribes

(i) Adults

1. Fourth longitudinal vein (M_1) with a distinct stump vein (vestige of M_2) (Fig. 11); facial carina gradually but strongly lowered in the middle; Africa south of the Sahara . . . . . KIRKIOESTRINI, new
   M_1 without a stump vein (Figs. 12-14); facial carina variable 2
2. Facial carina interrupted between the third antennal segments . . . 3
   Facial carina of even height, uninterrupted, dividing the antennal fovea into two symmetrical depressions, where the antennae are located 4
3. Parafrontalia with pustules; Africa, Eurasia RHINOESTRINI, new
   Parafrontalia with pits; Eurasia and Africa . . . . . OESTRINI Leach
4. Inferior transverse vein of wing (tp) ending near the curve of M_1 (Fig. 13); abdomen with strongly erect longitudinal callosities, very evident (Fig. 43); Africa south of the Sahara . . . .
   . . . . GEDOELSTIINI, new
   Tp ending almost in the middle of the first posterior cell (Fig. 14); abdomen with transverse or oblique callosities, not very evident; Palaearctic . . . . CEPHALOPININI, new

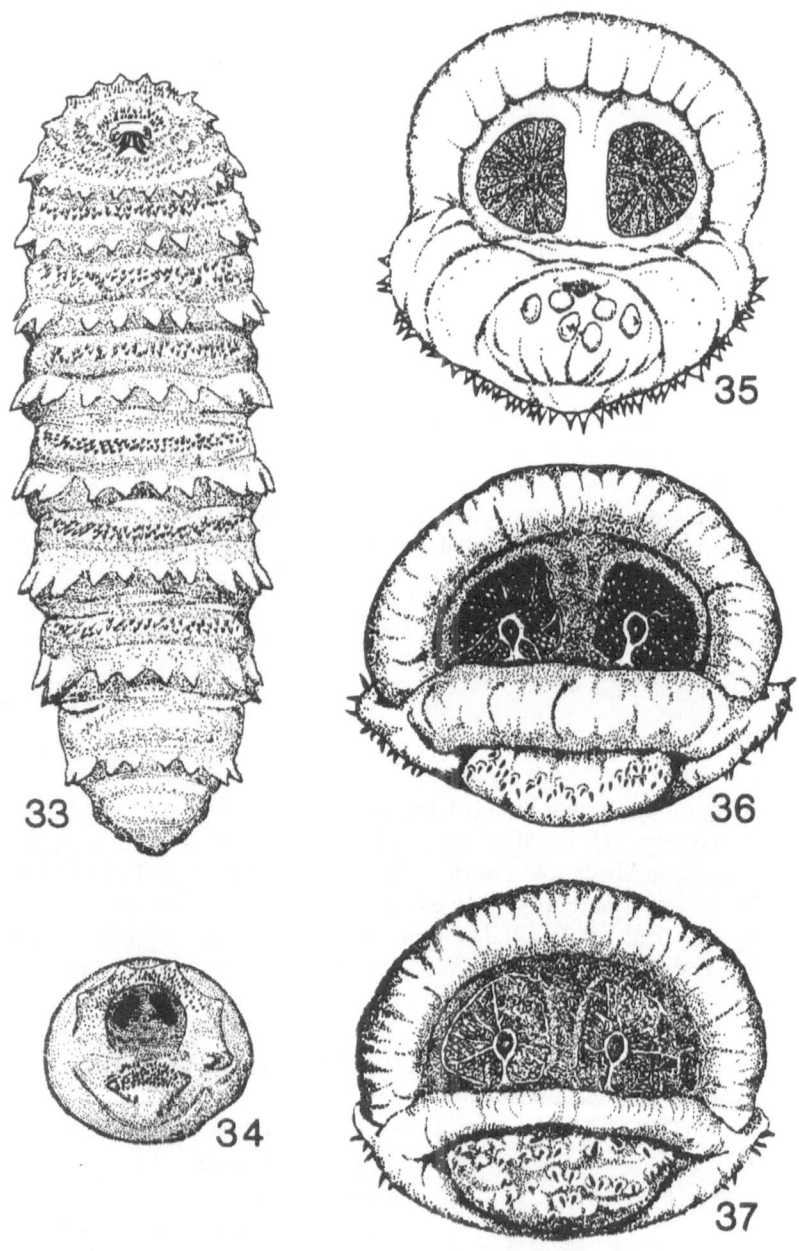

Figs. 33-37. Fig. 33. Larva of *Cephalopina*; Figs. 34-37. Posterior spiracles; Fig. 34. *Cephalopina*; Fig. 35. *Kirkioestrus*; Fig. 36. *Gedoelstia cristata*; Fig. 37. *G. haessleri*.

(ii) 3rd stage larvae (adapted from LAURENCE, *1961)*

1. Posterior spiracles distinctly and markedly emarginate along internal margin, crescent-shaped; antennal lozes with 2 pseudocelli; parasites of Suidae, Hippopotamidae, Giraffidae, and a few groups of Bovidae, and of Equidae (genus *Equus*); Africa and Eurasia . . . . RHINOESTRINI, new
   Posterior spiracles rounded, totally encircling the button of the pseudospiracle, or never obviously emarginate along the internal margin 2
2. Larvae with smooth, warty, fleshy processes forming complete rows on segments 3-11; posterior spiracles as in Fig. 33; antennal lobes with 1-3 pseudocelli, generally 2; parasites of Camelidae (genus *Camelus*); Palaearctic . . . . CEPHALOPININI, new
   Larvae never as above 3
3. Antennal lobes with 3 pseudocelli each; posterior spiracles nearly circular, totally encircling the button of the pseudospiracle (Fig. 35); parasites of Hippotraginae (Bovidae); Africa south of the Sahara . . . . KIRKIOESTRINI, new
   Antennal lobes with 2 pseudocelli each; posterior spiracles with different shapes 4
4. Posterior spiracles with dorsoventral line from centre to lower margin, sometimes open and emarginate (Figs. 36-37); parasites of Antilopinae (Bovidae); Africa south of the Sahara . . . .
   . . . . GEDOELSTIINI, new
   Posterior spiracle with horizontal line from centre to inner edge (Fig. 39), usually not emarginate; or hind spiracles completely rounded with no trace of line (Fig. 38); parasites of Caprinae (Bovidae) and Hippotraginae (Bovidae); Africa and Eurasia . . . .
   . . . . OESTRINI Leach

2. Tribe Kirkioestrini, new

Length, 11-14 mm. Ocelli present; facial carina consisting of a narrow keel gradually but strongly lowered in the middle; cell R_5 closed and petiolate; M_1 with stump vein; larvae parasitic on Hippotraginae (Bovidae).

DISTRIBUTION: Africa. Type-genus, *Kirkioestrus* Rodhain & Bequaert.

Genus *Kirkioestrus* Rodhain & Bequaert

Kirkia Gedoelst, 1914: 210 (preocc. Pollonera, 1909). Type-species, *blanchardi* Gedoelst (mon.).

Kirkioestrus Rodhain & Bequaert, 1915: 694 (nom. nov. for *Kirkia* Gedoelst). Type-species, *Kirkia blanchardi* Gedoelst (aut.).

Neokirkia Townsend, 1918: 153 (unnecessary nom. nov. for *Kirkia* Gedoelst). Type-species, *Kirkia blanchardi* Gedoelst (aut.). (Townsend's des. of *Kirkia minuta* Gedoelst as the type-species of *Neokirkia* invalid, as it is a replacement name).

Key to species:

(i) Adults (after Zumpt, *1965: 157)*

1. Terminal portion of third longitudinal vein (R_{4+5}) not recurrent; legs brownish-yellow, with tibiae and femora partially blackish; length, 11-12 mm . . . . *minutus* (Rodhain & Bequaert)
   Terminal portion of R_{4+5} recurrent; legs entirely brownish-red; length, 13-14 mm . . . . *blanchardi* (Gedoelst)

(ii) 3rd stage larvae (after Zumpt, *1965: 157)*

1. Ventral segments with short rows of spines posterolaterally; inferior lip of the last segment (12) with about half a dozen spines . . .
   . . . . *minutus* (Rodhain & Bequaert)
   Segments 5-11 ventrally with short rows of spines posterolaterally; inferior lip of the last segment (12) with about a dozen large spines . . . . *blanchardi* (Gedoelst)

Kirkioestrus blanchardi (Gedoelst)

'Larve de Kirk en *Buselaphus lichtensteinii*' Blanchard, 1893: 132, fig. 5.
Kirkia blanchardi Gedoelst, 1914: 211 (larva).
Kirkia surcoufi Gedoelst, 1914: 212 (larva); Roubaud, 1914a: 215 (larva); 1914b: 194, figs. 55-56, 57 (distr.).
Kirkia sp. (?*blanchardi*); Rodhain & Bequaert, 1915: 454 (larva; descr. ♂♀).
Kirkioestrus blanchardi; Rodhain & Bequaert, 1916: 158; Zumpt, 1962; 1965: 158, figs. 215-217; Emden, 1945; Zumpt, 1962 (in *Tragelaphus strepsiceros*).
Kirkioestrus surcoufi; Rodhain & Bequaert, 1915: 694; 1916: 153 (synonymy); Emden, 1945; Laurence, 1961: 597; Zumpt, 1962 (in *Tragelaphus strepsiceros*).
Distribution: Africa south of the Sahara–Volta (Boromo); Ivory Coast (Gouro, Odienné); Chad (Chari River); Zaire (Dungu River);

Uganda; Mozambique (Zambezi River, Beira District). This seems to be a species of the northern savannahs, overlapping with *minutus* in the Great Lakes and Mozambique areas.

HOSTS: *Alcelaphus buselaphus* (Common Hartebeest), and *A. lichtensteinii* (Lichtenstein's Hartebeest) (Bovidae, Hippotraginae: Alcelaphini).

Kirkioestrus minutus (Rodhain & Bequaert)

Kirkia minuta Rodhain & Bequaert, 1915: 456 (in 'Hartebeest' and 'Korrigum').

Kirkioestrus minutus; RODHAIN & BEQUAERT, 1916: 159 (larva III, imago ♀); EMDEN, 1945 (in 'Blue Wildebeest'); LAURENCE, 1961: 597; ZUMPT, 1965: 157, figs. 211-214.

Oestrus compositus Gedoelst, 1916: 260.

DISTRIBUTION: Africa south of the Sahara – southern Sudan; Ethiopia (south); Zaire (Dungu River, Kabare, south of Lake Kiwu); Tanzania; Ruanda; Burundi; Kenya (Mombasa); Zambia (Kapiri Mposhi); Mozambique; Namibia; South Africa (Natal, Transvaal), Botswana. This seems to be a species predominating in the southern savannahs, with great overlap with *blanchardi* to the north (region of the Great Lakes, etc.).

HOSTS: *Connochaetes taurinus* ('Blue Wildebeest'), *Alcelaphus buselaphus* ('Common Hartebeest'), *A. lichtensteinii* ('Lichtenstein's Hartebeest'), and *Damaliscus korrigum* ('Korrigum'); all Bovidae, Hippotraginae: Alcelaphini.

3. Tribe Rhinoestrini, new

Length, 9-13 mm. Ocelli present; M_1 without stump vein; R_5 closed and petiolate; facial carina variable; parafacialia with or without dark spots; parafrontalia with small or large pustules, isolated or confluent; abdomen with or without brown or black, shining tubercles. Larvae variable, parasites of certain groups of Artiodactyla (except Cervidae) and *Equus*.

DISTRIBUTION: Africa and Eurasia. Type-genus, *Rhinoestrus* Brauer.

Key to subtribes and genera (adapted from ZUMPT, 1965: 159-160)

1. Wings with cross-vein r-m situated opposite or slightly before level of apex of Sc; parafacialia without dark spots, the setae located in small crater-like pits which are yellow or brownish-yellow;

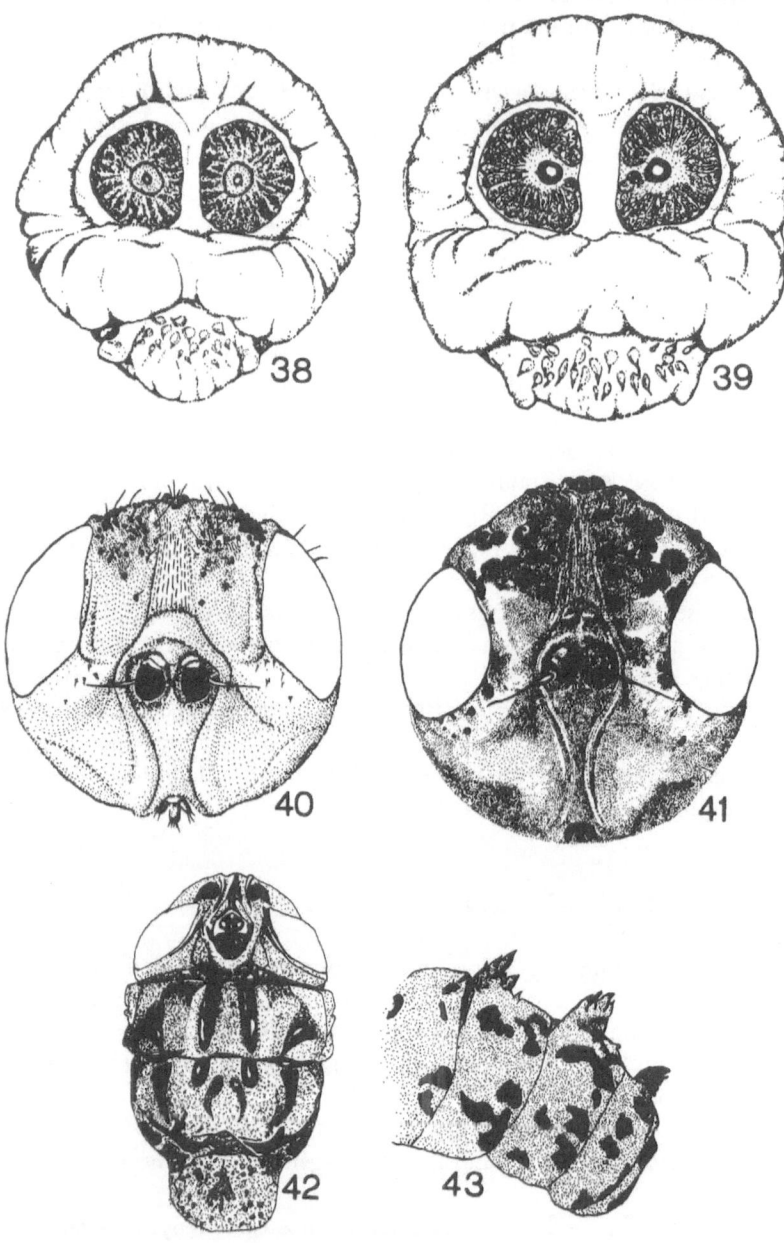

Figs. 38-43. Fig. 38. Posterior spiracle of larva of *Oestrus ovis*; Fig. 39. Same, *Loewioestrus variolosus*; Fig. 40. Head, *Gruninia tshernyshevi*; Fig. 41. Same, *Rhinoestrus purpureus*; Fig. 42. Head and thorax of *Gedoelstia cristata*; Fig. 43. Abdomen of *G. cristata*.

parafrontalia with relatively small and low tubercles, partially or predominantly confluent; mesonotum without weals, but only with fine and dense granules, each of which carries a black seta; scutellum also with granules, but never with tubercles; abdomen without tubercles, only with setiferous granules similar to the thorax; length, 11-13 mm; Africa (Zaire) (Subtribe SUINO-ESTRINA, new) . . . . *Suinoestrus*, gen. n.

Cross-vein r-m situated beyond level of apex of Sc; other combinations of characters 2

2. Lower part of the parafrontalia with a characteristic, well developed, longitudinal keel; parafacialia without dark spots, only with crater-like, yellow or brownish-yellow pits as the remainder of the parafrontalia (Fig. 40); abdomen without tubercles, only with small setiferous granules; mesonotum without weals; wings without basal spots; length, 10-11 mm; Asia (Tadzhikistan and Kazakhstan) (Subtribe GRUNINIINA, new) *Gruninia*, gen. n.

Lower part of the parafrontalia without keel; parafacialia with dark, black or brown spots, frequently developed as papulae (Fig. 41); abdomen with shining black or brown tubercles, each of which carries a reclinate seta; mesonotum with weals always present, although sometimes not very well defined; wings always with basal spots; length, 9-13 mm; Africa, Eurasia (Subtribe RHINOESTRINA, new) . . . . *Rhinoestrus* Brauer

Key to the genera and species of Rhinoestrini, based on the 3rd stage larvae (adapted from ZUMPT, 1965: 160-161).

1. Ventral armature of segments consists of quadrangular scales, forming 2-4 uninterrupted rows on the anterior part of each segment; dorsal armature absent, except on the pseudocephalon; parasites of *Antidorcas marsupialis*; South Africa and Namibia . . . .
. . . . *Rhinoestrus vanzyli* Zumpt & Bauristhene

Ventral armature of segments consists of spines; a dorsal spinulation present or absent 2

2. Dorsal armature absent on segments 3-12 3

Dorsal armature present at least on some anterior segments 5

3. Latero-ventral bulges of segments with spines; parasites of *Giraffa camelopardalis*; Tanzania . . . . *Rhinoestrus giraffae* Zumpt

Latero-ventral bulges of segments without spines 4

4. Last segment with 2-4 slightly irregular rows of spines on the anterior margin of the ventral surface; parasites of *Procapra gutturosa*; Tadzhikistan and Kazakhstan . . . .
. . . . *Gruninia tshernyshevi* (Grunin)

Last segment ventrally without spines; parasites of *Phacochoerus aethiopicus*; Zaire, Central African Republic.................
.... *Rhinoestrus phacochoeri* Rodhain & Bequaert

5. Posterior margin of latero-ventral bulges of segments without groups of spines; parasites of *Hippopotamus amphibius*; Cameroons, Zaire .... *Rhinoestrus hippopotami* Grünberg
Posterior margin of latero-ventral bulges of at least some of the segments with small groups of spines.................... 6

6. Segment 11, dorsally, completely bare; parasites of *Potamochoerus porcus*; Zaire .... *Suinoestrus nivarleti* (Rodhain & Bequaert)
Segment 11, dorsally, with several rows of anteriorly directed spines on the posterior margin 7

7. Last segment bare ventrally, except on the terminal projections, which are always spinous; parasites of *Antidorcas marsupialis*; South Africa, Namibia ....
.... *Rhinoestrus antidorcitis* Zumpt & Bauristhene
Last segment with 4-7 rows of spines ventrally 8

8. Posterior spiracles with a typical channel in the internal margin . 9
Posterior spiracles largely excavated on the inner ventral side, the margins forming a straight angle 10

9. Posterior spiracle higher than wide; parasites of *Equus*; Palaearctic, introduced with horses in the Oriental and Ethiopian regions ...
.... *Rhinoestrus purpureus* (Brauer)
Posterior spiracles as high as wide; parasites of *Equus*; Central Asia, Middle East, Africa .... *Rhinoestrus usbekistanicus* Gan

10. Dorsal surface of segments 3-4 with 2-3 complete rows of spines anteriorly; on the 5th segment the rows are interrupted medianly, and segment 6 has at most lateral groups of spines; parasites of *Equus*; European part of USSR, Kazakhstan, Uzbekistan, Mongolia, China .... *Rhinoestrus latifrons* Gan
Dorsal surface of segments 3-4 with 3-4 complete rows of spines anteriorly; segment 5 with 3-4 rows interrupted medianly, and 6-10 with lateral groups of spines; parasites of *Equus*; South Africa, Namibia .... *Rhinoestrus steyni* Zumpt

Rhinoestrus giraffae is only known from the larvae; *Rh. phacochoeri* is known in the adult phase, but the original description by RODHAIN & BEQUAERT is so inadequate that this species cannot be included in the key. Knowledge of the adults of these species will be extremely important for the establishment of the phylogeny and evolutionary history of the Rhinoestrini. Adults of the known species were described by GRUNIN (1957, 1966) and ZUMPT (1965).

3.1. Subtribe SUINOESTRINA, **new**

Characterized especially by the position of cross-vein r-m, situated opposite or before the level of the apex of Sc; other characters as in the above keys. Type-genus, *Suinoestrus,* gen. n.

Genus *Suinoestrus,* **new**

Rhinoestrini; parafacialia without dark spots, the setae located in pits similar to footprints, crater-like; abdomen without tubercles, only with small setiferous granules. Type-species *Rhinoestrus nivarleti* Rodhain & Bequaert.

Suinoestrus nivarleti (Rodhain & Bequaert), **n. comb.**

'Larves des narines d'un sanglier du Congo français' Blanchard, 1896: 668, pl. 18, figs. 7-11 (in the explanation of the plates, p. 677, the larva is erroneously designated as *Oestrus ovis*).

Rhinoestrus nivarleti Rodhain & Bequaert, 1912: 370, figs. 14-16.

REFS. – RODHAIN & BEQUAERT, 1916: 119; ZUMPT, 1958: 57; 1965: 167, figs. 225-227.

DISTRIBUTION: Africa – Zaire (Poste Yongama, between rivers Lomami and Lualaba; Stanley Pool, Kunzulu, Kwamouth, along Congo River, between 3°-4°30'S and 15°-16°30'W; Yambuya, Aruwini River; Kisangani [ex Stanleyville]; Bambili, Uelé River; Bili, Bili River; Litaki). This species is probably associated with the Congo forest belt; ZUMPT (1965: 168) examined specimens of the genus *Potamochoerus* from Northern Rhodesia (near Monze) and Mozambique (Beira District) and found them free from the parasite.

HOST: *Potamochoerus porcus* (Suidae).

BIOLOGY: Adults of *S. nivarleti* are found in May, November, and December, and it is possible that they occur throughout the year; the pupal period is from 28-35 days.

3.2. Subtribe GRUNINIINA, **new**

Length, 9-11 mm; mesonotum without weals; r-m situated beyond apex of Sc; thorax and abdomen with granules; wings without basal spots; inferior part of parafrontalia with well-developed, characteristic longitudinal keel. Type-genus, *Gruninia,* gen. n.

Genus *Gruninia*, **new**

CHARACTERS: the same of the subtribe. Type-species, *Rhinoestrus tshernyshevi* Grunin.

Gruninia tshernyshevi (Grunin), **n. comb.**

Rhinoestrus tshernyshevi Grunin, 1951: 467.
REFS. – GRUNIN, 1957: 134; GRUNIN & SLUDSKII, 1960: 210; ZUMPT, 1965: 170.
DISTRIBUTION: Central Asia (Tadzhikistan and Kazakhstan).
HOSTS: *Ovis orientalis* and *Ovis ammon* (Bovidae, Caprinae).
BIOLOGY: Adults reared by GRUNIN & SLUDSKII emerged in June, after a pupal period of 3 weeks.

3.3. Subtribe RHINOESTRINA, **new**

Length, 9-13 mm. Parafacialia with dark spots, black or brown; para-frontalia without longitudinal keel; abdomen with well developed, black or brown tubercles; wings with well evident basal spots; r-m situated beyond the level of the apex of Sc. Type-genus, *Rhinoestrus* Brauer.

Genus *Rhinoestrus* Brauer

Rhinoestrus Brauer, 1886: 289. Type-species, *Cephalomyia purpurea* Brauer (orig. des.).
Hippooestrus Townsend, 1932: 447. Type-species, *Rhinoestrus hippopotami* Grünberg (orig. des.).
REFS. – ABDUSSALAM, 1940; AKCHURIN, 1945; ALEKSANDROV, 1945; BOUR-GELAT, 1960; BIL'DUSHNIKOV, 1949; BRAUER, 1886; CANNON, 1917; CHAMS & MOHSÉNINE, 1956; CLARK, 1815; GAN, 1947a-b; GASCOU-GNOLLE, 1961; GRÜNBERG, 1904, 1906; GRUNIN, 1951b, 1953, 1957, 1966 (rev. Palaearctic spp.); GRUNIN & SLUDSKII, 1960; GRYUNER, 1929a; HAESELBARTH *et al.*, 1968; HENNIG, 1952 (larvae); HUTCHEON, 1891-1914; KARPENKO, 1947; KHOLODKOWSKI, 1907; KOLOMIETS, 1951a-b, 1952a-b, 1953, 1954, 1955; KRIVKO, 1957a; LAURENCE, 1961; LAR-ROUSSE, 1926; LAVROV, 1952; NIKITIN, 1962; PAULL.(?), 1844; PINKER-TON, 1971; PORTSCHINSKY, 1906-1915 (biol.); RAILLIET, 1918; RONDANI, 1854; SACHS, 1970; SAYIN, 1967; SALAMATIN, 1930; SIBONI, 1952; STEEL,

1887; Suvortsev, 1889; Wetzell, 1970b, 1971a; Zumpt, 1958a, 1959, 1962, 1965 (rev.).

The genus *Hippooestrus* Townsend may be revalidated in the future, when the Rhinoestrina are better known; the species-group centred about *Rhinoestrus hippopotami* seems to be well defined, but at present it is unsatisfactory to segregate it only on the basis of the size of adults.

Key to species, based on adults (modified from Zumpt, 1965: 159-160)

1. Larger species, length 11-15 mm (*hippopotami*-group) 2
   Smaller species, length 8-11 mm (*purpureus*-group) (*R. phacochoeri* also seems to belong to this group, but the adults are insufficiently characterized) 5
2. The pair of longitudinal, pre-sutural stripes of the mesonotum fuscous and ill-defined, due to an intensive wrinkling, at most shining only anteriorly 3
   The longitudinal pair of pre-sutural stripes of the mesonotum narrow, well-defined, and almost reaching the transverse suture, shining all over their extension; thorax predominantly dark brown, abdomen black; South Africa and Namibia . . . .
   . . . . *antidorcitis* Zumpt & Bauristhene
3. Body predominantly dark brown or brownish-yellow; the pair of mesonotal stripes is black and almost entirely dull; Ethiopian species 4
   Body predominantly reddish-brown; the pair of mesonotal stripes ill-defined, reddish, and shining on the anterior portion; Palaearctic species . . . . *latifrons* Gan
4. Tubercles of parafrontalia isolated from each other; only a few approximated; Zaire . . . .   *hippopotami* Grünberg
   Tubercles of parafrontalia confluent, only a few on the upper part more or less isolated (only males considered); South Africa, Namibia . . . . *steyni* Zumpt
5. Mesonotum with densely rugose, flattened, fuscous or in part shining weals 6
   Mesonotum with slightly raised, shining, black or blackish-brown weals, the presutural pair of internal stripes is smooth and without a dense wrinkling; Palaearctic, Oriental, Ethiopian . . . .
   . . . . *purpureus* (Brauer)
6. Parafrontalia with relatively flattened and largely confluent tubercles; abdominal tubercles flattened, of small and moderate size; South Africa and Namibia . . . . *vanzyli* Zumpt & Bauristhene
   Parafrontalia with large and voluminous tubercles, mostly isolated, only a few near the eye margins confluent; abdominal tubercles

conical, shining, the median ones very large; Central Asia, Middle East, Africa .... *usbekistanicus* Gan

Rhinoestrus antidorcitis Zumpt & Bauristhene

Rhinoestrus antidorcitis Zumpt & Bauristhene, 1962: 11, figs. 9-13.
REF. – ZUMPT, 1965: 171, figs. 235-239.
DISTRIBUTION: South Africa and Namibia, wherever the host occurs.
HOST: 'Springbuck', *Antidorcas marsupialis* (Bovidae, Antilopinae, Antilopini).
BIOLOGY: *Antidorcas* is also parasitized by *Rhinoestrus vanzyli*. In western Transvaal the mature larvae of *antidorcitis* were caught in June, and adults have emerged from the end of July to the middle of August, after a pupal period of 49-56 days. The number of larvae of this species, as well as the number of larvae of *vanzyli*, found in the same host, is low.

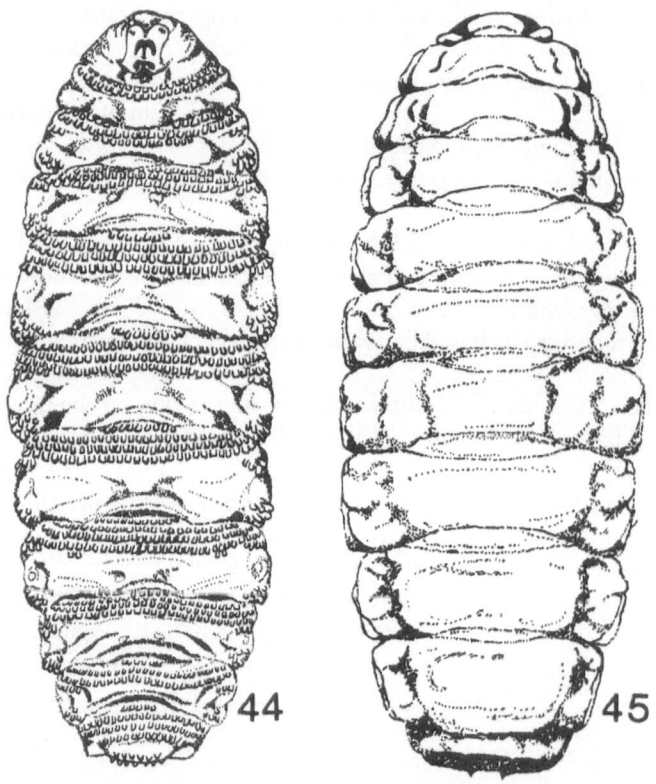

Figs. 44-45. Larva of *Rhinoestrus vanzyli*; Fig. 44. Ventral; Fig. 45. Dorsal.

Rhinoestrus giraffae Zumpt

Rhinoestrus giraffae Zumpt, 1965: 169, figs. 230-231.
REFS. – LAURENCE, 1961: 595; WETZELL, 1970b. (This species is only known from the larvae).
DISTRIBUTION: Tanzania (Old Shinyanga).
HOST: Giraffe, *Giraffa camelopardalis* (Giraffidae).

Rhinoestrus hippopotami Grünberg

Rhinoestrus hippopotami Grünberg, 1904: 37; SURCOUF & GEDOELST, 1909: 615; RODHAIN & BEQUAERT, 1916: 217; ZUMPT & BAURISTHENE, 1962: 19, 21; ZUMPT, 1965: 166, fig. 224;   WETZELL, 1973: 5-14 (all larval stages, adult, distr.).
Hippooestrus hippopotami; Townsend, 1932: 447.
DISTRIBUTION: Cameroons (Adamoua plateau, N'Gaounderé); Sudan (Mudiriyat el Istwa'ya, Yei River, near Wayo in Moro; high White Nile); Zaire (Ango, Ueré River; Stanley Pool, Congo River; Kongolo, Lualaba River); Uganda. According to ZUMPT (1965: 167): 'A hippo shot in Northern Rhodesia (nr. Monze) proved to be free of larvae, and it is possible that the parasite does not extend as far south as its host.'
HOST: *Hippopotamus amphibius* (Hippopotamidae).

Rhinoestrus latifrons Gan

Rhinoestrus latifrons Gan, 1947: 24.
REFS. – RUBTSOV, 1948: 138; 1950: 248; 1953: 48; GRUNIN, 1957: 130 (1966: 78); ZUMPT, 1965: 164.
DISTRIBUTION: European USSR, Kazakhstan, Uzbekistan, Mongolia, China.
HOST: *Equus caballus* L. (Equidae).
BIOLOGY: Similar to *purpureus* (Brauer).

Rhinoestrus phacochoeri Rodhain & Bequaert

Rhinoestrus phacochoeri Rodhain & Bequaert, 1915: 452.
REFS. – RODHAIN & BEQUAERT, 1916: 136; ZUMPT & BAURISTHENE, 1962: 21; ZUMPT, 1965: 168, figs. 228-229.
DISTRIBUTION: Zaire, Cameroons.
HOST: 'Warthog', *Phacochoerus aethiopicus* (Suidae).

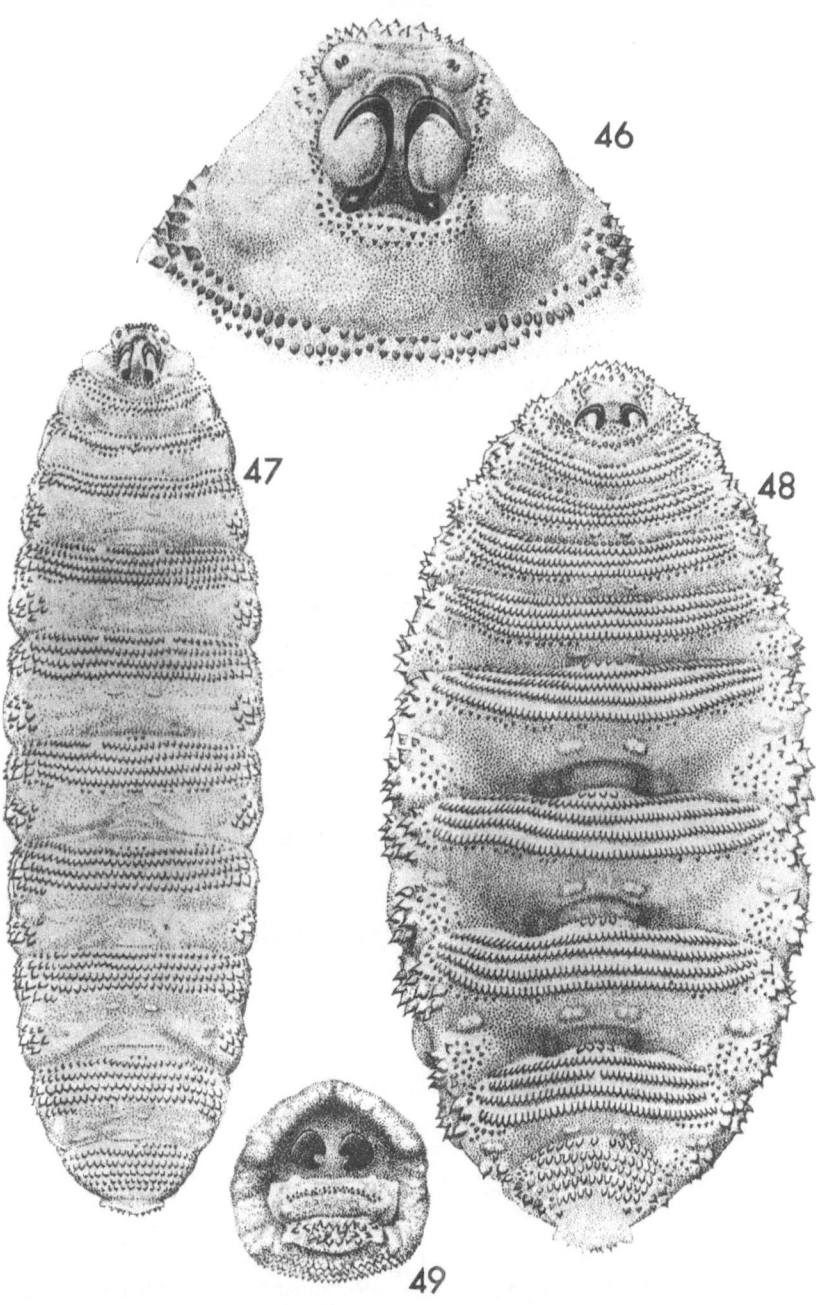

Figs. 46-47, 49. Larva of *Rhinoestrus purpureus*; Fig. 46. Ventral view of pseudocephalon; Fig. 47. Ventral view of larva; Fig. 49. Posterior spiracle; Fig. 48. Ventral view of larva of *Rhinoestrus latifrons*.

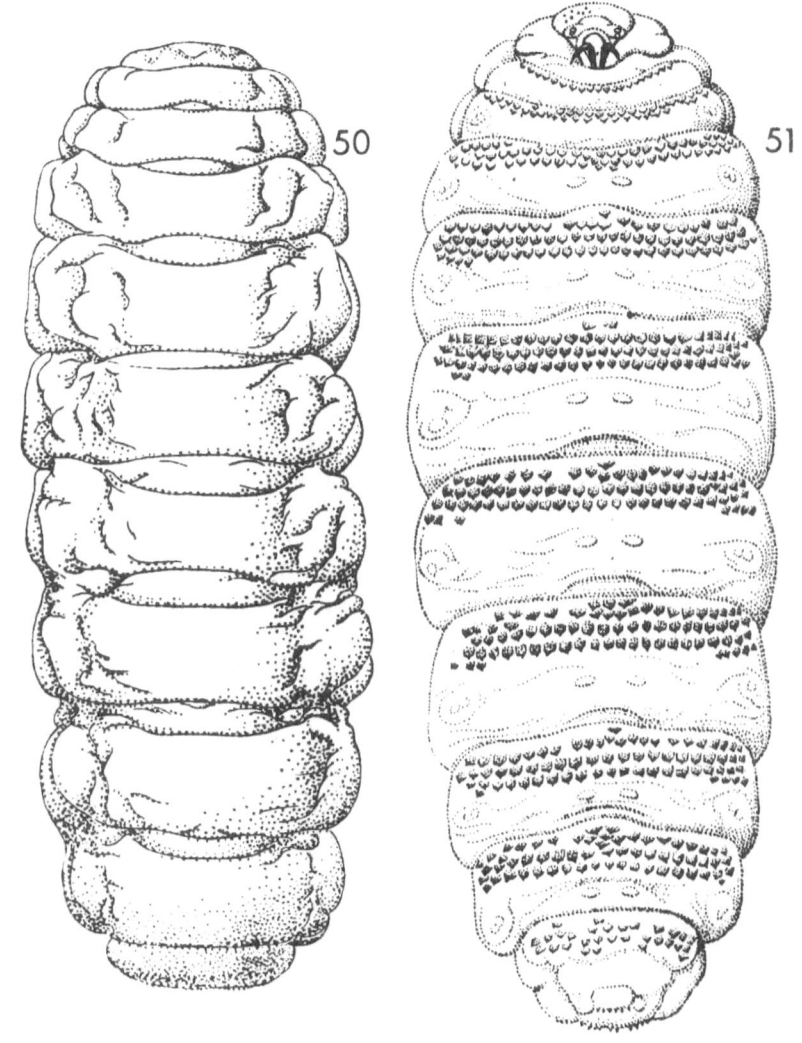

Figs. 50-51. Larva of *Gruninia tshernyshevi*; Fig. 50. Dorsal; Fig. 51. Ventral.

Rhinoestrus purpureus (Brauer)

Cephalomyia purpurea Brauer, 1858: 457.
Oestrus purpureus; BRAUER, 1863: 158.
Rhinoestrus purpureus; BRAUER, 1886: 289; PORTSCHINSKY, 1915: 1-47;
    PAVLOVSKII, 1929: 60; SLAVIN, 1934: 30; GAN, 1947: 122; GRUNIN, 1950:
    244; KOLOMIETS, 1951: 550; LAVROV, 1952: 277; KOLOMIETS, 1952a: 20,

1952b: 208; GRUNIN, 1953: 42; KOLOMIETS, 1953: 188, 1955: 213; GRUNIN, 1957: 121, figs. [1966: 74, figs. 5, 9, 26, 37, 59, 69, 82, 193, 203-219]; ZUMPT & BAURISTHENE, 1960: 20-22; ZUMPT, 1965: 161, figs.

Rhinoestrus nasalis (Linnaeus) of BRUMPT, 1913: 700 (misident.).

DISTRIBUTION: Palaearctic; introduced with horses in the Oriental and Ethiopian regions.

HOSTS: *Equus caballus* and *E. asinus* and their hybrids.

BIOLOGY: 'The female produces between 700 and 800 larvae, which are expelled in batches of 8-40 into the nostrils and sometimes also the orbits of the hosts. The first instar larvae are found in the nasal cavities, where they remain in this stage for a few weeks or even months. In southern Russia, for instance, the first instars are present from September until June, steadily decreasing in number. The speed of development of the same batch of larvae varies considerably. Some of them moult to the second stage in March, others later, and in July second instar larvae are still found. The same is true for the third stage; some already appear at the end of March, others only in July and August. The Russian authors believe that there is only one generation in the Ukraine. The older first instar larvae move further up to the posterior parts of the nasal passages, and the second and third instar larvae are also found in the pharyngeal area.'

'In Usbekistan, however, and other warmer parts of Central Asia, two generations are thought to be accomplished, and the flies are on the wing from March to mid-June and again in September and October, and second larval stages are found as early as January. The explanation of this phenomenon is probably that in areas with a cold winter, the development of the first instar larva is considerably retarded during the unfavourable season.'

'Female flies were kept in captivity for 25 days on the average, but a few lived up to 38 days. The life-span of the males is only about two weeks. The pupal period is given as from 15 to 32 days.' (ZUMPT, 1965: 162).

Rhinoestrus steyni Zumpt

Rhinoestrus steyni Zumpt, 1958: 56, figs.

REFS. – ZUMPT & BAURISTHENE, 1962: 19, 21; ZUMPT, 1965: 165.

DISTRIBUTION: South Africa (Natal, Transvaal), Namibia.

HOSTS: *Equus burchelli* (Burchell's zebra) and *Equus zebra* (Mountain zebra).

BIOLOGY: 3rd stage larvae were collected in May and June; the pupal period of one specimen lasted almost a month.

Rhinoestrus usbekistanicus Gan

Rhinoestrus usbekistanicus Gan, 1947a: 24; GAN, 1947b: 122; GRUNIN, 1953: 47; 1957: 47 [1966: 72; figs. 70, 190, 191]; ZUMPT, 1965: 163, figs. (synonymy).

Rhinoestrus szlampi Zumpt, 1959: 4, fig.

REF. – ZUMPT & BAURISTHENE, 1962: 20, 22.

DISTRIBUTION: Central Asia, Middle East, Africa.

HOSTS: *Equus burchelli* in Africa south of the Sahara; horse, ass and donkey in the Palaearctic region.

BIOLOGY: In the semi-arid areas of Central Asia *usbekistanicus* predominates over *latifrons* and *purpureus*. In Uzbekistan and Tadzhikistan there are two generations a year: the adults of the first generation are on the wing from May to June, those of the second from September to October. In southern Africa the adults emerged in August and October, and the pupal period was of 3 weeks; a female was captured in September, in Tanzania (ZUMPT, 1965: 164).

Rhinoestrus vanzyli Zumpt & Bauristhene

Rhinoestrus vanzyli Zumpt & Bauristhene, 1964: 4, figs. 1-8.

REF. – ZUMPT, 1965: 173, figs. 240-245.

DISTRIBUTION: Same as *antidorcitis*.

HOST: 'Springbuck', *Antidorcas marsupialis*.

BIOLOGY: *R. vanzyli* is more common in *Antidorcas* than *antidorcitis*; the mature larva abandons the host in July, the adults emerging after 30-50 days.

4. Tribe Oestrini Leach

M_1 without a stump vein; median carina of antennal fovea interrupted between the third antennal segments; parafrontalia with pits.

DISTRIBUTION: Eurasia and Africa; introduced with sheep in most parts of the world. Type-genus, *Oestrus* Linnaeus.

Key to subtribes, genera and species

(i) Adults (adapted from ZUMPT, 1965: 174)

1. Wings short and broad; r-m situated in the middle of or beyond the discal cell (Subtribe OESTRINA Leach) 2

Wings long and narrow; r-m situated before the middle of the discal cell (Subtribe LOEWIOESTRINA, new) . . . .
. . . . *Loewioestrus variolosus* (Loew)
2. Thorax with large, flattened tubercles of a black colour; mesonotum with a well-defined pattern of weals, as in certain Rhinoestrini; length, 12-15 mm . . . . *Oestroides macdonaldi* (Gedoelst)
Thorax with small black granular tubercles; mesonotum without weals (Genus *Oestrus* Linnaeus) 3
3. Wing veins black; length, 12-13 mm . . . . *caucasicus* Grunin
Wing veins yellow 4
4. Mesonotum with black hairs; tubercles of the 3 brownish pollinose presutural stripes toothed and stronger than those of the inter-mediary areas; when the same sexes are compared, the para-facialia are more strongly rugose and with less and larger pits then in the species below; length, 11-16 mm . . . .
. . . . *aureoargentatus* Rodhain & Bequaert
Mesonotum with yellow hairs; tubercles of the brownish and yel-lowish pollinose presutural stripes rounded, and not, or very little, stronger than those of the intermediary areas; parafacialia more finely rugose, with more and smaller pits; length, 10-12 mm
. . . . *ovis* Linnaeus

(ii) 3rd stage larvae (adapted from ZUMPT, 1965: 175)

1. Dorsal surface of segments 3-6 with rows or groups of spines ante-riorly 2
Dorsal surface of segments without spines, except segment 2, which may present a few spines 3
2. Segments 3-5 dorsally with 2-3 regular rows of spines; the following 3 segments with lateral groups of spines . *Oestrus caucasicus* Grunin
Segments 3-4 dorsally with irregular rows of spines; the following 2 or 3 segments only with lateral groups of spines; rarely segment 5 with an almost complete row of spines . . . .
. . . . *Oestrus aureoargentatus* Rodhain & Bequaert
3. Dorsal surface of segment 2 without spines . . . .
. . . . *Loewioestrus variolosus* (Loew)
Dorsal surface of segment 2 with 1-2 irregular rows of spines, some-times interrupted in the middle 4
4. Segments 6-8 ventrally with 5-6 rows of spines . . . .
. . . . *Oestroides macdonaldi* (Gedoelst)
Segments 6-8 ventrally with 3-4 rows of spines . . . . . . *Oestrus ovis* L.

4.1. Subtribe LOEWIOESTRINA, new

Wings long and narrow, r-m situated before the middle of the discal cell.

DISTRIBUTION: Africa south of the Sahara. Type-genus, *Loewioestrus* Townsend.

Genus *Loewioestrus* Townsend

Loewioestrus Townsend, 1918: 152. Type-species, *Cephalomyia variolosa* Loew (orig. des.).

Loewioestrus variolosus (Loew)

Cephalomyia variolosa Loew, 1863: 15.
Oestrus variolosus; BRAUER, 1863: 156, fig.; RODHAIN & BEQUAERT, 1916: 105, figs.; ZUMPT, 1961: 10; 1965: 181, figs.
Oestrus bertrandi Rodhain & Bequaert, 1915: 453, 692.
Oestrus disjunctus Gedoelst, 1916: 259.
Oestrus interruptus Gedoelst, 1919: 333.
DISTRIBUTION: Africa south of the Sahara.
HOSTS: Antelopes of the tribes Alcelaphini and Hippotragini (Bovidae): 'Blue Wildebeest', *Connochaetes taurinus*; 'Red and Lichtenstein's Hartebeest', *Alcelaphus buselaphus* and *A. lichtensteinii*; 'Korrigum', 'Tsesseby', and 'Blesbok', *Damaliscus korrigum*, *D. lunatus*, and *D. dorcas*; 'Roan Antelope', *Hippotragus niger*; and 'Gemsbok', *Oryx gazella*.
BIOLOGY: Breeds throughout the year; pupal period: 23-24 days.

4.2. Subtribe OESTRINA Leach

Wings short and broad; r-m situated at middle of, or beyond middle of discal cell.

DISTRIBUTION: Ethiopian and Palaeartic regions; introduced with sheep in several parts of the world. Type-genus, *Oestrus* Linnaeus.

Genus *Oestroides* Gedoelst, **stat. n.**

Oestrus, subg. *Oestroides* Gedoelst, 1912: 431. Type-species, *macdonaldi* Gedoelst (by pres. design.).

Characters as in the above keys.

Oestroides macdonaldi (Gedoelst)

Oestrus (Oestroides) macdonaldi Gedoelst, 1912: 430.
Oestrus macdonaldi; RODHAIN & BEQUAERT, 1916: 114; LAURENCE, 1961:
599, fig.; ZUMPT, 1965: 182; 1968: 99-100 (synonymy).
Oestrus bassoni Zumpt, 1961: 3, figs.; 1965: 182, fig.; 1969: 99-100 (synonymy).
DISTRIBUTION: Africa south of the Sahara.
HOSTS: 'Lichtenstein's Hartebeest', *Alcelaphus lichtensteinii*; 'Red Hartebeest', *Alcelaphus buselaphus*; 'Korrigum', *Damaliscus korrigum*; 'Blesbok', *Damaliscus dorcas*.

Genus *Oestrus* Linnaeus

Oestrus Linnaeus, 1758: 584. Type-species, *ovis* Linnaeus (CURTIS, 1826:
pl. 106).
Cephalemyia Latreille, 1818: 273. Type-species, *Oestrus ovis* Linnaeus
(mon.).

Oestrus aureoargentatus Rodhain & Bequaert

Oestrus aureoargentatus Rodhain & Bequaert, 1912: 381, figs.; RODHAIN &
BEQUAERT, 1916: 98, figs.; ZUMPT, 1961: 11; 1965: 180, figs.
Oestrus regalis Austen, 1934: 248.
DISTRIBUTION: Africa south of the Sahara.
HOSTS: 'Roan Antelope', *Hippotragus equinus*; 'Sable Antelope', *Hippotragus niger*; 'Korrigum', *Damaliscus korrigum*; 'Tsesseby', *Alcelaphus
lunatus*; 'Common and Lichtenstein's Hartebeest', *Alcelaphus buselaphus* and *A. lichtensteinii*; and 'Blue Wildebeest', *Connochaetes
taurinus*.
BIOLOGY: Infestation of the Hippotragini and Alcelaphini by this
species is very heavy, more than 100 larvae been found in a single host.
It seems that this species breeds all the year round, as all larval stages
are found together in the same host in several months of the year.

Oestrus caucasicus Grunin

Oestrus caucasicus Grunin, 1948: 1124, figs.; GRUNIN, 1957: 105, figs.
[1966: 65, figs. 58, 168-169]; GRUNIN, GREBENYUK & SARTBAEV, 1955:
365 (adult); GREBENYUK & SARTBAEV, 1955: 89 (biol.).

Oestrus caucasicus gvozdeni Grunin, 1950: 88, figs. (3rd stage larva).

DISTRIBUTION: Central Asia, Caucasus.

HOSTS: 'Tur', *Capra caucasica*; 'Siberian ibex', *Capra ibex*; *Capra cylindricornis*; and *Capra sibirica*.

BIOLOGY: There are probably two generations a year; the mature larvae abandon the host in May and September.

Oestrus ovis Linnaeus

'Vers qui naissent dans le nez des moutons' RÉAUMUR, 1738: 552, pl. 35, figs. 8-25.

Oestrus ovis Linnaeus, 1758: 430. (For references see below)

Cephalomyia ovis; LATREILLE, 1818: 273; 1829: 503; MACQUART, 1833: 195; 1835: 51; ROBINEAU-DESVOIDY, 1863: 57; JOLY, 1846: 212, 283; RONDANI, 1857: 23.

Oestrus nasalis ovinus Numan, 1851: 132-152, 2 pls.

Oestrus perplexus Hudson, 1892: 63 (*nomen nudum*), pl. 7, fig. 12.

Oestrus ovis, var. *corsicae* Gómez-Fernández, 1946: 58, fig.

Oestrus ovis, var. *granatae* (Gomez-Fernández, 1946: 58, fig.

REFS. – REDI, 1686; VALLISNIERI, 1710; 1733; RÉAUMUR, 1734; LINNAEUS, 1737; SCHREBER, 1759; LINNAEUS, 1761; GEOFFROY, 1763; LINNAEUS, 1767; FABRICIUS, 1775; DE GEER, 1776; FABRICIUS, 1781; SCHRANK, 1781; MODEER, 1787; HERBST, 1787; FISCHER, 1788; GMELIN, 1788; FABRICIUS, 1794; CLARK, 1797; CEDERHJELM, 1798; DWIGUBSKY, 1802; SCHRANK, 1803; FABRICIUS, 1805; SHAW, 1806; LATREILLE, 1809; BAYLE-BARELLE, 1809; OLIVIER, 1811; DONOVAN, 1813; CLARK, 1815; LATREILLE, 1816; LEACH, 1817; FALLÉN, 1818; SALM, 1818; MEIGEN, 1824; MACQUART, 1833; GENÈ, 1835; DAVIES, 1835; BARTON, 1837; DAHLBOM, 1837; GUÉRIN-MÉNEVILLE, 1837; BLANCHARD, 1840; WESTWOOD, 1840; MALAGUTI, 1842; ZETTERSTEDT, 1844; JOLY, 1846; FITCH, 1849; WALKER, 1849; SAXESEN, 1850; NUMAN, 1850; 1851; WALKER, 1853; LÖWE, 1854; RONDANI, 1854; HENNING, 1855; SIMONDS, 1856; RONDANI, 1857; LOEW, 1857; BRAUER, 1858; SCHINER, 1862; ROBINEAU-DESVOIDY, 1863; BRAUER, 1863; KOWALSKI, 1863; RONDANI, 1865; GOUREAU, 1866; HERING, 1866; FEDTSCHENKO, 1868; RILEY, 1869; LUCAS, 1876; MAIR, 1880; MÉGNIN, 1880; MÉGNIN & BOUTHERY, 1880; WHITWORTH, 1880; WINCHESTER, 1882; PORTSCHINSKY, 1884; MERTON, 1884; ORMEROD, 1885; 1888; CURTICE, 1890; HUDSON, 1892; BLANCHARD, 1892; SANGAGLI, 1892; BAKER, 1895; GILRUTH, 1895; OSBORN, 1896; LUGGER, 1897; HUTCHEON, 1899; GRIMSHAW, 1900; LEON, 1900; BUTTERFIELD, 1900; LEONARDI & LUNARDONI, 1900; THIERRY, 1900; HUTTON, 1901; YERBURY, 1901; FELT, 1903; JOHNSON, 1903; MÜLLER,

1903; SAITTA, 1903; THEOBALD, 1903; FROGGATT, 1905; THIERRY, 1905; BAU, 1906; COLLINGE, 1906; BEZZI & STEIN, 1907; SPEISER, 1907; GROSSO, 1907; SERGENT & SERGENT, 1907a-b; BRÈTHES, 1908; SERGENT & SERGENT, 1908a-b; PORTIER, 1909; CHRISTY, 1909; SERGENT & SERGENT, 1910; THEOBALD, 1910; KING, 1911; FROGGATT, 1911; BOUET & ROUBAUD, 1912; BRIOT, 1912; MIDDLETON, 1912; SERGENT & SERGENT, 1912; ADER, 1913; FARMAKOVSKII, 1913; PORTSCHINSKY, 1913; RUPPERT, 1913; SERGENT & SERGENT, 1913; SERRES, 1913; VAINSHTEIN, 1913; BROWN, 1914; FRANCAVIGLIA, 1914; HUTCHEON, 1914; ROUBAUD, 1914a-b; KUHN, 1914; PERKINS, 1914; STEFANI, 1915; LOCHHEAD, 1916; LUTZ, 1917; SPÄRCK, 1918; FÜLLEBORN, 1919; PATTON, 1920; LARROUSSE, 1921; SHANNON, 1922; GABRIELIDIS, 1922; MILLER, 1922; STARCK, 1923; BAKER, 1924; BUEN, 1924; LARROUSSE, 1924; GAMINARA, 1925; HERMS, 1925; LARROUSSE, 1926; PLESKE, 1926; SHANNON & DEL PONTE, 1926; FELT, 1928 (as *Wohlfartia vigil*, error); SÉGUY, 1928; IHERING, 1929; ICZN, 1929 (fixation of type-species); PATTON & EVANS, 1929; PAVLOVSKII, 1929; FLETCHER & SEN, 1929-31; DUPUIS D'UBY, 1930; GARUDACHAR, 1930; IHERING, 1930; NEIVA, 1930; PAINTER, 1930; PLESKE, 1930; STOUDER, 1930; BAIRD, 1931; BISSET, 1931; COULON & DINULESCU, 1931; DUPUIS D'UBY, 1931; PARADOKSOV, 1931; GILDOW & HICKMAN, 1931; STOUDER, 1931; TRABUT, 1931; STEWART (J. R.), 1932; STOUDER, 1932; DRENSKY, 1933; MITCHELL & COBBETT, 1933; PETTIT, 1933; TRABUT, 1933; BHATIA, 1934; CORNELL & DAUBNEY, 1934; GALLIARD, 1934; SHUMAKOVITCH, 1934; LYON, 1934; SABBAGH, 1934; KALMYKOV, 1935; CRAWFORD, 1935; DUTOIT, 1935; DUTOIT & CLARK, 1935; LYON, 1935; RODRÍGUEZ NOVOA, 1935; TOULAND & MÉDINGER, 1935; BASKAKOV, 1936; BOLDYREV & USPENSKII, 1936; DOTEN et al., 1936; KULIEVA, 1936; PATTON, 1937; PASCHEFF, 1937; DUTOIT, 1938; DEL PONTE, 1939; DOTEN et al., 1939; SNIDERMAN, 1939; TEMPLE et al., 1939; COBBETT, 1940a-b; FALLIS, 1940; STOUDER, 1940; COBBETT & MITCHELL, 1941; SULTANOV, 1941; MANINE, 1941; MELLO, 1941; GAN, 1942; HEDGES, 1942; HONESS & FROST, 1942; SCOTT, 1942; RODRÍGUEZ NOVOA, 1942; UNSWORTH, 1943; JAMES, 1944; ÁLVAREZ & MATÉ, 1944; ISOLA & OSIMANI, 1944; OTTEN, 1944; OSIMANI & SALSAMENDI, 1945; SALCÉS FERMÍN & CALVO, 1945; BASKAKOV, 1946; GÓMEZ-FERNÁNDEZ, 1946; HANDSCHIN, 1946; LOTIN, 1946; SAPOGOV, 1947; ABIDZHANOV, 1948; JAMES, 1948; KEISER, 1948; LECLERCQ, 1948; MAZINA, 1948; BUCK, 1949; SULTANOV & BEKUZIN, 1949; UNSWORTH, 1949; COUEY, 1950; GAN, 1950; GIL COLLADO, 1950; KORNILOVA, 1950; LECLERCQ, 1950; LIZCANO HERRERA, 1950; ABRAMOV, 1951; KLËNIN, 1950a-b; HENNIG, 1952; FETHERS, 1952; GUERRA GRANDE, 1952; NIKOLIĆ, 1952; SERGENT, 1952a-b; SOFRONOV & LYSOV, 1952; BASU et al., 1953; CROWTHER, 1953; GAN, 1953; GRUNIN, 1953; KLËNIN, 1953; PANICK, 1953; REYNON,

1953; Schneider, 1953a-b; Brauns, 1954; Gan, 1954; Klёnin, 1954a-b; Kuklin, 1954; Lapierre & Pette, 1954; Rabello & Malheiro, 1954; Schneider, 1954; Vermeil, 1954; Bobokhodzhaeva, 1955; Chereshnev, 1955; Gil Collado, 1955; Gómez-Fernández, 1955a-b; Grebenyuk & Sartbaev, 1955; Klёnin, 1955; Krivko, 1955; León & Andrade, 1955a-b; Grzywiński & Madej, 1955; Koide, 1955; Lumbreras Cruz & Polack, 1955; Miloshev, 1955; Bugaeva, 1956; Cobbett, 1956; DuToit & Fiedler, 1956; Koide, 1956; Kolomiets & Alfimova, 1956; Kolomiets, Alfimova & Kapustin, 1956; Krivko, 1956; Simison, 1956; Suvortsev, 1956; Grunin, 1957; Doury, 1957; Dukalov, 1957; Kononyuk, 1957; Krivko, 1957a-b; Kulakov, 1957; Meira et al., 1957; Merdivenci, 1957; Pampiglione, 1957; Pereira, 1957; Dinulescu, 1958; Eydan, 1958; Favier, 1958; Spencer, 1958; Wilkinson, 1958; Pampiglione, 1958a-d; Sicart, Ruffié & Meira, 1958; Kolomiets, Alfimova, Kapustin & Krivko, 1958; Bacigalupo & Villamil, 1959; Peterson et al., 1959; Chavarría Ch. & Ávila Carrillo, 1959; Stampa, 1959; Yanovich, 1959; Atías M. et al., 1960; Chereshnev, 1960a-b; Deduit et al., 1960; Dinulescu et al., 1960; DuToit & Meyer, 1960; Guennec & Robineau, 1960; Itúrbide, 1960; Kabos, 1960; Krivko & Redko, 1960; Miré et al., 1960; Nosik & Goncharov, 1960; Scherf, 1960; Alexander, 1961; Chebotarev, 1961; Correa, 1961; Damian, 1961; Dinulescu, 1961; Kato & Murakami, 1961; Machida, 1961; Miller et al., 1961; Stampa & Pols, 1961; Tsai, 1961; Aspinall, 1962; Atencio León & Ramírez, 1962; Basson, 1962; Burford, 1962; Drummond, 1962; Gukassian, 1962; Kato & Murakami, 1962; Meleney et al., 1962; Negru et al., 1962; Semenov, 1962; Sharma & Verma, 1962; Nicoli et al., 1962; Nikitin, 1962; Semenov, 1962; Apodaca et al., 1963; Meleney et al., 1963; Petrov & Bratanov, 1963; Pfadt & Campbell, 1963; Roberts & Colbenson, 1963; Aukhadiev, 1964; Boyes, 1964; Knapp & Drudge, 1964; Liu Kao & Ts'ui, 1964; Negru & May, 1964; Negru, May & Sirbu, 1964; Scott, 1964; Semenov, 1964; Theodor, 1964; Pęadt, 1964; Drummond & Graham, 1965; Graber & Gruvel, 1965; Knapp & Drudge, 1965; Vartic, Suteu & Tricǎ, 1965; Kal'kis, 1965a-b; Kamarli, 1965; Mazzoleni, 1965; Saccà et al., 1965; Capelle, 1966; Chaudhuri, 1966; Drummond, 1966; Grunin, 1966; Lora D. et al., 1966; Ortecho & Marble, 1966; Ivanov & Petrov, 1966; Nitzulescu, 1966; Petrov et al., 1966; Guimarães, 1967; Knapp & Rogers, 1967; Lyons et al., 1967; Vartic, Suteu & Tricǎ, 1967; Colin & Moss, 1968; Knapp & Rogers, 1968; Haeselbarth et al., 1968; Segerman & Zumpt, 1968; Pasǎre & Vicea, 1968; Rogers et al., 1968; Schirre, 1968; Salvador Yépez et al., 1968; Basson, 1969; Brown, 1969; Hitchcock & Foos, 1969; Buchanan, 1969; Buchanan et al., 1969; Pasǎre & Vicea, 1969;

Ware, 1969; Dewhirst & Echeverria, 1969; Meleney & Apodaca, 1969; Ware *et al.*, 1969; Abul-Hab, 1970; Hoffmann & Goldsmid, 1970; Rakusin, 1970; Wetzell & Bauristhene, 1970; Vilenberg *et al.*, 1971; Dorzh & Minař, 1971; Guevara Benítez *et al.*, 1971; Horack *et al.*, 1971; Pinkerton, 1971; Salvador Yépez & Gallardo Z., 1971; Knapp, 1972; Nepoklënov *et al.*, 1972; Nepoklënov, 1972; Ranatunga & Rajamahendran, 1972; Ranatunga & Weilgama, 1972; Sayin *et al.*, 1972; Tonkozhenko *et al.*, 1972; Kettle, 1973; Pokidov, 1973; Shcherban, 1973; Ternovoi, 1973.

DISTRIBUTION: Palaearctic region; introduced with sheep in most parts of the world.

HOSTS: *Ovis ammon, Ovis aries, Capra ibex, Capra sp.* (Caucasus).

BIOLOGY: See above references.

Unplaced Oestrini

1. *Oestrus dubitatus* Basson & Zumpt, 1969: 57, figs. 1-3.

TYPE-LOCALITY: S.W. Africa; Ovamboland, Etosha National Park.

HOST: *Connochaetes taurinus* Burchell. Known only from larva.

5. Tribe Gedoelstiini, new

M_1 without stump vein; median carina of antennal fovea uninterrupted, of even height; tp ending near the curve of M_1; abdomen with longitudinal, strongly erect callosities (Fig. 43).

DISTRIBUTION: Africa south of the Sahara.

TYPE-GENUS, *Gedoelstia* Rodhain & Bequaert.

Genus *Gedoelstia* Rodhain & Bequaert

Gedoelstia Rodhain & Bequaert, 1913: 173. Type-species, *cristata* Rodhain & Bequaert (orig. des.).

Key to species, based on adults (after Zumpt, 1965: 183)

1. Mesonotum with shining black weals; abdomen with large paired tubercles on tergites 3-5; length, 13-16 mm . . . .
   . . . . *cristata* Rodhain & Bequaert
   Mesonotum without shining weals, only with an ill-defined, dull black or brown pattern; abdomen with smaller tubercles; length,

12-14 mm . . . . *haessleri* Gedoelst
The two species of *Gedoelstia* hybridize naturally in the Kalahari region (Namibia and Botswana) (BASSON, ZUMPT & BAURISTHENE, 1963).

Gedoelstia cristata Rodhain & Bequaert

Gedoelstia cristata Rodhain & Bequaert, 1913: 176, figs.
REFS. – RODHAIN & BEQUAERT, 1916: 144, figs.; EMDEN, 1944: 426; ZUMPT, 1965: 183, figs.
DISTRIBUTION: Africa south of the Sahara, wherever antelopes of the tribe Alcelaphini occur.
HOSTS: 'Black and blue Wildebeest', *Connochaetes gnou, C. taurinus*; 'common and Lichtenstein's Hartebeest', *Alcelaphus buselaphus, A. lichtensteinii*; 'korrigum', *Damaliscus korrigum*; 'gemsbok', *Oryx gazella*.
BIOLOGY: 'In the normal hosts all three larval stages are found in the nasal cavities, the second and third mainly in the frontal sinuses. The first instar larvae were also recovered from the cardio-vascular system of Wildebeest and Hartebeest. BASSON (1962) believes that the first instar larvae are normally deposited into the orbits, enter a vein for completion of at least a part of the life-cycle in the cardiovascular system, and then migrate up the trachea to the nasal cavities, where they moult to the second stage.'
'The Gemsbok and domestic animals are abnormal hosts, in which the larvae do not reach the nasal cavities and never develop beyond the first stage.'
'In Southern Africa, the imagines hatched from June to August and from December to April. One female of the hybrid type was found to harbour over 2,000 larvae.' (ZUMPT, 1965: 185).

Gedoelstia haessleri Gedoelst

Gedoelstia haessleri Gedoelst, 1915: 148; EMDEN, 1944: 426; ZUMPT, 1965: 185, figs.
Gedoelstia paradoxa Rodhain & Bequaert, 1915: 453; RODHAIN & BEQUAERT, 1920: 177.
Gedoelstia impolita Austen, 1934: 243.
DISTRIBUTION: Probably the same as *cristata*.
HOSTS: *Connochaetes taurinus* and *gnou*; *Alcelaphus buselaphus* and *lichtensteini*; *Damaliscus dorcas, lunatus* and *korrigum*. First stage larvae were also found in cattle, goats, horses and man.
BIOLOGY: Similar to that of *cristata*. Larvae of *haessleri* are mainly found

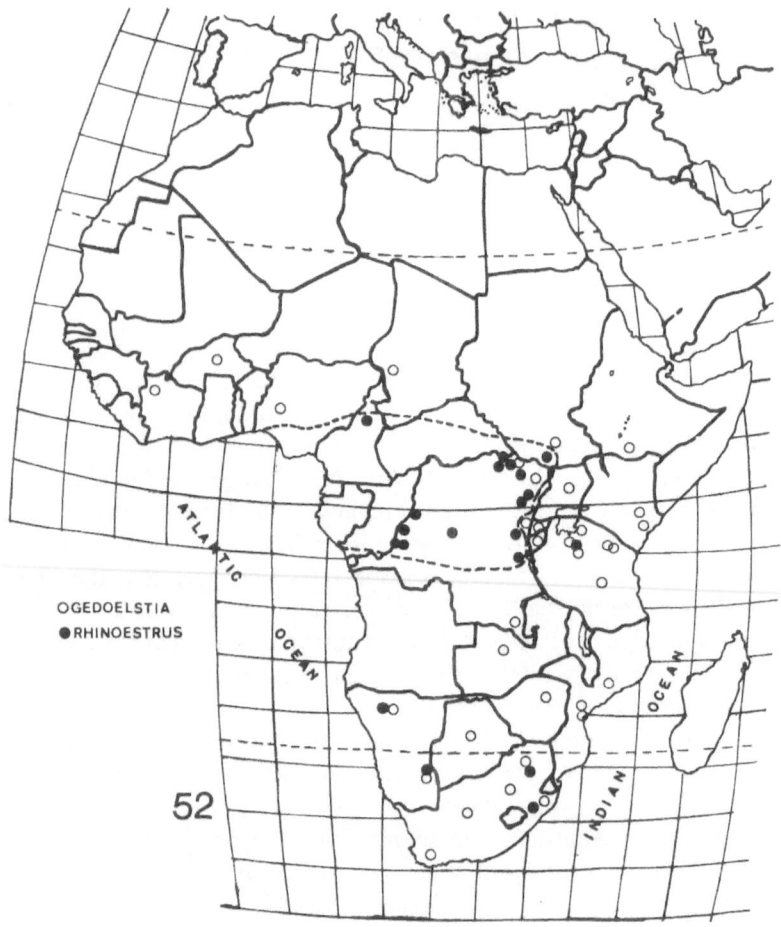

Fig. 52. Geographical distribution of *Gedoelstia* and *Rhinoestrus* in Africa.

in the subdural cavity of the 'blue wildebeest', their preferred host. BASSON (1962) believes that the larvae are deposited in the host's eye orbits, then entering a vein and completing a part of their first stage in the cardio-vascular system; eventually they migrate to the subdural cavity through the foramens of the cribriform plate and ethmoid bone, then to the nasal cavities, where they moult to the second stage.

6. Tribe Cephalopinini, new

M_1 without stump vein; median carina of antennal fovea uninter-

rupted, of even height; tp ending almost in the middle of the first
posterior cell; abdomen with oblique or transverse, not very pro-
nounced, callosities.

DISTRIBUTION: Palaearctic region (introduced with the camel in the
Oriental region).

TYPE-GENUS, *Cephalopina* Strand.

Genus *Cephalopina* Strand

Cephalopsis Townsend, 1912: 53 (preocc. Rafinesque, 1815). Type-
species, *Oestrus maculatus* Wiedemann (orig. des.) = *titillator* (Clark).
Cephalopina Strand, 1928: 48 (nom. nov. for *Cephalopsis* Townsend).
Type-species, *Oestrus maculatus* Wiedemann (aut.) = *titillator* (Clark).

Cephalopina titillator (Clark)

Oestrus titillator Clark, 1816: 4, pl. 2, fig. 22.
Oestrus maculatus Wiedemann, 1830: 256; SCHINER, 1862: 393.
Oestrus libycus Clark, 1843: 93, pl. 2.
Cephalomyia maculata; MACQUART, 1843: 25.
Cephalomyia maculata; BRAUER, 1860: 557; 1863: 163, pl. 3, fig. 4, pl. 7,
    figs. 3, 9; PORTSCHINSKY, 1884: 122; BAU, 1906: 14, pl. 2, figs. 9, 9a;
    BEZZI & STEIN, 1907: 590; KING, 1911: 127; SHARP, 1909: 545, fig. 245;
    ROUBAUD, 1914: 187, fig. 51.
Pharyngobolus cameli Steel, 1887: 27, pl. (erroneously labelled as 'Oestrus
    cameli').
Cephalopsis maculata; TOWNSEND, 1912: 53.
Cephalopsis titillator; RODHAIN & BEQUAERT, 1915: 691; 1916: 72.
Cephalopina titillator; ZUMPT, 1965: 187, figs.
REFS. – ABIDZHANOV, 1950; BLAGOVESHCHENSKII, ORLOV & KRASNUSOV,
    1937; SÉGUY, 1923; SELYAEV, 1933; STEEL, 1887; TSAPRUN, 1935;
    HAESELBARTH *et al.*, 1966; GABER, TABO & SERVICE, 1967; LODHA &
    CHAUDHURI, 1962; LANDOIS, 1903; PATTON, 1920; MIDDLETON, 1907;
    SOLIMAN & EL-HINAIDY, 1966.
DISTRIBUTION: Palaearctic region (introduced with the camel in the
Oriental region).
HOSTS: camel and dromedary, *Camelus bactrianus* and *C. dromedarius*
(Camelidae).
BIOLOGY: There are two generations per year; larvae of the first gener-
ation begin to abandon the host in the middle of March and continue
for 3 weeks; larvae collected by PATTON (1920) pupated between 23

and 27 March and emerged on 10 April. The females deposit their larvae in mid-April; these larvae leave the nasal cavities of the host in mid-August, and reappear in September; these flies begin larviposition in mid-September. It is common to see around dromedaries from 40-50 male and female flies, which can be collected around the head of the animal.

Tsaprun (1935) investigated the biology of this species in Central Asia. According to this author, larvae remain in the host for 10-11 months, therefore only one generation per year exists. Up to 165 larvae were found in a single camel, and 64 blocked the naso-pharyngeal cavities. The pupal stage lasted on average 25 days; captive females lived from 4-16 days. The females had from 800-900 eggs.

Grunin (1957) agrees with Patton, admitting two generations per year: one from May to June, the other from September to October; the larval period of the first generation lasts 3 to $3^1/_2$ months, the larvae dropping from the host from mid-August to the end of September; the larvae of the second generation remain in the host during winter, and drop in great numbers from mid-August to the end of April, some remaining in the host until June.

It is interesting to note that Grunin remarks that the (Bactrian) camel is normally three times more infested than the dromedary.

VI. PHYLOGENY OF THE OESTRIDAE

1. Morphological trends of the Oestridae

It may be assumed that the ancestral stock from which the Oestridae evolved should present the following characters: pilosity undeveloped; reduced chaetotaxy; eyes and antennae of normal size; antennal fovea shallow, without a median carina; reduced mouthparts; postscutellum undeveloped; wings with cell R_5 open and M_2 present as a stump vein. As to the larvae, the 1st stage should be spindle-shaped, the body segments covered by several rows of spines, the arrangement of these rows being almost identical on the dorsal and ventral surfaces of the body; the posterior spiracle was a porous plate.

The highly specialized larval life, as parasites of internal cavities of certain groups of mammals, and the consequent decrease in the life span of adults, resulted in several morphological adaptations: 1. The mouthparts became gradually more reduced, almost absent; all the energy of the adult comes from the immature stages; as already seen, adults at most lick a few drops of water; 2. The eyes tended to increase, with a consequent decrease of the frontal stripe; a more or less accentuated sexual dimorphism involving the width of the front developed; the sight power in these flies seems to be fundamental for the localization of aggregation sites and hosts; 3. With predominance of vision, the antennae became smaller with a tendency to be concealed inside the foveae; these tend to become deeper, and finally a median carina was developed, first an incomplete one and finally a fully developed one, which separates the fovea into two symmetrical depressions; 4. The wings became strengthened to assure a greater efficiency and rapidity of flight, necessary for the search of aggregation sites and hosts; this is achieved by the closing of cell R_5, the modification in wing shape, modification in the position of crossveins and the elimination of vestigial veins such as M_2; 5. The pilosity underwent two different pathways: in one group it increased, giving the fly the appearance of bombids; in the other group it became almost entirely absent; 6. The body wall developed certain structures, such as granules, pits, pustules, weals, tubercles, etc., of probable mimetic function and also for inter- and intraspecific recognition.

The larvae underwent several modifications in their morphology, as already seen above.

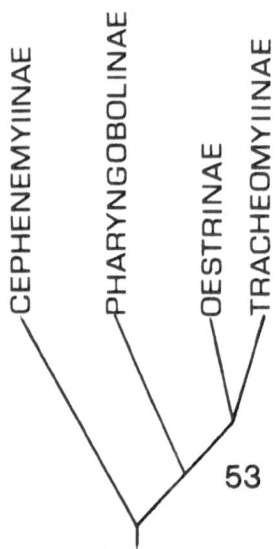

Fig. 53. Phylogeny of Oestridae subfamilies.

2. Phylogeny of the subfamilies (Fig. 53)

The ancestral stock, as characterized above, became fragmented into four main groups: 1. The Cephenemyiinae, which have maintained a great part of the primitive characters but which have acquired a more or less dense pilosity; 2. The Pharyngobolinae, where cell R_5 became closed at the wing margin and where a complete facial carina was developed; 3. The Oestrinae, with cell R_5 closed and petiolate, with a tendency to lose M_2; they have primitively kept the pointed spines of the larvae and a carina was developed only in some groups; 4. The Tracheomyiinae, probably derived from the same basic source as the Oestrinae, with a closed and petiolate R_5, ocelli absent and a medially interrupted carina; the spines of the larvae became quadrangular, with a sinuous border.

The Pharyngobolinae and Tracheomyiinae are currently represented by only one genus each and will not be discussed further here.

3. Phylogeny of the Cephenemyiinae (Fig. 54)

The Cephenemyiinae divided into two branches: 1. The Cephenemyiini, with a tendency to increase of the body pilosity, becoming similar to certain Hymenoptera; M_2 remained relatively long and the

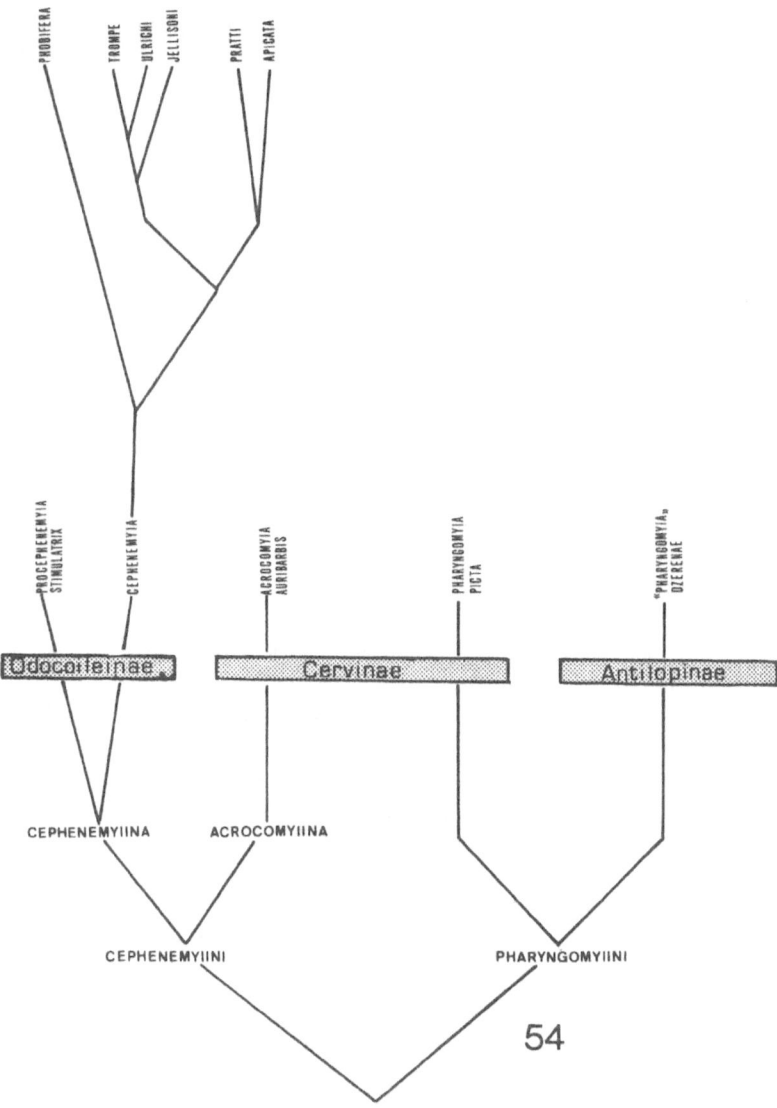

Fig. 54. Phylogeny of the Cephenemyiinae.

6th abdominal tergite almost as long as the 5th; the 1st stage larvae remained spindle-shaped, but developed long fringes of elongate spines on the posterior margin of the abdominal segments; 2. The Pharyngomyiini, which have acquired only a moderate pilosity (which does not conceal the integument) but which present several derived characters

– M_2 became shorter; the 6th abdominal tergite became reduced to almost half the length of the 5th; the larvae became greatly modified, clavate, due to an expansion of thoracic segments II-III and abdominal segment I, but did not develop long fringes on the posterior margin of the abdominal segments.

The Cephenemyiini also gave rise to two groups: (i) the Cephenemyiina, characterized by the absence of a facial carina or, if the carina is present, it is weakly developed and without hairs; and (ii) the Acrocomyiina, with a well-developed facial carina covered with hairs. The Cephenemyiina have two genera – one without carina (*Procephenemyia*, exclusively Palaearctic), and the other with an incipient carina devoid of hairs (*Cephenemyia*, Nearctic and Palaearctic). The Acrocomyiina have only one genus (*Acrocomyia*, exclusively Palaearctic). It is worth noting that the Cephenemyiina in their larval stages are primarily parasites of Cervidae of the subfamily Odocoileinae, while the Acromyiina specialized in parasitizing Cervinae.

The Pharyngomyiini also seem to have given rise to two groups; one, represented by *Pharyngomyia picta*, is parasitic on Cervinae; the other, represented by '*Pharyngomyia*' *dzerenae*, parasitizes Antilopinae (Bovidae). There are significant differences between these two groups in the larvae; however, only discovery of the adult of *dzerenae* can confirm the separation of these two groups into two genera.

With regard to the species of *Cephenemyia*, I consider *phobifera* from the eastern United States, characterized by the uniform yellow colour of the mesonotal pilosity, as the most primitive; all the other five species of this genus show transverse bands of black hairs on the mesonotum. This second group of species may be divided as follows: (i) species with abdominal hairs predominantly yellow or reddish-yellow – *trompe* (circumpolar), *jellisoni* (western United States) and *ulrichi* (temperate zone of Eurasia); it seems that both *jellisoni* and *ulrichi* are descendants from the basic stock that gave rise to *trompe*; *jellisoni* differs by the narrower front; *ulrichi* by the presence of a larger dark spot on the wing around the crossvein r-m; (ii) species with abdominal pilosity bi- or tricolorous, with black and white mixed to yellow; this group has two species – one in the northwestern (*apicata*) and the other in the southwestern (*pratti*) United States.

4. Phylogeny of the Oestrinae (Fig. 55)

The Oestrinae can be immediately separated into two groups: one (Kirkioestrini) with M_2 present as a stump vein; the other (4 other tribes) with M_2 completely absent. Among these other four tribes, two

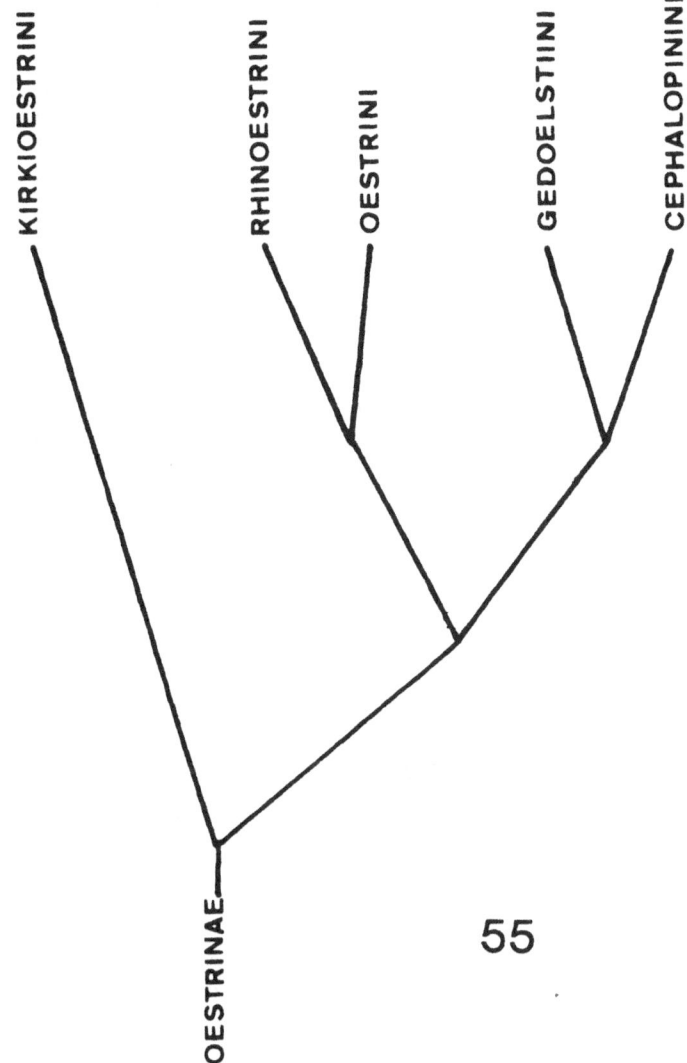

Fig. 55. Phylogeny of Oestrinae tribes.

(Rhinoestrini and Oestrini) possess an interrupted facial carina, while the others (Gedoelstinii and Cephalopinini) show a complete, uninterrupted carina.

The Rhinoestrini developed pustules on the parafrontalia; the Oestrini pits. The Gedoelstiini developed longitudinal callosities on the abdomen, whereas in the Cephalopinini these callosities are transverse or oblique.

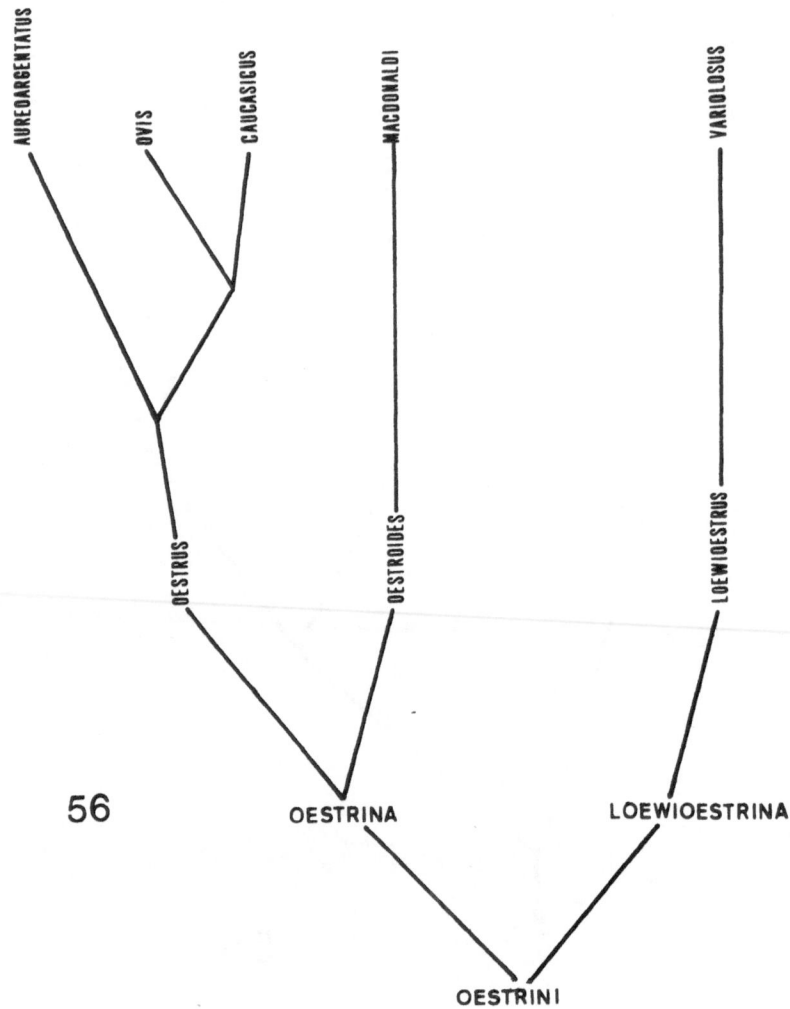

Fig. 56. Phylogeny of the Oestrini.

With the exception of the Rhinoestrini and Oestrini, the other two tribes are monobasic and will not be further discussed here.

The more primitive Rhinoestrini are represented by the subtribe Suinoestrina (only genus, *Suinoestrus*), characterized by the absence of spots on the parafacialia and by the position of r-m, situated opposite or slightly before the level of the apex of Sc. The two other subtribes, Rhinoestrina and Gruniniina, have spots on the parafacialia and r-m is located beyond the level of the apex of Sc; the Gruniniina, however, developed a well-developed longitudinal keel on the lower part of the parafrontalia, which is absent in the Rhinoestrina.

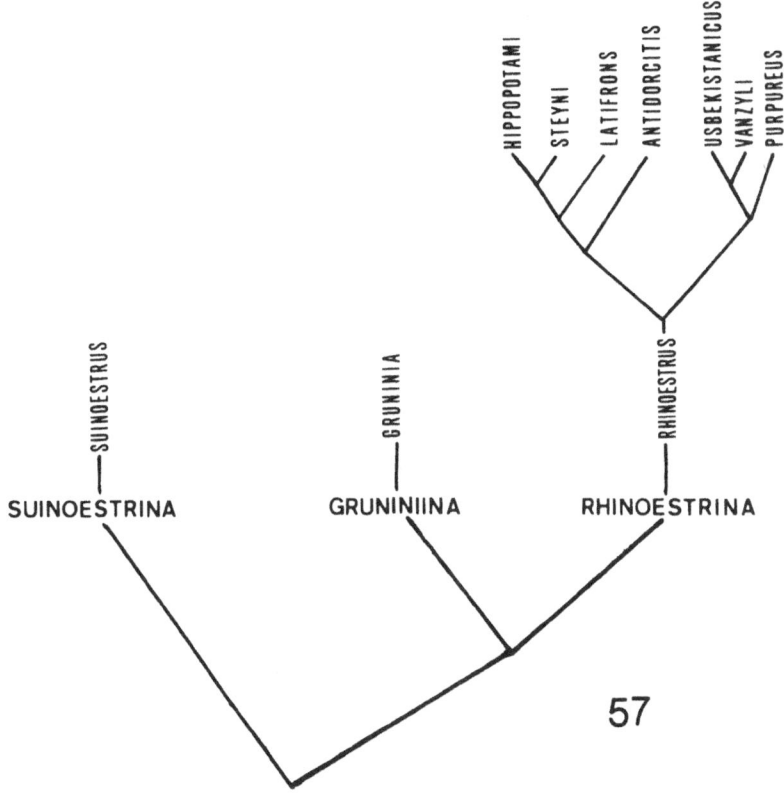

Fig. 57. Phylogeny of the Rhinoestrini.

The genus *Rhinoestrus* presents two species-groups: (i) *hippopotami-*group (with the species *hippopotami, antidorcitis, latifrons* and *steyni*); (ii) *purpureus*-group (*purpureus, vanzyli* and *usbekistanicus*). With only limited knowledge of the genus *Rhinoestrus* the construction of a plausible phylogeny is very difficult; a highly tentative attempt is shown in Fig. 57.

The Oestrini also may be separated into two subtribes, according to the position of cross-vein r-m – Oestrina and Loewioestrina. The Loewioestrina are monobasic; the Oestrinae have two genera: *Oestroides*, with flattened tubercles on the thorax similar to those of certain Rhinoestrina and *Oestrus*, with only granular tubercles. The probable phylogeny of the group is shown in Fig. 56.

Fig. 58. *Cephenemyia*, habitus.

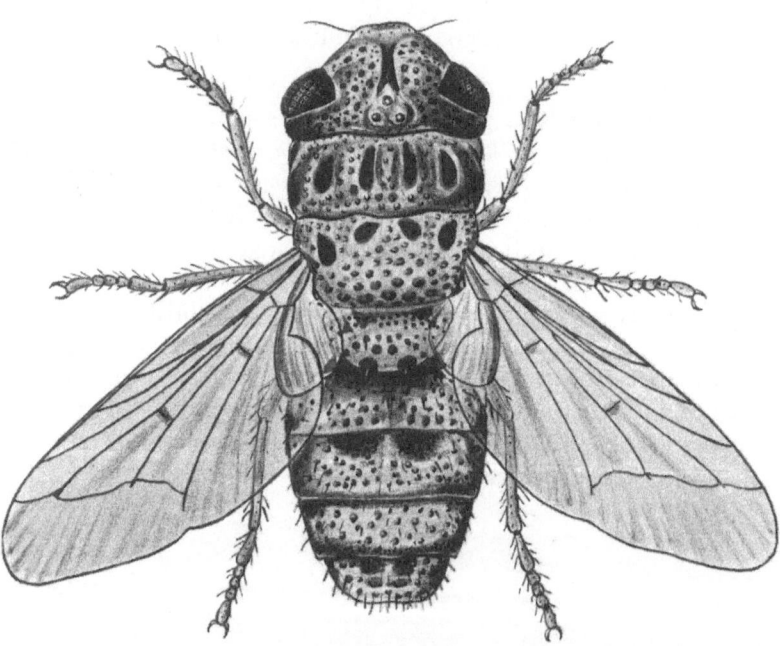

Fig. 59. *Oestrus ovis*, habitus.

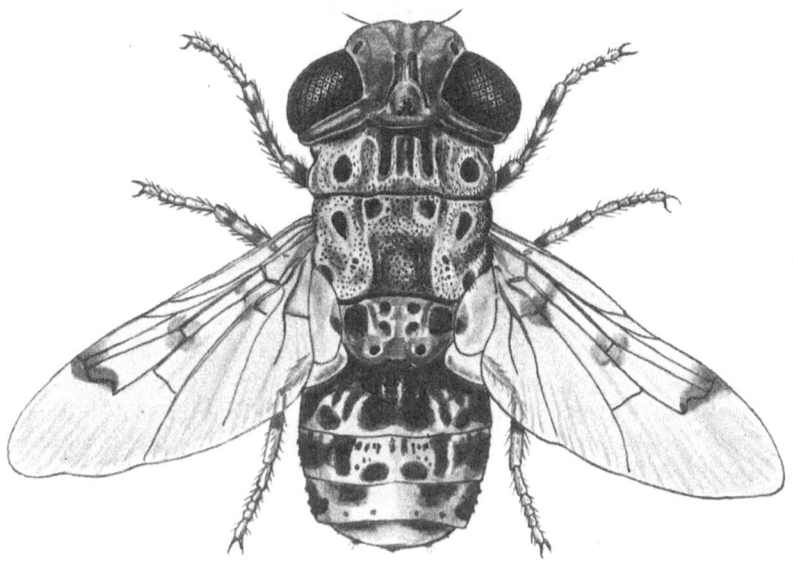

Fig. 60. *Cephalopina titillator*, habitus.

VII. BIBLIOGRAPHY OF OESTRIDAE

ABDURAKHMANOV, M. V., (1946): [Death of a horse due to stoppage of the bronchi by larvae of flies]. *Veterinariya* 23: 41.

ABDUSSALAM, M., (1940): The occurrence of equine nasal bot, *Rhinoestrus purpureus* (Brauer, 1858) in India. Proc. 26th Indian Sci. Congr. (Lahore, 1939) 4: 8.

ABIDZHANOV, A. A., (1948): Ekologiya lichinki i kukolki ovech'ego ovoda (*Oestrus ovis* L.). [Ecology of the larva and pupa of *O. ovis* L.]. *Trudy Inst. Bot. Zool. Akad. Nauk Uzbek. SSR, Sb. Zootekhn. Parazit.* 1: 144-164.

ABIDZHANOV, A. A., (1950): Biologiya polostnogo ovoda verblyuda *Cephalopsis titillator* Cl. v usloviyakh Srednei Azii. [Biology of the cavicole camel bot fly, *C. titillator*, in Central Asia]. *Ibid.* 2: 93-107.

ABRAMOV, I. V., (1951): Entomozy. Estroz (ovodovaya invaziya) ovets. [*Oestrus ovis* in sheep]. In: F. A. TERENT'EV & A. A. MARKOV, Infektsionnye i invazionnye bolezni ovets i koz: 475-484, figs. Moscow.

ABUL-HAB, J., (1970): Seasonal occurrence of sheep bot fly, *Oestrus ovis* L. (Diptera, Oestridae) in Central Iraq. *J. med. Ent.* 7(1): 111-115, figs.

ADER, W. M., (1913): [Note on *O. ovis*). Zanzibar Protect. Med. Sanit, Rep. for 1913.

ADYUSHEV, KH., (1957): Metody likvidatsii nosovogo ovoda ovets. [Control methods for sheep nasal bot]. *Sel'.Khoz. Bashkir.* 7: 47.

AGASSIZ, L., (1846): Nomenclator zoologici index universalis, continens nomina systematica classium, ordinum, familiarum et generum animalium omnium, tam viventium quam fossilium, 393 pp. Soloduri.

AKCHURIN, B. S., (1945): Rinestroz loshadei v Bashkirskoi ASSR. [Rhinoestrosis of horses in the Bashkir ASSR]. *Veterinariya* 22(6): 21-22. [Abstr. in *Rev. appl. Ent.* (B) 34(1): 14, 1946].

AKCHURIN, B. S. & K. V. AYUPOV, (1957): K voprosu biologii ovech'ego ovoda *Oestrus ovis* v Bashkirskoi ASSR. [Biology of *O. ovis* in the Bashkir ASSR]. *Byull. nauchno-tekh. Inf. Kazakh. nauchno-issled. vet. Inst.* 1: 33-34.

ALDRICH, J. M., (1915): The deer bot-flies (genus *Cephenomyia* Latr.). *J. N.Y. ent. Soc.* 23: 144-150.

ALEKSANDROV, N. A., (1945): Rinestroz loshadei. [Rhinoestrosis of horses]. *Veterinariya* 22(6): 19-20. [Abstr. in *Rev. appl. Ent.* (B) 34(1): 13, 1946].

ALEXANDER, A. S., (1931): Rout the bots and grubs. *Rur. N. Yorker* 90: 310.

ALEXANDER, R. A., (1964): Uitpeuloog (bulging eye disease): A recently described oculo-vascular myiasis of domestic animals in southern Africa. *Advanc. vet. Sci.* 9: 35-60.

ALLEN, (-), (1872): Larvae of oestris [sic] or bot-fly. *Boston med. surg. J.* (n.s.) 10(18)[= vol. 87]: 306.

ALTUM, (-), (1881): Die Rehrachenbremse (*Cephenomyia stimulator* F.). *Z. Forst- u. Jagdw.* 13(3): 153-156, figs.

ALVAREZ, (-) & (-) MATÉ, (1944): Un caso de miasis estrosa nasal. *Med. esp.* 12: 295-296.

ANACKER, H., (1890-91): [Articles] Nasenbremse, *Oestrus*, Oestrusbeulen, Oestruslarvenkrankheiten, Rinderbremse. *Tierarzt* 7: 85, 338, 348-349, 1890; 8: 450-451, 1891.

ANDERSON, J. R. & W. OLKOWSKI, (1968): Carbon dioxide as an attractant for host-seeking *Cephenemyia* females. *Nature, Lond.* 220(5163): 190-191.

ANDERSON, W. B., (1935): Ophthalmomyiasis. A review of the literature and a report of a case of ophthalmomyiasis interna posterior. *Am. J. Ophthal.* 18: 699.

APODACA, S. A., W. P. MELENEY & H. O. PETERSON, (1963): Ecdysis observed in second instar of *Oestrus ovis* Linné, 1761 [sic] in vitro. *J. Parasit.* 49(4): 659.

ASHIKHMIN, A. D. & G. M. BOLTACHEV, (1959): Lechenie ovets pri polostnom ovode. [Treatment of sheep against cavicole bots]. *Veterinariya* 36(1): 43-44.

ASPINALL, K. W., (1962): Annual report of the Department of Veterinary Services and Animal Industry 1962: 1-62. Malawi.

ATENCIO LEÓN, A. & A. J. RAMÍREZ, (1972): Miasis cavitaria de las ovejas. *Revta vet. venezol.* 32(188): 164-168, figs.

ATÍAS, M. A., R. DONCKASTER RODRÍGUEZ, H. SCHENONE F. & M. OLIVARES, (1960): Miasis ocular producida por larvas de *Oestrus ovis. Boln. chil. Parasit.* 15(2): 37-38, figs.

AUDOUIN, J. V., (1823): Céphalémye. *Cephalemyia*; Céphénemyie. *Cephenemyia*. In: Nouveau Dictionnaire Classique des Sciences Naturelles, 3: 329-330, 336-337. Paris.

AUKHADIEV, T., (1964): K voprosu lecheniya estroza ovets. [On the treatment of ovine oestrosis]. *Vest. sel'.-khoz. Nauki, Alma-Ata* 7(10): 63-66, figs.

AUSTEN, E. E., (1898): Notes on the oestrine parasites of British deer. *Entomologist's mon. Mag.* (2) 9[= vol. 34]: 8-13.

AUSTEN, E. E., (1930): On a new dipterous parasite (Family Calliphoridae, subfamily Calliphorinae) of the Indian Elephant, with notes on other dipterous parasites of elephants. *Proc. zool. Soc. Lond.* 1930: 677-688, 2 figs.

AUSTEN, E. E., (1934): Two new Oestridae (Diptera) parasitic on African antelopes. *Ann. Mag. nat. Hist.* (14) 10: 242-250, 2 figs.

BACIGALUPO, J. & C. F. VILLAMIL, (1959): Miasis humana por *Oestrus ovis* Linneo, 1761 [sic]. Primeras Jornadas Entomoepidemiologicas Argentinas, B. Aires: 833-836, 1 fig.

BAIRD, W. H. W., (1931): Observations on *Oestrus ovis* L. Rep. Dep. vet. Sci. anim. Husbandry Tanganyika 1930: 46-47.

BAKER, C. F., (1895): Biological notes on some Colorado Diptera. *Ent. News* 6: 173-174.

BAKER, E. T., (1924): *Oestrus ovis* in sheep. *Vet. Med.* 14(8): 400-404.

BANZHAFF, W., (1928a): Biologische Beobachtungen an Rachenbremsen. *Int. ent. Z* 22: 85-91, 1 pl., 1 fig.

BANZHAFF, W., (1928b): Wenig bekannte Parasiten unseres Wildes. *Natur Mus., Frankf.* 58(6): 281-284, figs. 1-4.

BARRETT, R. E. & D. E. WORLEY, (1966): The incidence of *Dictyocaulus* sp. in three populations of elk in southcentral Montana. *Bull. Wildl. Dis. Assoc.* 2(1): 5-6.

BARTON, T. T., (1837): The *Oestrus* in the frontal sinuses of the sheep at an unusually early period. *Veterinarian, Lond.* 10: 281.

BASKAKOV, V. P., (1936): Rezultaty opytov lecheniya ovets ot polostnogo ovoda. [Results of sheep protection against cavicole bot flies]. *Trudy nauchno-issled. vet. Opytn. St. NKZ Uzbek SSR, Parazit. Bolezni ovets Uzbek.* 7(11): 68-75.

BASKAKOV, V. P., (1946): Novoe v biologii polostnogo ovoda ovets i metody bor'by s nim. [News on the biology of the sheep bot fly and measures for its control]. *Veterinariya* 23(4): 15-16.

BASSON, P. A., (1962): Studies on specific oculo-vascular myiasis of domestic animals (Uitpeuloog). I. Historical review. II. Experimental transmission. III. Symptomatology, pathology, aethiology and epizootology. *Onderstepoort J. vet. Res.* 29: 81-87, 203-210, 211-216.

BASSON, P. A., (1966): Idem. IV. Chemotherapy. *Ibid.* 33(2): 297-303.

BASSON, P. A., (1969): Idem. V. Histopathology. *Ibid.* 36(2): 217-231, figs.

BASSON, P. A. & F. ZUMPT, (1969): *Oestrus dubitatus* n. sp. from the nasal cavities of the blue wildebeest (*Connochaetes taurinus* Burchell, 1823) in the Etosha National Park, South West Africa. *Madoqua* 1: 57-60, 3 figs.

BASSON, P. A., F. ZUMPT & E. BAURISTHENE, (1963): Is there a species hybridization in the genus *Gedoelstia*? (Diptera, Oestridae). *Z. ParasitKde* 23: 348-353, 9 figs.

BASU, B. C., P. B. MENON & C. M. SEN GUPTA, (1953): Cases of myiasis in man due to *Oestrus ovis* Linnaeus (sheep bot). *Indian med. Gaz.* 88(6): 321-322.

BAU, A., (1906): Diptera, Fam. Muscaridae, Subfam. Oestrinae. *Genera Insect.* 43: 1-31, pls.

BAU, A., (1920): Die Elchranchenbremse *Cephenomyia ulrichi* Brauer und ihre Larven-stadien. *Zentbl. Bakt. ParasitKde* (1) 84: 541-554.

BAU, A., (1928): Die Gattung *Cephenomyia* Latreille (Diptera, olim Oestridae). *Ibid.* (2) 75: 458-459.

BAYLE-BARELLE, G., (1809): Saggio intorno agli insetti nocivi ai vegetabili economici, agli animali utili, all'agricoltura ed ai prodotti dell'economia rurale, 180 pp., 2 pls. Milano. [Also publ. in *Giorn. Soc. Incor. Sci. Milano*, vols. 5-7].

BAYLE-BARELLE, G., (1824): Degli insetti nocivi all'uomo, alle bestie e all'agricoltura, alle ortaglie, ecc., x + (1) + 143 pp., 4 pls. Milano.

BECHER, E., (1882): Zur Kenntniss der Mundtheile der Dipteren. *Denkschr. Akad. Wiss., Wien* 45(2): 123-162, pls. 1-4.

BEDFORD, G. A. H., (1926a): A check-list and host-list of the external parasites found on South African Mammalia, Aves, and Reptilia. 11th and 12th Rep. Vet. Educ. Res., Dep. Agric. Union S. Afr., Pretoria 1: 705-784.

BEDFORD, G. A. H., (1926b): Check-list of the Muscoidea and Oestridae which cause myiasis in man and animals in South Africa. *Ibid.* 785-817.

BENNETT, G. F., (1955): Studies on *Cuterebra emasculator* Fitch, 1856 (Diptera, Cutere-bridae) and a discussion of the status of the genus *Cephenemyia* Ltr., 1818. *Can. J. Zool.* 33: 75-98.

BENNETT, G. F., (1961): On the biology of *Cephenemyia phobifera* (Diptera: Oestridae), the pharyngeal bot of the white tailed deer, *Odocoileus virginianus. Ibid.* 40(7): 1195-1210, 9 figs.

BENNETT, G. F. & C. W. SABROSKY, (1962): The Nearctic species of the genus *Cephene-myia* (Diptera, Oestridae). *Ibid.* 40: 431-448, 7 figs., 1 pl., 1 table.

BERGMANN, A. M., (1899): Om oestriderna och deres ekonomiska betydelse. *Ent. Tidskr.* 20: 133-155, pls. 2-4. [Also publ. in *Svensk VetTidskr.* 4(9): 433-446, 1899].

BERGMANN, A. M., (1916): Om renens oestrider. *Skand. VetTidskr.* 6(12): 309-340, pls. 1-5, figs. 1-2 [cont.].

BERGMANN, A. M., (1917a): Idem [concl. of 1916]. *Ibid.* 7(1): 1-34, pls. 6-26, figs. 14-63.

BERGMANN, A. M., (1917b): Om renens oestrider [republ. of 1916 and 1917a]. *Meddr St. vet.-bakt. Anst.* 10: 1-66, pls. 1-26, figs. 1-63. [Also publ. in *Ent. Tidskr.* 38(1): 1-32, pls. 1-5, figs. 1-13; (2): 113-146, pls. 6-26, figs. 14-64, 1917; Abstr. in *Rev. appl. Ent.* (B) 19(2): 40-41, 1922].

BERGMANN, A. M., (1919): Ueber die Oestriden des Renntieres. *Z. InfektKrankh. parasit. Krankh. Hyg. Haustiere* 20(1): 65-91, pls. 1-11, figs. 1-39; (2): 97-116; (3): 179-201.

BERGMANN, A. M., (1932): Kozhnyi i nosovoi ovody severnogo olenya. [Warble and bot flies of the reindeer]. Sb. Olenev., tundr. Vet. Zootekhn. Kom. Severa pri Prezid. VTsIK, pp. 234-257. Ed. 'Vlast' Soveten', Moscow.

BEZZI, M. & P. STEIN, (1907): Schizophora, Eumyidae. Schizometopa, In: T. BECKER, M. BEZZI, K. KERTÉSZ & P. STEIN, Katalog der paläarktischen Dipteren 3: 189-828.

BHATIA, H. L., (1934): The bot flies of goats and sheep. *Agric. Live-Stk India* 4(5): 516-523, pls. 42-43.

BIL'DUSHNIKOV, A. P., (1949): Iz praktiki primeneniya skipidara pri rinestroze, para-skaridoze i parafilyarioze loshadei. [The use of turpentine in rhinoestrosis, para-scaridiosis and parafilariosis in horses]. *Veterinariya* 26(4): 42. [Abstr. in *Helminth. Abstr.* 18(3): 118, 120, 1950].

BISSET, N., (1931): The possible association of the sheep nostril fly, Oestrus ovis, with pneumonia in sheep. Welsh J. Agric. 7: 363-367. [Abstr. in Vet. J. 88(5): 229-233, 1932].

BLAGOVESHCHENSKII, D. I., N. P. ORLOV & G. N. KRASNUSOV, (1937): K biologii verblyuzh'ego ovoda Cephalopina titillator Cl. i bor'be s nim ('Kumyr' verblyudov). Mat. po vredit. zhivotnov i faune preismushch. yuzhn. Kazakhstana. [On the biology of the camel bot fly and its control ('Kumyr' of the camel)]. Trudy kazan. Fil. Akad. Nauk SSSR 2: 101-121.

BLANCHARD, R., (1892): Sur la présence de la larve d'Oestrus ovis chez la chèvre. Bull. Soc. ent. Fr. 1892: cclvi-ccxlvii.

BLANCHARD, R., (1893): Contribution à l'étude des diptères parasites, IV. Sur une larve extraite du sinus frontal d'un antilope. Ibid. 1893: cxxxii-cxxxiv.

BLANCHARD, R., (1896): Idem. XIII. Sur un oestre du Congo. Annls Soc. ent. Fr. 65: 668, pl. 18.

BLICKLE, R. L., (1956): Notes on the life history of Cephenomyia phobifer Clark (Diptera). Ent. News 67(1): 13-14.

BLOCH, M. E., (1779): Beiträge zur Naturgeschichte der Würmer, welche in andern Thieren leben. Beschäft. Berl. Ges. naturf. Freunde 4: 534-561.

BLOOMFIELD, E. N., (1898): Cephenomyia auribarbis, Mg.: Larvae, etc. Entomologist's mon. Mag. (2) 9 (= vol. 34): 7-8.

BLUMENBACH, J. F., (1830): Oestrus bovis, equi, ovis. Die Ochsen-, Pferde- und Schaf-Bremse, in his Abbildungen Naturh. Gegenstände (n. ed.), 5: ?pp., 1 pl.

BOBOKHODZHAEVA, M. YA., (1955): K voprosu ob oftalmomiaze ob lichinok polostnykh ovodov v Tadzhikistane. [On the question of ophthalmomyiases by larvae of cavicole bot flies in Tadzhikistan]. Trudy Akad. Nauk tadzhik. SSR 40: 147-151.

BOLDYREV, V. I. & V. S. USPENSKII, (1936): Kozhnyi i nosovoi ovod severnogo olenya, pp. 1-51. [Warble and bot flies of the reindeer]. Ed. KOIZ, Moscow-Leningrad.

BONSDORFF, E. J., (1866): Finnlands tvåvingade Insekter (Diptera). Bidr. Känn. Finl. Nat. Folk 7 (1864): vi + 306 pp.

BORTHEN, (-), (1879): [Østruslarven]. Norsk Mag. Laegevidensk. (3) 8: ? [Also in Forhandl. norsk. med. Selskab 1879: 140].

BOUET, G. & E. ROUBAUD, (1912): L'oestre des moutons au Sénégal. Bull. Soc. Path. exot. 5(9): 733-736.

BOURGELAT, C., (1760): Sur les vers trouvés dans les sinus frontaux, dans le ventricule, et sur la surface extérieure des intestins d'un cheval. Mém. Math. Savants étrangers, Paris 3: 409-432.

BOUVIER, G., H. BURGISSER & P. A. SCHNEIDER, (1952): Développement des larves de Cephenomyia stimulator Clark (Dipt. Calliphoridae) du chevreuil en Suisse. Mitt. schweiz. ent. Ges. 25(3): 265-267.

BOYES, J. W., (1964): Somatic chromosomes of higher Diptera. VIII. Karyotypes of species of Oestridae, Hypodermatidae and Cuterebridae. Can. J. Zool. 42: 599-604, 12 figs., 1 table.

BRANDT, E., (1881): Mittheilungen über das Nervensystems der Oestriden. Horae Soc. ent. ross. 16: vi-vii.

BRAUER, F. M., (1858): Neue Beiträge zur Kenntniss der europäischen Oestriden. Verh. zool.-bot. Ges. Wien 8: 449-470, pl. 10, figs. 1-8, pl. 11, figs. 1-7.

BRAUER, F. M., (1860): Neue Beiträge zur Kenntniss der europäischen Oestriden. Ibid. 10: 641-658.

BRAUER, F. M., (1862): Cephenomyia Ulrichi, die Rachenbremse des Elennthieres. Ibid. 12: 973-976.

BRAUER, F. M., (1863): Monographie der Oestriden, 292 pp., illus. Wien.

BRAUER, F. M., (1865): Ueber Oestriden-Larven. Zool. Gart., Frankf. 6(11): 410-413.

BRAUER, F. M., (1866): Pharyngobolus africanus, m. Ein Oestride aus dem Rachen des afrikanischen Elephanten. Nachtrag zur Monographie der Oestriden. *Verh. zool.-bot. Ges. Wien* 16: 879-883, pl. 19, figs. 1-2.

BRAUER, F. M., (1867a): Beobachtungen von Oestriden auf neuangekommenen Thieren in zoologischen Gärten [Correspondence dated 12 Dec. 1866]. *Zool. Gart., Frankf.* 8(2): 76.

BRAUER, F. M., (1867b): [Ueber Oestriden, Briefe von 14 Dez. 1866]. *Ibid.* 8(3): 113-114.

BRAUER, F. M., (1876): Beschreibung neuer und ungenügend bekannten Phryganiden und Oestriden. *Verh. zool.-bot. Ges. Wien* 25: 69-73, pl. 4, figs. 1-5. [Also sep. publ., same pp. and date].

BRAUER, F. M., (1883): Zweiflügler des kaiserlichen Museums zu Wien. III. *Denkschr. Akad. Wiss., Wien* 47(1): 1-100.

BRAUER, F. M., (1886a): The larvae of Oestridae (tranl. by B. P. Mann from the 'Monographie'). *Psyche, Cambr., Mass.* 4 (1883-85): 305-310.

BRAUER, F. M., (1886b): Nachträge zur Monographie der Oestriden. I. Ueber die von Frau A. ZUGMAYER und Hrn. F. WOLF entdeckte Lebensweise des *Oestrus purpureus. Wien. ent Ztg* 5: 289-304.

BRAUER, F. M., (1887): Idem. II. Zur Charakteristik und Verwandtschaft der Oestriden-Gruppen in Larven und vollkommenen Zustande. *Ibid.* 6: 4-16.

BRAUER, F. M., (1892): Ueber die aus Afrika bekannt gewordenen Oestriden und insbesondere über zwei neue von Dr. HOLUB aus Südafrika mitgebrachte Larven aus dieser Gruppe. *Ber. Akad. Wiss., Wien* 101(1): 4-16, pl. 1.

BRAUER, F. M., (1893): Vorarbeiten zu einer Monographie der Muscaria Schizometopa (exclusive Anthomyidae). *Verh. zool.-bot. Ges. Wien* 43: 447-525.

BRAUER, F. M., (1898): Beiträge zur Kenntniss aussereuropäische Oestriden und parasitischer Muscarien. *Denkschr. Akad. Wiss., Wien* 44: 259-282, pl. 1.

BRAUER, F. M. & J. VON BERGENSTAMM, (1889): Die Zweiflügler des kaiserlichen Museums zu Wien. IV. Vorarbeiten zu einer Monographie der Muscaria Schizometopa (exclusive Anthomyidae). Pars I. *Ibid.* 56(1): 69-180, 11 pls.

BRAUNS, A., (1954): Untersuchungen zur angewandten Bodenbiologie. I. Terricole Dipterenlarven, 179 pp., 96 figs., 6 pls; II. Puppen terricole Dipterenlarven, 156 pp., 75 figs. Göttingen, Frankfurt & Berlin.

BREEV, K. A., (1950): O povedenin krovososushchikh dvukrylykh i ovodov pri napadenin ikh na severnogo olenya i otvetnykh reaktsiyakh olenei. I. Povedenie krovososushchikh dvukrylykh i ovodov pri napadeniya ikh na severnogo olenya. [Ueber das Verhalten der blutsaugende Dipteren und Bremsen bei ihren Angriffen auf das Renntier und die Reaktionen des Renntiers. I. Das Verhalten der blutsaugenden Dipteren und Bremsen bei Angriffen auf das Renntier]. *Parazit. Sb.* 12: 167-198.

BREEV, K. A., (1965): Aktivnost' napadeniya kozhnogo (*Oedemagena tarandi* L.) i nosovogo (*Cephenomyia trompe* L. [sic]) ovodov na severnogo olenya i faktory eë reguliaruyushchnie. [Force of attack of *O. tarandi* and *C. trompe* on the reindeer and factors regulating it]. *Ibid.* 16: 155-183.

BREEV, K. A., (1965): [Present means of combat against bot flies and perspectives of development]. *Trudy zool. Inst., Leningr.* 34: 308-318.

BRÈTHES, J., (1908): El estro de los ovinos. *Agricultura Nac., B. Aires* 1: 287-288.

BRIOT, A., (1912): La biologie et la destruction des larves d'oestrides. *Cosmos, Paris* 66: 40-41.

BROWN, H. S., Jr., J. C. HITCHCOCK, Jr. & R. Y. FOOS, (1969): Larval conjunctivitis in California caused by *Oestrus ovis. Calif. Med.* 111(4): 272-274.

BROWN, W. G., (1914): The nasal fly (*Oestrus ovis*). *Qd agr. J.* (n.s.) 2(3): 220-221.

BROWNING, B. M. & E. M. LAUPPE, (1964): A deer study in a redwood-Douglas fir forest type. *Calif. Fish Game* 50(3): 132-147, figs.

BRUNETTI, E. A., (1923): Pipunculidae, Syrphidae, Conopidae, Oestridae. In: Fauna of

British India, including Ceylon and Burma. *Diptera* 3: 424 pp., figs. 1-85, pls. 1-6. London.

BRUNETTI, E. A., (1925): Some notes on Indian Syrphidae, Conopidae and Oestridae. *Rec. Indian Mus.* 27(2): 75-79.

BUCHANAN, R. S., (1969): Sheep parasites. Their importance and control with systemic insecticides in Arizona. *Diss. Abstr.* (B) 29(11): 3984B-3985B.

BUCHANAN, R. S., L. W. DEWHIRST & G. W. WARE, (1969): The importance of sheep bot fly larvae and their control with systemic insecticides in Arizona. *J. econ. Ent.* 62(3): 675-677. 1 fig.

BUCK, M., (1949): Sur un cas d'infestation massive à *Oestrus ovis* avec localization inhabituelle sur une antenaise mérinos de Sud-Afrique. *Bull. Soc. Path. exot.* 42(5-6): 326.

BUEN, S. DE, (1924): Un caso de miasis ocular por *Oestrus ovis* L. *Archos Inst. nac. Hig., Madr.* 3(3): 219-220, figs.

BUGAEVA, I. V., (1956): K voprosu ob oftalmomiaze. [On the question of ophthalmomyiasis]. *Oftal. Zh.* 11(1): 59-60.

BURFORD, M. R., (1962): Restless sneezing sheep? The trouble may be nasal bot. *J. Dep. Agric. S. Aust.* 65(9): 399, 403.

BURENSUND, A. S., (1750): Om Braemse-Bylder i Lapmarken, in his Extract af Kunglika Svenska Vetenskaps-Akademien Afhandlingar fra 1739 til og med 1750, 2: 360-363.

BUTTERFIELD, J. F., (1900): *Oestrus ovis. J. comp. Med. vet. Arch.* 21(1): 23-24.

CAMERON, A. E., (1932a): The nasal bot fly, *Cephenomyia auribarbis* MEIGEN (Diptera, Tachinidae) of the red deer, *Cervus elaphus* L. *Parasitology* 24: 185-195.

CAMERON, A. E., (1932b): Arthropod parasites of the red deer (*Cervus elaphus* L.) in Scotland. *Proc. r. phys. Soc. Edinburgh* 22: 81-89.

CAMERON, A. E., (1937): Insects and other pests of 1936. Parasites of the red deer (*Cervus elaphus*). *Trans. Highl. agric. Soc. Scot.* (5) 49: 130-147.

CANNON, G. T., (1917): Notes on development of *Oestrus* larvae in the pharynx of the horse. *Vet. Rec.* 30: 107-109, 1 pl.

CAPELLE, K. J., (1966): The occurrence of *Oestrus ovis* L. (Diptera, Oestridae) in the bighorn sheep from Wyoming and Montana. *J. Parasit.* 52(3): 618-621, 3 figs.

CAPELLE, K. J. & C. M. SENGER, (1959): Occurrence of *Cephenemyia jellisoni* in a sample of Montana mule deer. *Ibid.* 45(4): 32.

CATTS, E. P., (1964): Field behavior of adult *Cephenemyia* (Diptera: Oestridae). *Can. Ent.* 96: 579-585, 7 figs.

CATTS, E. P. & R. GARCIA, (1963): Drinking by adult *Cephenemyia* (Diptera: Oestridae). *Ann. ent. Soc. Am.* 56(5): 660-663.

CEDERHJELM, (-), (1798): Faunae Ingricae prodromus, exhibens methodicam descriptionem insectorum, agri Petropolensis praemissa Mammalium, Avium, Amphibiorum et Piscium enumeratione, 18 + 348 pp., 3 pls. Lipsiae.

ČEPELÁK, J., (1972): Prve poznatky o vyskyte a ekologii kuklic a střečkov Cherchovskego pohoria (Tachinidae, Oestridae-Diptera). [First observations on the occurrence and ecology of maggots and bot flies of the Cherchovsky Mountains]. *Ent. Problemy* 10: 106-124, figs.

CHAMS, G. & H. MOHSÉNINE, (1956): Un cas de myiase ophthalmique par la larve de *Rhinoestrus purpureus. Acta med. Iran.* 1: 22.

CHAUDHURI, R. P., (1966): Insect tormentors of livestock. II. Warble flies and bot flies. *Indian Fmr* 15(11): 26-28, 33, figs.

CHAVARRÍA CHAVARRÍA, M. & R. ÁVILA CARRILLO, (1959): Eine neuartige wirksame Behandlung der durch *Oestrus ovis* Linn. ausgelösten Myiasis cavitaria. *Zbl. vet.-Med.* 6(9): 816-824, illus.

CHAVARRÍA CHAVARRÍA, M. & R. ÁVILA CARRILLO, (1960): Nuevo tratamiento efectivo y practico de la miasis cavitaria ocasionada por Oestrus ovis Linn. *Cienc. vet., México* 5(2): 167-173.

CHEBOTAREV, R. S., (1961): Metody likvidatsii estroza (ovodovoi bolezin) ovots. [Methods of control of oestrosis (bot fly disease) of sheep], pp. 298-299. In: A. P. MARKEVICH, Metody izucheniya parazitologicheskoi situatsii i bor'ba s parazitami sel'skokhozyaistvennikh zhivotnikh. [Methods of studying parasitological situations and eradication of parasitic diseases of domestic animals], 350 pp., illus. Kiev.

CHERESHNEV, N. A., (1955): Mekhanizirovannyi sposob obrabotki ovets protiv estroza. [Mechanical method of treating sheep for oestrosis]. Tezisy i Dokl. pyatoi i-proizv. Konf. Vet. i-issled. Uchrezhd. Sibiri, Posv. 25-letiyu Sib. Zonal'n. i Issled. Vet. Inst. (20-23 Iyuliya 1955 g.): 178-170. Omsk.

CHERESHNEV, N. A., (1960a): [Verwendung der Maschinen OZhU-5 und einiger wirksamen Mittel zur Abwehr der cavicole Bremsen der Tiere]. *Veterinariya* 3: 66-69.

CHERESHNEV, N. A., (1960b): [Ein tragbares Gerät zur Erzeugung von Aerosolen und die Anwendung des letzteren gegen cavicole Bremsen und Ektoparasiten der landwirtschaftlichen Haustiere]. (*Bull. d. wiss.-techn. Inform. d. Kirgis. wiss. Forschungsinst. f. Tierzucht u. Tierheilkunde*) 5: 77-79.

CHERESHNEV, N. A. & A. M. KRIVKO, (1957): Primenenie aerosolei protiv polostnykh ovodov ovets i loshadei. [Aerosols against sheep and horse bot flies]. *Trudy Inst. Vet. kazak. Fil. vses. Akad. sel'.-khoz. Nauk* 8: 316-326.

CHILLCOTT, J. G., (1965): Family Oestridae (including Hypodermatidae), pp. 1111-1113. In: A. STONE *et al.*, A Catalog of the Diptera of America north of Mexico. U.S. Dep. Agric., *Agric. Handb.* 276: 1-1696. Washington, D.C.

CHRISTY, J. McN., (1909): Maggots in lamb's nose. *Transv. agric. J.* 2(5): 135-136.

CLARK, B., (1797): Observations on the genus Oestrus. *Trans. Linn. Soc. Lond.* 3: 289-329, pl. 23.

CLARK, B., (1798): Observations sur les oestres. *J. Phys., Paris* 3 [= vol. 46]: 329-337.

CLARK, B., (1815): An essay on the bots of horses and other animals, 72 pp., 2 col. pls. London.

CLARK, B., (1815? or 1816): Supplementary sheet. Discovery of the fly of the white bot, [4 pp.]. London.

CLARK, B., (1827): On the insect called Oistros by the ancients, and of the true species intended by them under this appelation: In reply to the observations of W. S. MAC-LEAY, Esq., and the French naturalists. To which is added, a description of a new species of Cuterebra. *Trans. Linn. Soc. Lond.* (Zool.) 15(2): 402-411. [Also publ. in *Phil. Mag.* (n.s.) 3(16): 283-289, 1828].

CLARK, B., (1841): An appendix or supplement to a treatise on the Oestri and Cuterebrae of various animals [Secretary's summary]. *Proc. Linn. Soc. Lond.* (Zool.) 1: 99, 100 [Meetings of April 6th and 20th, 1841].

CLARK, B., (1843): An appendix or supplement to a treatise on the Oestri and Cuterebrae of various animals. *Trans. Linn. Soc. Lond.* (Zool.) 19(2): 81-94.

CLARK, B., (1847): Note on the bot infesting the stag. *Zoologist* 5: 1569-1570, 3 figs.

CLARK, B., (1848): Addenda, 1848 [to CLARK, 1843], 5 figs. London.

CLARK, B., (1857a): The supposed new Oestrus. *Zoologist* 15: 5542.

CLARK, B., (1857b): Further note on the supposed new Oestrus. *Ibid.* 15: 5630.

CLARK, B., (1857c): The tzetze fly of Africa identified with Oestrus ovis. *Ibid.* 15: 5720-5721.

COBBETT, N. G., (1940a): An effective treatment for the control of the sheep head grub, Oestrus ovis, in areas where the winters are cold. *J. Am. vet. med. Ass.* 97: 565-570, illus.

COBBETT, N. G., (1940b): A method of large-scale treatment of sheep for the destruction of the head grub (= Oestrus ovis). *Ibid.* 97(765): 571-575, figs.

COBBETT, N. G., (1956): Head grub of sheep. U.S. Dep. Agr. Yearb. 1956: 407-411.

COBBETT, N. G. & W. C. MITCHELL, (1941): Further observations on the life cycle and incidence of the sheep bot, *Oestrus ovis*, in New Mexico and Texas. *Am. J. vet. Res.* 2: 358-366.

COLIN, J. S. & C. S. MOSS, (1968): El problema de *Oestrus ovis* L. (Diptera, Oestridae) en hatos de ovinos del Edo. de México. *Folia ent. mex.* 18-19: 79.

COLLART, A., (1952): *Cephenomyia auribarbis* (Meigen) en Belgique (Diptera: Calliphoridae). *Bull. Annls Soc. r. ent. Belg.* 88(3-4): 66.

COLLINGE, W. E., (1906): Note on the deposition of the eggs and larvae of *Oestrus ovis* L. *J. econ. Biol.* 1: 72-73.

COOKE, (-), (1857): Description of an *Oestrus* new to Britain. *Zoologist* 15: 5438. [See CLARK, 1857a-b].

COOKE, (-), (1862): Note on a species of *Oestrus. Ibid.* 20: 8023.

COQUEBERT DE MONTBRET, A. J., (1799-1804): Illustratio iconographica insectorum, quae in Musaeis Parisinis observavit et in lucem edidit Joh. Christ. Fabricius, praemissis eiusdem descriptionibus. Accedunt species plurimae, vel minus vel nondum cognitae. Decas Primas. ['An VII' (= 1799)]: 1-44 pp.; Decas Secunda ['An X' (= 1801)]: 1-90 pp.; Decas Tertia ['An XII' (= 1804)]: 1-142 pp., 30 pls. Didot, Paris.

CORNELL, R. L. & R. DAUBNEY, (1934): *Oestrus ovis*, p. 44, in Conference of Governors of British East African Territories. Research Conferences. Conference on coordination of veterinary research, held at Kabete, 6th to 10th January, Nairobi.

CORREA, O., (1961): Oestrose [sic] ovina. *Granja (Rev. agropec. Sul-Brasil)* 17: 21-22.

COUEY, F. M., (1950): Rocky Mountain bighorn sheep of Montana. *Montana Fish Game Comm. Bull.* 2: 1-90.

COULON, G. & G. DINULESCU, (1931): Un cas de myiase oculaire à *Oestrus ovis* L. en Corse. *Annls Parasit. hum. comp.* 9(2): 140-143.

COWAN, I. McT., (1943): Notes on the life history and morphology of *Cephenemyia jellisoni* Townsend and *Lipoptena* [sic] *depressa* Say, two dipterous parasites of the Columbian black-tailed deer [*Odocoileus hemionus columbianus* (Richardson)]. *Can. J. Res.* (D) 21: 171-187, 9 figs.

CRAWFORD, M., (1935): Nasal myiasis in sheep and goats. Ceylon Admin. Rep., Pt. 4 (Educ., Sci & Art) (H) Vet.: 48.

CROSS, H. E., (1925): Bot flies of the Punjab. *Bull. agric. Res. Inst. Pusa* 160: 1-16, 7 pls.

CROWTHER, R. W., (1953): The sheep nostril fly. *Countryman, Nicosia* 7(2): 10-12.

CURRAN, C. H., (1934): The families and genera of North American Diptera, 512 pp., 235 figs., 2 pls. N. York.

CURTICE, C., (1890): The animal parasites of sheep, 322 pp. Washington.

CURTIS, J., (1828): *British Entomology*: Being illustrations and descriptions of the genera of insects found in Great Britain and Ireland 5: 195-241. London.

DAHLBOM, A. G., (1837): Kort underrättelser om Scandinaviska insekters allmännare skåda och nytta i huskällingen. En handbok för landbrukare och naturforskare, 370 + 36 + 10 pp., 2 pls. Lund.

DAHLGRÜN, W., (1930): Die Oestriden. *Ther. Mh. VetMed.* 1-2: 3-12, 17 figs.; 3-4: 45-53, 20 figs.

DAMIAN, T., (1961): Un foyer de myiases des sinus chez les moutons. *Probleme Zool. vet.* 11(2): 72.

DAVIES, A. H., (1835): Vitality of *Oestrus ovis. Entomologist's mon. Mag.* 3(1): 103-104. [Also publ. in *Arcana Sci. & Art.* 9: 200-201, 1836].

DEDUIT, Y., J. CALLOT & C. STEINER, (1960): Un nouveau cas de thimni, myiase oculaire externe à *Oestrus ovis*, en France. *Strasb. méd.* (n.s.) 11(5): 328-331.

DE GEER, C., (1776): Mémoires pour servir à l'histoire des insectes, 6: 523 pp., 30 pls. Stockholm.

DEICH, A. B., (1909): Oestruslarven beim Reh. *Ber. Vet Wes. Sachs.* 53 (1908): 77.

DEICH, A. B., (1912): Oestruslarven. *Ibid.* 56 (1911): 63.

DEL PONTE, E., (1939): Revisión de los 'Oestridae' argentinos. *Physis, B. Aires* 17: 525-534.

DEL PONTE, E., (1959): Diptera, Oestridae, in Primeras Jornadas Entomo-epidemiologicas Argentinas, B. Aires: 575.

DEMOUSSY, (-), (1819): Mémoire sur les oestres. *Mém. Soc. cent. r. Agric., Paris* 1819: 339-376.

DINULESCU, G., (1958): Noi date asupra dezvoltării lui *Oestrus ovis* Linné la ovine. *Studii Cerc. Inframicrobiol.* 9(2): 275-278.

DINULESCU, G., (1960): [Observations sur la classification des Oestridae]. *Studii Cerc. Biol.* (Ser. Biol. anim.) 12(1): 7-20, 6 figs.

DINULESCU, G., (1961): Diptera, Fam. Oestridae. *Fauna Rep. Pop. Rom. Insecta* 11(4): 165 + (3) pp., 4 col. pls., 76 figs. Bucharest.

DINULESCU, G., H. ALMAŞEANU, V. NESTOROV & D. CERNI, (1960): Contribution à l'étude des Oestridae de la R.P.R. *Trav. Mus. Hist. nat. Gr. Antipa* 2: 279-285, 5 figs.

DONNOVAN, E., (1813): The natural history of British insects, explaining them in their several states, with the periods of their transformations, their food, economy, etc., together with the history of such minute insects as require investigation by the microscope, 16: 91 + 10 pp. London.

DORŽ, C. & J. MINÁRŽ, (1971): Warble flies of the families Oestridae and Gasterophilidae (Diptera) found in the Mongolian People's Republic. *Folia Parasit., Praha* 18(2): 161-164.

DOTEN, S. B., C. E. FLEMING & L. R. VAWTER, (1936): The relation of methods of herding sheep on the open to the prevalence of grub in the head, Oestrus ovis, 1934-1935. *An. Rep. Nevada agric. exp. Sta.* 1934-35: 22.

DOTEN, S. B., C. E. FLEMING & L. R. VAWTER, (1938): The relation of methods of herding sheep on the open range to the prevalence of grub in the head (*Oestrus ovis*), 1934-1937. *Ibid.* 1936-37: 33.

DOURY, P., (1957): Un cas de myiase oculaire à *Oestrus ovis* Linné observé à Tamanrasset (Hoggar). *Archs Inst. Pasteur Algér.* 35(2): 76-77.

DRÓZDZ, J., (1961a): Nowe dane o biologii larw *Pharyngomyia picta* Meig. (Diptera, Oestridae), parazyta *Cervus elaphus* L. [New data on the biology of *P. picta*, parasite of *P. picta*, parasite of *C. elaphus*]. *Wiad. parazyt.* 7(2; Suppl.): 373-379.

DRÓZDZ, J., (1961b): Cephenomyinae (Diptera: Oestridae) jeleniowatych w Polsce. [Cephenemyiinae of Cervidae in Poland]. *Ibid.* 7(2; Suppl.): 381-382.

DRUMMOND, R. O., (1961): A new organophosphorous systemic insecticide for the control of larvae of *Oestrus ovis* L. in sheep. *J. Parasit.* 47(4; Suppl.): 36.

DRUMMOND, R. O., (1962): Control of larvae of *Oestrus ovis* in sheep with systemic insecticides. *Ibid.* 48(2): 211-214.

DRUMMOND, R. O., (1966): Systematic insecticides to control larvae of *Oestrus ovis* in sheep. *Ibid.* 52(1): 192-195.

DRUMMOND, R. O. & O. W. GRAHAM, (1965): Systemic insecticides in livestock insect control. *Vet. Rec.* 77 (48; = n° 4566): 1418-1420.

DRYNSKI, P. S., (1933): Parazitni mukhi ot' semeistvo Oestridae v' Bulgariya. [Parasitic flies of the family Oestridae in Bulgaria]. *Izv. tsarsk. prirodonauch. Inst. Sof.* 6: 125-149, 15 figs.

DUDZINSKY, W., (1964): Studies on *Cephenomyia stimulator* Cl. (Diptera, Oestridae) parasitizing the European roe deer (*Capreolus capreolus* L.). *Wiad. parazyt.* 10(4-5): 615-616.

DUDZINSKY, W., (1970a): Studies on *Cephenomyia stimulator* (Clark) (Diptera, Oestridae), the parasite of the European roe deer, *Capreolus capreolus* (L.). I. Biology. *Acta parasit. pol.* 18(42-50): 555-572, figs.

DUDZINSKY, W., (1970b): Idem. II. Invasiology. *Ibid.* 18(42-50): 573-592.

DUKALOV, I. A., (1957): Estroz ovets v Rostovskoi oblasti. [Ovine oestrosis in the Rostov region]. *Veterinariya* 34(6): 23-28.

DUPUIS D'UBY, P., (1930): Un cas de myiase oculaire. *Algér. méd.* (4) 34(27): ? (Abstr. in *Medna Paises cálid.* 4(3): 266, 1931).

DUPUIS D'UBY, P., (1931): A propos d'un nouveau cas de myiase oculaire à *Oestrus ovis* observé à Alger. *Archs Inst. Pasteur Algér.* 9(4): 630-637, figs. 1-2.

DUTOIT, R., (1935): The external parasites of sheep. I. Fly pests. *Fmg S. Afr.* 10: 321-324, figs. 1-7.

DUTOIT, R., (1938): Idem. *Ibid.* 13: 401-407, figs. 1-3.

DUTOIT, R. & R. CLARK, (1935): The sheep nasal fly. A method of treatment for sheep infested with larvae of *Oestrus ovis*. *Jl. S. Afr. vet. med. Ass.* 6(1): 25-32, figs. 7-8.

DUTOIT, R. & G. H. FIEDLER, (1956): A new method of treatment for sheep infested with the larvae of the sheep nasal-fly, *Oestrus ovis* L., in the Union of South Africa. *Onderstepoort J. vet. Res.* 27: 67-75.

DUTOIT, R. & N. MEYER, (1960): A case in South Africa of ocular myiasis in man due to the first-stage larvae of the nasal bot fly of the sheep (*Oestrus ovis* L.). *S. Afr. med. J.* 34(28): 581-582.

DWIGUBSKY, J. A., (1802): Primitiae faunae Mosquensis, seu enumeratio animalium, quae sponte circa Mosquam vivunt, 215 pp. (Diss. Inaug., 14 Juni 1802), Mosquae.

DYK, V., (1962): On the invasion of the naso-pharyngeal cavity of roe deer with *Cephenomyia* larvae. Summ. Papers Sci. Sess. 175th Anniv. Coll. vet. Sci., Budapest: 26-27.

DYK, V. & I. DYKOVÁ, (1962): Oblastní rosdíly ve výskytu a uvolňváni larv střečka *Cephenemyia stimulator* v hostilele. *Sb. čsl. Akad. zeměd. Věd* (vet. Med.) 7(3): 193-206.

DYK, V. & R. ZAVADIL, (1968): Pozdní výskyt larev střečka hltanového u srnců. [Occurrence of larvae of the genus *Cephenemyia* in roebucks]. *Veterinářství* 18(11): 486-487.

EICHHORN, G., G. GRÄFNER, T. HIEPE & H. RIBBECK, (1970): Untersuchungen über das Vorkommen von Rachenbremsenlarven beim Rehwild. *Mh. VetMed.* 25(7): 278-282, 3 figs.

EMDEN, F. I. VAN, (1945): Keys to Ethiopian Tachinidae. I. Phasiinae. *Proc. zool. Soc. Lond.* 114: 389-436, 3 pls.

ENDERLEIN, G., (1911): Neue Gattungen und Arten aussereuropäischen Fliegen. *Stettin. ent. Ztg* 72: 135-209.

ENDERLEIN, G., (1936): 22. Ordnung. Zweiflügler, Diptera, Abt. 16, 259 pp., 317 figs. (= Lfg. 2, part). In: P. BOHMER, P. EHRMANN & F. ULMER (eds.), Die Tierwelt Mitteleuropas, 6 (Insekten, III. Teil). Leipzig.

EREMIN, V. M., (1950): Lokalizatsiya lichinok polostnykh ovodov v oblasti glotki. [Localization of larvae of the botfly in the region of the pharynx]. *Veterinariya* 27(7): 54-55.

ESPMARCK, Y.. (1968): Observations of defence reactions to oestrid flies by semi-domestic forest reindeer (*Rangifer tarandus* L.) in Swedish Lapland. *Zool. Beitr.* 14: 155-167, 2 figs.

EYDAN, R., (1958): Un cas de miase oculaire à *Oestrus ovis* L. observé à Guerrara (Mzab). *Archs Inst. Pasteur Algér.* 36(4): 485-486.

FABRICIUS, J. C., (1775): Systema entomologiae, sistens insectorum classes, ordines, genera, species, adiectis synonymis, locis, descriptionibus, observationibus, 832 pp. Flensburgi et Lipsiae.

FABRICIUS, J. C., (1781): Species insectorum exhibentes eorum differentias specificas, synonyma, auctorum, loca natalia, metamorphosin, 2: 517 pp. Hamburgii et Kilonii.

FABRICIUS, J. C., (1794): Entomologia systematica emendata et aucta, 4: 472 pp. Hafniae.

FABRICIUS, J. C., (1805): Systema antliatorum secundum ordines, genera, species, 373 + 30 pp. Brunsvigae.

FALLIS, A. M., (1940): Studies on *Oestrus ovis* L. *Can. J. Res.* (D) 18(12): 442-446, pls.

FARMAKOVSKII, N. V., (1913): Lichinki ovech'ego ovoda v glazy cheloveka. [Larvae of *O. ovis* in the human eye]. *Vrach. Gaz.* 10: 345-347.

FAVIER, G., (1958): Un cas de myiase oculo-nasale à *Oestrus ovis* observé dans l'annexe de la Saoura (Sud Oranais). *Archs Inst. Pasteur Algér.* 36(2): 182-183.

FEDCHENKO, A. P., (1868): Materialy dlya entomologii gubernii Moskovskogo okruga. Spisok dvukrylykh nasekomykh. [Materialen zur Entomologie des Moskauer Gouvernments. Diptera]. *Izv. Obshch. Lyubit. Estest. Antrop. Etnogr.* 1868.

FELT, E. P., (1903): Seventeenth report of the State Entomologist on injurious and other insects in the State of New York. *Bull. N.Y. St. Mus.* 53 (= Ent. 14): 699-925, pls. 1-4.

FELT, E. P., (1928): Observations and notes on injurious and other insects of New York State. *Ibid.* 274: 145-176.

FETHERS, G., (1952): Sheep nasal botfly. *Pastoral Rev. Graz. Rec.* 62(6): 589.

FISCHER, J. L., (1787): Observationes de oestro ovino atque bovino factae, 69 pp., 4 col. pls. Lipsiae.

FITCH, A., (1849): [Note on *Oestrus ovis*]. *Trans. N.Y. agric. Soc.* 9: 800.

FITCH, C. P., (1928): A preliminary note on the occurrence of a head and throat bot in the wild deer (*Cervus virginianus*) of Minnesota. *Cornell Vet.* 18(4): 353-357, fig. 1.

FLETCHER, T. B. & S. K. SEN, (1929-31): A veterinary entomology for India. *J. cent. Bur. Anim. Husb. Dairy. India* 3(2-3): 50-57, 95-100, 3 pls., 1929; 4(1): 1-5, 3 pls., (3): 90-104, 2 pls., 1930; (4): 127-138, 5 pls., 1931.

FRANCAVIGLIA, M. C., (1914): Larva di *Oestrus ovis* L., per la prima volta rinvenuta nell'orecchio umano. *Boll. Sed. Accad. gioenia Sci. nat.* 31: 23-27.

FRAUENFELD, G. V., (1864): Das Vorkommen des Parasitismus im Thier- und Pflanzenreiche. Als Festschrift zur 50-jährigen Jubelfeier des naturforschenden Gesellschaft in Emden, 32 pp. Vienna.

FREY, R., (1914): *Cephenomyia ulrichi* Brauer en på älg lefvande, för landet ny oestrid. *Meddn Soc. Fauna Flora fenn.* 40 (1913-14): 117-119, 307, figs.

FRINGS, C. & M. NUSSBAUM, (1909): Ein Vorschlag zur Feststellung des Geschlechtszahlen bei den Rachenbremsen. *Bonn. Sber. Ges. Naturk.* (Naturw. Abt.) 1908: 25-31.

FROGGATT, W. W., (1905): Sheep infested with the larvae of the nasal fly (*Oestrus ovis*) at Megalong. *Agric. Gaz. N.S.W.* 16(4): 342.

FROGGATT, W. W., (1911): The nasal fly of sheep (*Oestrus ovis*) in Australia. *Ibid.* 22(3): 223-227, pl.

FROGGATT, W. W., (1913): The kangaroo bot fly, *Oestrus macropi*, sp. n. *Ibid.* 24(7): 565-568, 1 pl.

FÜLLEBORN, F., (1919): Ueber Ophthalmomyiasis und einen solchen Fall aus Nord-Frankreich. *Arch. Schiffs- u. Tropenhyg.* 23: 349-359.

GABRIELIDIS, A. & J. GUIART, (1922): Myiase oculaire à *Oestrus ovis* à Constantinople. *Paris méd.* 43(12): 249-250, figs. [Also in *Bull. Acad. Méd.* (3) 87(9): 253-255, 1922].

GALLIARD, H., (1934): Un nouveau cas de myiase oculaire à *Oestrus ovis* en France. *Annls Parasit. hum. comp.* 12(3): 177-181, figs. 1-2.

GAMINARA, A., (1925): Un caso de miasis ocular por *Oestrus ovis*. *Boln Cons. nac. Hig., Montev.* 19: 73-78.

GAN, Z. I., (1942): Biologiya ovech'ego ovoda (*Oestrus ovis* L.). [Biology of the sheep bot fly, *O. ovis*]. *Trudy uzbek. Fil. Akad. Nauk SSSR* 12 (Zool., 1): 27-99.

GAN, Z. I., (1947a): Polostnye ovody loshadei Uzbekistana. [Cavicole bot flies of horses in Uzbekistan]. *Byull. Akad. Nauk uzbek. SSR* 7: 24-28.

GAN, Z. I., (1947b): K biologii nekotorykh vidov roda *Rhinoestrus*. [On the biology of some species of the genus *Rhinoestrus*]. *Izv. Akad. Nauk uzbek. SSR* 5: 122-131.

GAN, Z. I., (1950): Novoe v merakh bor'by s ovech'im ovodom. [News on the control measures against sheep bot fly]. *Dokl. Akad. Nauk uzbek. SSR* 12: 40-43.

GAN, Z. I., (1953): Ovechii ovod *Oestrus ovis* L. [The sheep bot fly, *O. ovis*], 160 pp. Akad. Nauk uzbek. SSR, Tashkent.

GAN, Z. I., (1954a): Novoe v bor-be lichinkami ovech'ego ovoda. [News on the control of the sheep bot fly larvae]. *Veterinariya* 31(3): 47-52, illus.

GAN, Z. I., (1954b): Estroz ovets i mery bor'by s nim. [Ovine oestrosis and its control]. Tezisy Dokl. l. vses. Konf. Probl. vet. Dermat., Arakhn. Ent.: 192-194.

GARUDACHAR, M. K., (1930): The occurrence of the larvae of the sheep nasal bot fly (*Oestrus ovis*) in the trachaea. *Vet. Rec.* 10(15): 329-330.

GASCOUGNOLLE, R., (1961): Observation concernant un cas de myiase oculaire. *Méd. Afr. noire* 8(10): 209.

GEDOELST, L., (1912): Contribution à la faune des oestrides du Congo Belge. *Rev. Zool. afr.* 1: 426-432, 2 figs.

GEDOELST, L., (1914): Note sur un genre nouveau d'oestride. *Bull. Soc. Path. exot.* 7(3): 210-212.

GEDOELST, L., (1916): Note sur les oestrides. *Rev. Zool. afr.* 4: 144-161, 259-264.

GEDOELST, L., (1919): Inventaire d'une collection d'oestrides africains. *Bull. ent. Res.* 9: 333-340.

GEDOELST, L., (1922): Le trimorphisme larvaire des oestridés. *C. r. Séanc. Soc. Biol.* 86: 501-504.

GENÉ, C. G., (1827): Saggio su gli insetti nocivi all'agricoltura, agli animali domestici ed ai prodotti dell'economia rurale, 236 pp., 3 col. pls. Milano. [In MORETTI (ed.), Biblioteca Agraria, 2nd ed., 1836].

GEOFFROY, E. L., (1762): Histoire abrégée des insectes qui se trouvent aux environs de Paris, 2: 690 pp., 11 pls. Paris.

GERASIMOV, A. N., (1959): Sluchai popadaniya lichinok ovoda v kon'yunktival'niyu polost'. [On a case of bot fly larva in the conjunctival cavity]. *Trudy saratov. med. Inst.* 24: 178-179.

GIL COLLADO, J., (1950): Los estros de las ovejas. *Ganadería, Madr.* 7: 590-591, figs.

GIL COLLADO, J., (1955): Las especies españolas de estrideos (sensu lato). *Revta ibér. Parasit.* (Tomo Extraord.): 411-420.

GILDOW, E. M. & C. W. HICKMAN, (1931): A new treatment for *Oestrus ovis* larvae in the head of sheep. *J. Am. vet. med. Ass.* 32(2): 210-216, figs. 1-2.

GILRUTH, J. A., (1895): Nasal bot in sheep: grub in the head (*Oestrus ovis*). *Leafl. Farmers* (N.Z. Dep. Agr.) 23: 1-3.

GLÄSER, H. & A. STRÖSE, (1963): Ueber die Rachenbremsenkrankheit des Wildes, nebst Bemerkungen über die Dasselfliegen. *Dt. Jäger-Ztg* 60(49): 761-764, pl. [Abstr. in *Zentbl. Zool. allg. exp. Biol.* 4(10): 382, 1964].

GLAZENER, W. C. & F. F. KNOWLTON, (1967): Some endoparasites found in the Welder Refuge deer. *J. Wildl. Mgmt* 81(3): 595-597.

GMELIN, J. F., (1790): Caroli a Linné, Systema naturae per regna tria naturae (Ed. 13) 1. *Regnum Animale* (5): 2225-3220. Lipsiae.

GOLINI, V. I., S. M. SMITH & D. M. DAVIS, (1968): Probable larviposition by *Cephenemyia phobifer* (Clark) (Diptera: Oestridae). *Can. J. Zool.* 46(5): 809-814, fig.

GÓMEZ-FERNÁNDEZ, L., (1946): Revision critica de los casos de oftalmomíases españolas. *Revta. ibér. Parasit.* 6: 51-73.

GÓMEZ-FERNÁNDEZ, L., (1955a): La estrosis oculo-nasal (Thimni) del hombre en España, causada por el *Oestrus ovis* L. del carnero. *Ibid.* 15(2): 135-172.

GÓMEZ-FERNÁNDEZ, L., (1955b): La muy interesante morfología de la segunda larva del

Oestrus ovis L. *Ibid.* (Tomo Extraord.): 921-926, 2 figs.

GOMOYUNOVA, N. P. & T. K. KALVISH, (1971): [Biology and fungus diseases of *Oedemagena tarandi* L. and *Cephenemyia trompe* Modeer pupae in Chukotka]. *Izv. sib. Otdel. Akad. Nauk SSSR (Biol.)* 2(10): 93-102, 4 figs.

GOUREAU, C. C., (1966): Les insectes nuisibles à l'homme, aux animaux et à l'économie domestiques. *Bull. Soc. Sci. hist. nat. Yonne* 20(2): 3-258.

GRABER, M. & J. GRUVEL, (1965): Étude des agents des myiases des animaux domestiques et sauvages de l'Afrique équatoriale. *Revue Élev. Méd. vét. Pays trop.* (n.s.) 17(3): 535-554, maps.

GRABER, M. & J. GRUVEL, (1966): A study of species causing myiasis in wild and domestic animals in Equatorial Africa. Proc. 1st int. Congr. Parasit., Rome: 944-945.

GREBENYUK, R. V. & S. K. SARTBAEV, (1955): Polostnye ovody dikikh parnokopytnykh (*Capra sibirica* Meyer i *Ovis ammon* L.). [The cavicole bot flies of the wild goats *C. sibirica* and of the sheep *O. ammon*]. *Trudy Inst. Zool. Parazit., Frunze* 4: 89-94.

GREENE, C. T., (1956): Dipterous larvae parasitic on animals and man and some dipterous larvae causing myiasis in man. *Trans. Am. ent. Soc.* 82: 17-34, 21 figs.

GRIMSHAW, P. H., (1895): On the occurrence in Rossshire of *Cephenomyia rufibarbis*, a new British bot fly parasitic on the red deer. *Ann. Scot. nat. Hist.* 15: 155-158.

GRIMSHAW, P. H., (1900): Diptera Scotica. I. – Perthshire. *Ibid.* 23: 18-30.

GROSSO, G., (1907): Le larve d'estro ovino nei seni frontali ed aiacenti. *Moderno Zooiatro* 18: 841-845. [Abstr. in *Zentbl. Bakt. ParasitKde* (1) 41(19-21): 687, 1908].

GRÜNBERG, K., (1904): Ueber eine neue Oestridenlarve (*Rhinoestrus hippopotami* n. sp.) aus der Stirnhöhle des Nilpferdes. *Sber. Ges. naturf. Freunde Berl.* 1904: 35-39, pl.

GRÜNBERG, K., (1906): Einige Mitteilungen über afrikanische Oestriden. *Ibid.* 1906: 37-49.

GRUNIN, K. YA., (1947): Nosoglotochnyi ovod ussuriiskogo losya. [Nasal bot of the Ussuri elk (*Alces alces americanus bedfordi* Lyddeker)]. *Ent. Obozr.* 29(3-4): 224-231, figs.

GRUNIN, K. YA., (1948): Ovod (*Oestrus caucasicus*, sp.n.) parazitiruyushchnii na dagestanskom ture (*Capra cylindricornis* Blyth.). [A bot fly, *O. caucasicus*, parasitic on *C. cylindricornis*]. *Dokl. Akad. Nauk SSSR* (n.s.) 61(6): 1125-1127, fig.

GRUNIN, K. YA., (1950): Ovoda dzerena (*Procapra gutturosa* Gmel.) iz Mongol'skoi Narodnoi Respubliki. [Bot flies of *P. gutturosa* from the Mongolian People's Republic]. *Ibid.* 73(4): 861-864, figs.

GRUNIN, K. YA., (1951): O proiskhozdenii roda *Rhinoestrus* Br. (Diptera, Oestridae). [On the origin of the genus *Rhinoestrus* Br.]. *Ent. Obozr.* 31(3-4): 467-473, figs.

GRUNIN, K. YA., (1953): Lichinki ovodov domashnikh zhivotnykh SSSR. [Larvae of Oestridae from domestic animals in the USSR]. In: Akad. Nauk SSSR, *Oprideliteli po faune SSSR*, 51: 1-124, 139 figs. Moscow-Leningrad.

GRUNIN, K. YA., (1957): Nosoglotochnye ovoda (Oestridae). [Nasal bot flies (Oestridae)]. In: Zool. Inst. Akad. Nauk SSSR, *Fauna SSSR. Nasekomye dvukrylye*, 19(3): 1-145, 230 figs., 5 tables. [= Nov. Ser., n° 68]. Leningrad.

GRUNIN, K. YA., (1959): Otechestvennaya literatura po ovodam za 1957 g. [The Soviet literature on Oestridae s.l. published in the year 1957]. *Ent. Obozr.* 38(1): 265-270, 2 tables.

GRUNIN, K. YA., (1961a): Otechestvennaya literatura po ovodam za 1958 i 1959 gg. [The Soviet literature on Oestridae s.l. published in the years 1958 and 1959]. *Ibid.* 40(1): 240-246, 2 tables.

GRUNIN, K. YA., (1961b): O lichinke nosoglotochnogo ovoda kenguru (*Tracheomyia macropi* Frog.) (Diptera, Oestridae) iz Avstralii. [On the larva of the kangaroo nasal bot fly from Australia]. *Ibid.* 40(4): 929-933, 10 figs.

GRUNIN, K. YA., (1966): Oestridae [Fam. 64a']. In: E. LINDNER (ed.), *Die Fliegen der paläarktischen Region*, 8: 1-32, 90 figs., 4 tables [= Lfg. 264], 33-64, figs. 91-166 [= Lfg.

265], 65-96, figs. 167-240 [= Lfg. 266]. Stuttgart.

GRUNIN, K. YA., (1969): Calliphoridae, Gasterophilidae, Oestridae. Ergebnisse der zoologischen Forschungen von Dr. Z. KASZAB in der Mongolei (Diptera). *Faun. Abh. staatl. Mus. Tierk. Dresden* 3(2): 5-11.

GRUNIN, K. YA., R. V. GREBENYUK & S. K. SARTBAEV, (1965): Nosoglotochnyi ovod (*Oestrus caucasicus* Grunin) sibirskogo gormogo kozla. [*O. caucasicus* in *Capra sibirica*]. *Trudy zool. Inst. Akad. Nauk SSSR* 21:

GRUNIN, K. YA. & A. A. SLUDSKII, (1960): Novye dannie o nosoglotochkom ovode *Rhinoestrus tshernyshevi* Grunin (Diptera, Oestridae) arkhara (*Ovis ammon* L.). [New data on the nasal bot fly *Rh. tshernyshevi* of the wild sheep *O. ammon*]. *Ent. Obozr.* 39(1): 210-212, 2 figs.

GRYUNER, S. A., (1929a): Nosovoi ovod u loshadei i ego znachenie pri sape loshadei. [Nasal bot flies of horses and their significance in the glanders of horses]. *Trudy sib. vet. Inst.* 10: 303-322, 2 figs.

GRYUNER, S. A., (1929b): Nosovoi ovod u zhivotnykh semeistva Cervidae. [The bot flies of animals of the family Cervidae]. *Ibid.* 10: 323-328, figs. 1-5.

GRZYWIŃSKI, L. & Z. MADEJ, (1955): Estroza owiec. [Oestrosis in sheep]. *Medycyna wet.* 11(10): 593-597, figs.

GUENNEC, J. & G. M. ROBINEAU, (1960): Note sur trois cas de myiase conjonctivale. *Archs Ophthal., Paris* (n.s.) 20(6): 616-619, figs.

GUÉRIN-MÉNEVILLE, F. E., (1829-1844): Iconographie du règne animal de G. Cuvier, ou réprésentation d'après nature de l'une des espèces les plus remarquables et souvent non encore figurées, de chaque genre d'animaux, 3 vols. Paris.

GUERRA GRANDE, J. M., (1952): Consideraciones sobre un caso de oftalmomíasis externa por *Oestrus ovis. Archs Soc. oftal. hisp.-am.* 12(1): 86-89, illus.

GUEVARA BENÍTEZ, D., R. LOPEZ ROMAN & F. RAMOS ONTIVEROS, (1971): Nuevo caso de oftalmomíasis humana por *Oestrus ovis* L. en Granada. *Revta ibér. Parasit.* 31(3-4): 377-381, figs.

GUILLEBEAU, A., (1881): Ueber den Parasitismus einiger Oestriden. *Mitt. naturf. Ges. Bern* 1881(2): 7-11.

GUIMARÃES, J. H., (1967): Family Oestridae (including Hypodermatidae), in Dep. Zool., Secr. Agric., A Catalogue of the Diptera of the Americas south of the United States, 106: 1-4. S. Paulo.

GUKASSIAN, G. M., (1960): [Bremsenlarven auf der Schlundschleimhaut]. *Vest. Oto-rinolar.* 4: 99-100.

GUKASSIAN, G. M., (1962): [Bremsenlarven auf der Schleimhaut des Auges unter der Nase]. *Ibid.* 2: 101-102.

HADWEN, S., (1922): Parasites of reindeer and other animals in northern Alaska. *J. Parasit.* 9(1): 38-39.

HADWEN, S., (1923): Insects affecting live stock. *Bull. Can. Dep. Agric.* 29: 1-32, figs. 1-9.

HADWEN, S., (1926): Notes on the life history of *Oedemagena tarandi* L. and *Cephenemyia trompe* Modeer. *J. Parasit.* 13: 56-65.

HADWEN, S., (1932): Zametki ob istorii ovodov severnogo olenya *Oedemagena tarandi* L. i *Cephenomyia trompe* Modeer (referat M. LYUBIMOVA i S. AL'FA). *Sb. Olenev., tundr. Vet. Zootekhn.*: 296-298 (transl. of 1926).

HADWEN, S., (1940): *Cephenomyia* in Virginia deer. *J. Parasit.* 26(6); Suppl.: 16.

HAESELBARTH, E., J. SEGERMAN & F. ZUMPT, (1968): The arthropod parasites of vertebrates in Africa south of the Sahara (Ethiopian region), 3 (Insecta, excl. Phthiraptera). *Publs S. Afr. Inst. med. Res.* 13(52) (1966): 1-283, 136 figs., 15 pls.

HAIR, J. A., D. E. HOWELL, C. E. ROGERS & J. FLETCHER, (1969): Occurrence of the pharyngeal bot *Cephenemyia jellisoni* in Oklahoma white tailed deer, *Odocoileus virginianus. Ann. ent. Soc. Am.* 62(5): 1208-1210.

HANDSCHIN, E., (1946): Funde seltener Oestriden aus der Schweiz. *Mitt. schweiz. ent. Ges.* 20: 129-134, 3 figs.

HEDGES, H. S., (1942): Conjunctival myiasis due to *Oestrus ovis*. *Archs Ophthal., N.Y.* (n.s.) 28(2) [= vol. 85]: 251-253, figs.

HEINEMANN, E., (1936): Ein Opfer der Rachenbremse. *Dt. Jagd* (A) 6(31): 604.

HENNIG, W., (1952): *Die Larvenformen der Dipteren*, 3: 1-628, 338 figs. Berlin.

HENNING, C. F., (1855): Ueber *Oestrus equi*, *ovis* und *cervi capreoli*. *Allg. dt. naturh. Z.* 1:297-307, pl. 2.

HERBERG, M., (1938): Die Bekämpfung der Rachenbremse durch Vogelhege. *Dt. Jagd* (A) 9(4): 85-86.

HERBST, J. F. W., (1787): Kurze Einleitung zur Kenntniss der Insecten, für Ungeübte und Anfänger, 8: 1-200, index, 48 pls. Berlin & Stralsund. [Also publ. as Gemeinnütziger Naturgeschichte des Thierreichs, vols. 6-8, of the 'Naturgeschichte des Thierreichs' of BOROWSKI].

HERBST, J. F. W., (1790): Bremse, Oestrus. In: F. H. W. MARTINI, *Allgemeine Geschichte der Natur*, 9(12): 546-551, 1 pl., figs. 1-4. Berlin.

HERMAN, C. M., (1945): *Cephenemyia jellisoni* Townsend (Diptera Cuterebridae [sic]) reared from nasal bot of blacktailed deer. *Pan-Pacif. Ent.* 21(3): 120.

HERMAN, C. M., (1946): The nose bot fly of deer. *Calif. Fish Game* 32(1): 17-18, figs.

HERMS, W. B., (1925): Ophthalmomyiasis in man due to *Cephalomyia* (*Oestrus*) *ovis* (Linn.). *J. Parasit.* 12(1): 54-56, figs. A-C.

HERTING, B., (1955): Ein Beitrag zur Systematik der calyptraten Fliegen (kurze Mitteilung). *Mitt. schweiz. ent. Ges.* 28(2): 220-221.

HERTING, B., (1957): Das weibliche Postabdomen der calyptraten Fliegen (Diptera) und sein Merkmalswert über die Systematik der Gruppe. *Z. Morph. Ökol. Tiere* 45: 429-461, 21 figs.

HERTING, B., (1969): Records of Tachinidae (incl. Rhinophoridae) and Oestridae (Diptera) from Southern Spain, with descriptions of two new species. *Ent. Meddr* 37(3): 207-224.

HOFFMANN, B. L. & J. M. GOLDSMID, (1970): Ophthalmomyiasis caused by *Oestrus ovis* L. (Diptera: Oestridae) in Rhodesia. *S. Afr. med. J.* 44(22): 644-645, 4 figs.

HOFFMANN, (H.?), (1909): Ueber Rachenbremsen. *Bonn. Sber. Ges. naturk.* 1908-09 (naturw. Abt.): 21-25.

HOFFMANN, H., (1908): Rachenbremsen. *Z. Forst- u. Jagdw.* 40(12): 820-822.

HOFFMANN, H., (1938): Zur Bekämpfung der Rachenbremse. *Dt. Jagd* (A) 8(50): 906-908.

HOLMBOE, F. V., (1927): Forekommer bremselarver hos baeveren? *Norsk VetTidsskr.* 39 (9): 272-273.

HONESS, R. F. & N. Y. FROST, (1942): A Wyoming bighorn sheep study. *Bull. Wyo. Game Fish Dep.* 1: 1-126.

HORAK, I. G., J. P. LOUW & S. M. RAYMOND, (1971): Trials with rafoxanide: III. Efficacy of rafoxanide against the larvae of the sheep nasal bot fly *Oestrus ovis* Linné, 1761 [sic]. *J. S. Afr. med. Ass.* 42(4): 337-339.

HUBER, J. C., (1899): Bibliographie der klinischen Entomologie (Hexapoden, Acariden), 3 (Muscidae, Oestridae): 25 pp. Jena.

HUDSON, G. V., (1892): An elementary manual of New Zealand entomology: Being an introduction to the study of our native insects, (4) + 128 + (20) pp., 20 col. pls. London.

HUNTER, W. D., (1915): A new species of *Cephenomyia* from the United States (Diptera: Oestridae). *Proc. ent. Soc. Wash.* 17: 169-173.

HUTCHEON, D., (1891): Bots or paapjes in the horse. *Agric. J. Cape G.H.* 3(24): 227-238.

HUTCHEON, D., (1894): Idem. *Ibid.* 7(2): 37-39.

HUTCHEON, D., (1899a): Bots or worms in the heads of sheep and goats. *Ibid.* 14(10): 667-668.

HUTCHEON, D., (1899b): Bots or 'paapjes'. *Ibid.* 15(6): 400-407.

HUTCHEON, D., (1899c): Do bots kill horses? *Ibid.* 15(11): 759-760.

HUTCHEON, D., (1907): Bots or 'paapjes'. *Ibid.* 30(5): 676-683.

HUTCHEON, D., (1914): Idem. *Agric. J. Un. S. Afr.* 8(2): 194-200.

HUTSON, R., (1931): The nose-fly of deer, *Cephonomyia* [sic] *phobifer.* 70th An. Rep. Mich. St. Bd. Agric. (1930-31): 269-271, 2 figs. [Also in 44th An. Rep. Mich. agric. exp. St. 1930-31: 269-271, 2 figs., 1931; and in Rep. ent. Sect. Mich. St. Coll. Agric. 1930-31: 19-20, 2 figs., 1931].

HUTTON, F. W., (1901): Synopsis of the Diptera Brachycera of New Zealand. *Trans. Proc. N.Z. Inst.* (1900) 33 [= n.s., vol. 16]: 1-95.

IHERING, R. VON, (1929): Os oestrideos importados, seu papel como parasitas e em particular os *Gasterophilus* no Brasil. *Bolm Agric., S. Paulo* 30: 863-883.

IHERING, R. VON, (1930): Varios casos de *Oestrus* e *Gasterophilus* no Brasil. *Revta Soc. paul. Med. vet.* 1(2): 30-35.

INTERNATIONAL COMMISSION ON ZOOLOGICAL NOMENCLATURE, (1929): The type of *Oestrus* Linn., 1758, is *O. ovis* (Opinion 106). *Smithson. misc. Collns* 73(6) [= Publ. n° 3016]: 4-8.

JELLISON, W. L., (1935): *Cephenomyia pratti* (Diptera: Oestridae) reared from black-tailed deer. *Proc. helminth. Soc. Wash.* 2(2): 69.

JOLY, N., (1846): Recherches zoologiques, anatomiques, physiologiques et médicales sur les oestrides en général et particulièrement sur les oestres qui attaquent l'homme, le cheval, le boeuf et le mouton. *Annls Sci. phys., nat., Agric. Ind. Soc. r. Agric. Lyon* 9: 157-305, pls. 1-8.

JOLY, N., (1858): Sur l'hypermétamorphose des strepsiptères et des oestrides. *C. r. hebd. Séanc. Acad. Sci., Paris* 46(20): 942-944.

JOHNSTON, J. E., (1935?): Bot control. 61st An. Mt. Ontario vet. Ass. centr. Can. vet. Ass. 1935: 7-8.

KABOS, W. J., (1960): Tweevleugelike insekten–Diptera, VI. De nederlandsche dazen (Tabanidae) en horzels (Oestridae). *Wet. Meded. K. ned. Natuurh. Veren.* 38: 1-16, 21 figs.

KAL'KIS, YA. I., (1965a): Biologiya polostnogo ovoda ovets v lesostepnoi zone Altaiskogo kraya. [Biology of nasal bot flies of sheep in landscape zones in the Altai region]. *Trudy vses. nauchno-issled. Inst. vet. Sanit. Ektoparazit.* 26: 214-220.

KAL'KIS, YA. I., (1965b): Primenenie geksakhloranovogo dyma v bor'be s estrozom ovets. [Use of hexachlorane fumes in the control of *O. ovis* of sheep]. *Ibid.* 26: 221-227.

ISAICHIKOV, I. M., (1932): Bol'she vnimaniya bor'be s vragami olenevodstva, maralovodstva i karakulevodstva iz gruppy nasekomykh. [Greater attention to control of enemies of deer and sheep from the group of insects]. *Soyuzpushina, Vneshtogrizdat, Moscow* 7: 26-29.

ISOLA, W. & J. J. OSIMANI, (1944): Un nuevo caso de oftalmomíasis conjunctival produzida por *Oestrus ovis* en el Uruguay. *Archos urug. Med. cirurg.* 25(3): 260-264, figs.

ITURBIDE, A., (1960): El *Oestrus ovis* y la explotación ovina. *Investnes agropec.* 1(2): 105-107.

IVANOV, I. & D. PETROV, (1966): Nakhodka na *L. monocytogenes* u lichinki na *O. ovis* (Soobshchenie). [On the finding of *Listeria monocytogenes* in *O. ovis* larvae]. *Vet.-med. Nauk* 3(6): 641-643.

JAMES, M. T., (1944): Two erroneous records in American literature of the causative agents of myiasis. *J. Parasit.* 30(4): 273-274.

JAMES, M. T., (1948): The flies that cause myiasis in man. *U.S. Dep. Agric. misc. Publ.* 631 (1947): 1-175, figs. [For dating, see SABROSKY, 1950: 315].

JARRY, D. & NAIM, (1960): Contribution à l'étude des ophthalmomyiases à *Oestrus ovis* L. *Languedoc méd.-chir.* 6: 1-5.

KAMARLI, A. P., (1965): Mery bor'by s polostnym ovodom ovets. [Methods of sheep bot fly control]. *Veterinariya* 41[i.e., 42](2): 46-47.

KARPENKO, S. E., (1947): Zabolevanie loshadei rinestrozom. [Disease in horses by *Rhinoestrus*]. *Ibid.* 24(3): 42. [Abstr. in *Rev. appl. Ent.* (B) 36(10): 167-168, 1948].

KATO, H. & T. MURAKAMI, (1961): Infestation of sheep bot fly (*Oestrus ovis*) larvae in nasal sinuses of sheep observed in Iwate Pref., Japan. *J. Fac. Agric. Iwate Univ.* 5(2): 67-74, figs.

KATO, H. & T. MURAKAMI, (1962): Isolation of *Listeria monocytogenes* from an *Oestrus ovis* larva harvested from a sheep with *Listeria* encephalitis. *Jap. J. vet. Sci.* 24(1): 39-43, pl.

KEISER, F., (1948): Der erste Fall von Ophthalmomyiasis, hervorgerufen von *Oestrus ovis* L., aus der Schweiz. *Verh. naturf. Ges. Basel* 59: 29-44.

KELLNER, A., (1846-47): Bemerkungen über die als Larven im Rothwilde lebenden *Oestrus*. *Stettin. ent. Ztg* 7: 29-30, 1846; 8: 366-367, 1847.

KELLNER, A., (1853): Beobachtungen über die im Roth- und Rehwilde lebenden *Oestrus*-Arten. *Ibid.* 14: 89-93.

KETTLE, P. R., (1973): A study of the sheep botfly, *Oestrus ovis* (Diptera: Oestridae) in New Zealand. *N.Z. Ent.* 5(2): 185-191, 3 tables.

KHOLODKOVSKII, N. A., (1907): K' voprosu o razmnozhen'i i razvit'i zhivorodyashchikh mukh. [Zur Frage über die Fortpflanzungs- und Entwicklungsweise der vivipare Fliegen]. *Trudy imp. S-peterb. Obshch. Ent.* 38(1): 100-108.

KING, A., (1911): *Oestrus ovis* in sheep. *Rhodesia agric. J.* 9(1): 24-25.

KING, H., (1911): Report of the entomological section of the Wellcome Tropical Research Laboratories. 4th Rep. Wellcome trop. Res. Labs, Khartoum (B; gen. Sci.): figs. 12-16, pls. 1-9.

KIRKPATRICK, J., (1859): Ueber Engerlinge und Engerlingefliege [*Oestrus*]. 13. Jber. Ohio-Staats Landbaubehörde: 282-287, fig.

KLĖNIN, I. I., (1951a): Chislo generatsii *Oestrus ovis* na protyachenii goda. [Number of generations of *O. ovis* during the period of one year]. *Trudy chkalovsk. sel'.-khoz. Inst.* 4(2): 156-157.

KLĖNIN, I. I., (1951b): Nablyudeniya za vykhozhdeniem lichinok *Oestrus ovis* iz organizma ovets na protyazhenii sutoki. [Observations on the emergence of larvae of *O. ovis* in the organism of sheep in the twenty-four hour period]. *Ibid.* 4(2): 157.

KLĖNIN, I. I., (1953): Dinamika invazirovannosti ovets lichinkami *Oestrus ovis*. [Dynamics of the invasion of sheep by the larvae of *O. ovis*]. *Ibid.* 6: 137-141.

KLĖNIN, I. I., (1954a): K morfologii lichinok *Oestrus ovis*. [Morphology of the larva of *O. ovis*]. *Ibid.* 6: 143-153, figs.

KLĖNIN, I. I., (1954b): Materialy po klinike, terapii i profilaktive estroza ovets. [Materials for the clinics, therapy and prophylaxy of sheep oestrosis]. Tezisy Dokl. l. vses. Konf. Probl. vet. Dermat., Arakhn. Ent., Moscow: 189-191.

KLĖNIN, I. I., (1955): K epizootologii estroza ovets. [On the epizootology of ovine oestrosis]. *Trudy chkalovsk. sel'.-khoz. Inst.* 7: 273-280.

KLĖNIN, I. I., (1958): O merakh bor'by s ovodovoi bolezniyu ovets. [On measures for the control of bot fly on sheep]. *Sel'. Khoz. Povolzh'ya* 3(5): 75-77.

KLĖNIN, I. I., (1959a): Klinika i lechenie estroza ovets. [Clinics and treatment of ovine oestrosis]. *Trudy orenburgsk. sel.'-khoz. Inst.* 8: 195-200 (1958).

KLĖNIN, I. I., (1959b): Estroz ovets i mery bor'by s nim v usloviyakh Orenburgskoi oblasti. [Ovine oestrosis and control methods in the conditions of the Orenburg region], 28 pp. Diss. na Soisk. Uk. Step. Dokt. vet. Nauk, Khar'kov.

KNAB, F., (1913): A new bot-fly from reindeer (Diptera; Muscoidea). *Proc. biol. Soc. Wash.* 26: 155-156.

KNAPP, F. W., (1972): An ecological study of *Oestrus ovis* Linné. Abstr. 14th Int. Congr. Ent., Canberra: 291-292.

KNAPP, F. W. & J. H. DRUDGE, (1964): Efficacy of several organic phosphates against the bot-fly of sheep. *Am. J. vet. Res.* 25(109): 1686-1689.

KNAPP, F. W. & J. H. DRUDGE, (1965): Sheep nose bot. 77th An. Rep. Kentucky agric. exp. Sta. 1964: 53.

KNAPP, F. W. & C. E. ROGERS, (1967): Effects of sinus temperature upon overwintering of *Oestrus ovis. Proc. An. Mt. n-centr. Br. ent. Soc. Am.* (East Lansing, Mich.) 22: 170.

KNAPP, F. W. & C. E. ROGERS, (1968): A survey of sheep bot fly larva infestations in Kentucky. *J. econ. Ent.* 61(1): 23-25, 2 figs.

KNORRE, E. P., (1957): Materialy po biologii i znacheniyu nosoglotochnogo ovoda losya. [Materials on the biology and importance of the nasal bot of the elk]. *Zool. Zh.* 36(4): 563-574, figs.

KNOTT, J. J., (1877): A case of bots in the human subject. *Louisville med. News* 4(2): 15-16.

KNOWLTON, G. F. & J. A. ROWE, (1936): Deer botfly (*Cephenomyia pratti* Hunter). *Insect Pest Surv. Bull. U.S.* 16(3): 90.

KOCH, L. E., (1961): Kangaroo bot-fly larvae from Port Hedland. *West. Aust. Nat.* 7(7): 190-191.

KOIDE, S. S., (1955): *Oestrus ovis* (maggots in the eyes and nose). *Plantn Hlth* 20(4): 24.

KOIDE, S. S., (1956): Myiasis from *Oestrus ovis*. Report of three cases. *Hawaii med. J. inter-isl. Nurs. Bull.* 15(5): 460-461.

KOLOMIETS, YU. S., (1951a): Polostnoi ovod loshadi *Rhinoestrus purpureus* Br. na Ukranie. [The cavicole horse bot fly, *Rh. purpureus*, in Ukraine]. *Zool. Zh.* 30(6): 550-555.

KOLOMIETS, YU. S., (1951b): Polostnoi ovod loshadi *Rhinoestrus purpureus* na Ukranie i mery bor'by s nim. [Cavicole horse bot fly in Ukraine and measures for its control]. *Nauch. Trudy ukrainsk. Inst. eksp. Vet.* 18: 202-220, figs.

KOLOMIETS, YU. S., (1952a): Lechenie rinestroza loshadei. [Cure of equine rhinoestrosis]. *Ibid.* 19: 201-207.

KOLOMIETS, YU. S., (1952b): Allergicheskii metod diagnostiki pri rinestroza loshadei. [Allergic method for the diagnosis of rhinoestrosis in horses]. *Ibid.* 19: 208-216.

KOLOMIETS, YU. S., (1953): Izuchenie profilakticheskikh mery bor'by s rinestrozom loshadei. [Studies on prophylaxy of control of rhinoestrosis in horses]. *Ibid.* 20: 188-199.

KOLOMIETS, YU. S., (1954): Rinestroz loshadei i mery bor'by s nim. [Rhinoestrosis of horses and its control]. Tezisy Dokl. l. vses. Konf. Probl. vet. Dermat., Arakhn. Ent., Moscow: 186-188.

KOLOMIETS, YU. S., (1955): Shirokii opyt primeneniya profilakticheskoi i lechevno-profilakticheskoi obrabotki loshadei protiv rinestroza. [Experiment on the use of prophylactic and treatment-prophylactic dressings on horses agsinst rhinoestrosis]. *Nauch. Trudy ukrainsk. Inst. eksp. Vet.* 22: 213-217.

KOLOMIETS, YU. S. & A. V. ALFIMOVA, (1956): Biologicheskie osobennosti *Oestpus* [sic] *ovis* L. v usloviyakh Ukrainy. [Biologische Eigentümlichkeiten des *O. ovis* in der Ukraine]. *Ibid.* 23: 309-316.

KOLOMIETS, YU. S., A. V. ALFIMOVA & I. K. KAPUSTIN, (1956): Izyskanie profilakticheskikh mer bor'by s polostnym ovodom ovets. [Survey of prophylactic measures of control of sheep bot fly]. *Ibid.* 23: 303-308.

KOLOMIETS, YU. S., A. V. ALFIMOVA, I. K. KAPUSTIN & M. I. EMETS, (1958): Dal'neisheë izuchenie deistviya aerozolei geksakhlorana na organism ovets i primenenie ego v tselyakh profilaktiki estroza ovets. [Further study of the use of aerosol hexachlorane on the ovine organism and its use in complete prophylactic measures against ovine oestrosis]. *Ibid.* 24: 247-253.

KONONYUK, G. YA., (1957): Opyt bor'by s polostnym ovodom ovets. [Experiments on the control of the sheep bot fly]. *Trudy Inst. Vet. kazak. Fil. Akad. sel'.-khoz. Nauk* 8: 327-329.

Kornilova, A. F., (1950): Dva sluchaya popadaniya lichinok ovoda v kon'yunktival'nyi meshok glaza cheloveka. [Two cases of botfly larvae in the conjunctival sac of the human eye]. *Vest. Oftal.* 29(4): 41-42.

Kowalski, (-), (1863): Gehirn- und Darmaffection mit Oestruslarven bei Schafen. *Mitth. thierärztl. Praxis preuss. Staate* 10 (1861-62): 170-171.

Krembs, J., (1936-37): Folgen eines Rachenbremsenbefalls bei einer Rehgeiss. *Dt. Jagd* 18: ? [Abstr. in *Tierärztl. Rdsch.* 43(26): 439-440, 1937].

Krivko, A. M., (1955): Novye preparaty protiv polostnogo ovoda. [New preparations against cavicole bot flies]. *Sel.' Khoz. Kazak.* 11: 42-44.

Krivko, A. M., (1956): Estestvennoe razdvoenie vesennego pokoleniya ovechnego ovoda (*Oestrus ovis* L.) v Alma-Atinskoi oblasti. [Natural splitting of the vernal generation of *O. ovis* in the Alma-Ata region]. *Dokl. Akad. Nauk SSSR* 111(1): 248-249.

Krivko, A. M., (1957a): K morfologii i biologii polostnykh ovodov loshadei na yugo-vostoke Kazakhstana. [On the morphology and biology of the cavicole bot-flies of horses in south-eastern Kazakhstan]. *Trudy Inst. Vet. Kazakh. Fil. Akad. sel.'-khoz. Nauk* 8: 269-278, figs.

Krivko, A. M., (1957b): Morfologiya i biologiya polostnogo ovoda ovets (*Oestrus ovis* L.) na yugo-vostoke Kazakhstana. [Morphology and biology of the sheep bot fly (*O. ovis*) in south-eastern Kazakhstan]. *Ibid.* 8: 279-300, figs.

Krivko, A. M., (1957c): Ispytanie novykh insektitsidnykh preparatov protiv lichinok polostnogo ovoda ovets (*Oestrus ovis* L.). [Trials of new insecticide preparations against the cavicole sheep bot fly (*O. ovis*)]. *Ibid.* 8: 311-315.

Krivko, A. M., (1957d): Ispytanie bezapparatnykh geksakhloranovykh aerozolei v vide dyma na lichinok l stadii ovech'ego ovoda. [The use of hexachlorane aerosol without apparatus in form of smoke against 1st stage larvae of the sheep bot fly]. *Trudy kazakh. nauchno-issled. vet. Inst.* 9: 552-555.

Krivko, A. M., (1959): Materialy po izucheniyu polostnogo ovoda ovets v Kazakh-stane. [Materials for the study of the cavicole sheep bot fly in Kazakhstan], 14 pp. Diss. na Soisk. Uch. Step. Kand. vet. Nauk, Alma-Ata.

Krivko, A. M. & A. S. Red'ko, (1960): Estroz ovets i mery bor'by s nim. [Ovine oestrosis and measures for its control]. *Sel'. Khoz. Kazakh.* 7(5): 64-66.

Kuhn, W. G., (1914): Oestrus ovis. *Q. Bull. Kansas Cy vet. Coll.* 46: 1113.

Kuklin, C. H., (1954): Estroz ovets i metody khirurgicheskogo lecheniya. [Ovine oestrosis and methods of surgical treatment]. *Trudy omsk. vet. Inst.* 14: 153-162.

Kulakov, I. A., (1957): Estroz ovets v Rostovskoi oblasti. [Ovine oestrosis in the Rostov region]. *Veterinariya* 34(6): 23-28.

Kulieva, C. Ch., (1936): O miaze na eilagakh Eli-Su. [Ueber Myiase auf den Sommer-schafalmen (Eilagi) von Eli-Su]. *Trudy trop. Inst. azerb. SSR* 2: 136-137.

Landois, H., (1903): Die Kamel-Nasenbremse (*Oestrus maculatus* Wied.) im Westfäli-schen zoologischen Garten zu Münster. *Zool. Gart., Frankf.* 44: 53-55.

Langmuir, I., (1938): The speed of the deer fly. *Science, N.Y.* 87: 233-234.

Lapierre, J. & M. Pette, (1954): À propos d'un cas de myiase oculaire du à Oestrus ovis observé dans la région parisienne. *Bull. Soc. Path. exot.* 47(4): 561-566, pl.

Larrousse, F. L., (1921): La myiase oculaire à Oestrus ovis L. dans la région parisienne. *Ibid.* 14(9): 595-601, figs. 1-5.

Larrousse, F. L., (1924): Nouveau cas de myiase oculaire à Oestrus ovis L. en France. *Annls Parasit. hum. comp.* 2(3): 274.

Latreille, P. A., (1809): Genera crustaceorum et insectorum secundum ordinem natu-ralem in familias disposita, iconibus exemplisque plurimis explicata, 4: 399 pp., 4 pls. Parisiis et Argentorat.

Latreille, P. A., (1815): Rapport fait à l'Institut sur l'oeuvre de B. Clark. *Annls Agric. fr.* (1) 62: 266.

Latreille, P. A., (1818): Articles 'Oestre', 'Oestridés', and 'Oestrides', pp. 264-274, in Société de Naturalistes et d'Agriculteurs, Nouveau Dictionnaire d'Histoire Naturelle appliquée aux arts, à l'agriculture, à l'économie rurale et domestique, à la médecine, etc., 23: 1-612, pls. Deterville Libraire, Paris.

Latreille, P. A., (1825): Familles naturelles du règne animal, 570 pp. Paris.

Latreille, P. A., (1829): Les crustacés, les arachnides et les insectes. Tome second. In: [G. C. L. D. Cuvier], Le Règne Animal (Ed. 2), 5: 556 pp., 5 pls. Paris.

Laurence, B. R., (1961): On a collection of oestrid larvae (Diptera) from East African game animals. *Proc. zool. Soc. Lond.* 136(4): 593-601, 6 figs.

Laurence, B. R. & F. R. N. Pester, (1970): *Herpetomonas oestrorum* – a parasite within a parasite. *Trans. R. Soc. trop. Med. Hyg.* 64(1): 17-18.

Lavrov, P. A., (1952): Polostnoi ovod loshadi v Chkalovskoi oblasti. [Cavicole bot flies of horses in the Chkalov region]. *Trudy chkalovsk. sel'.-khoz. Inst.* 5: 277-281.

Leach, W. E., (1817): On the genera and species of eproboscideous insects, 20 pp., 3 pls. Edinburgh. [Also in *Mem. Wernerian nat. Hist. Soc.* 2: 547-566, 1818].

Leach, W. E., (1818): On the arrangement of oestrideous insects. *Mem. Wernerian nat. Hist. Soc.* 2(2)(1817): 567-568.

Leclercq, M., (1948): Revision des oestrides de Belgique (Diptera). *Bull. Mus. Hist. nat. Belg.* 24(41): 1-11.

Leclercq, M., (1950): Les 'oestrides' du Musée Zoologique de Strasbourg (Diptera). *Bull. Annls Soc. R. ent. Belg.* 86: 55-56.

Leclercq, M., (1955): À propos des oestrides de Belgique. *Revue Agric., Brux.* 8: 1-11.

Leclercq, M., (1971): Les mouches nuisibles aux animaux domestiques, un problème mondial, 199 pp., 78 figs. Les Presses Agronomiques de Gembloux, Bruxelles. [Pp. 126-147, figs. 67-72].

Lefèvre, E., (1890): Céphalémie (*Cephalemyia* Latr.). Grande Encyclopédie, Paris 9: 1164-1165, 1 fig.

Leidy, J., (1857): Remarks on the larva of a species of *Oestrus. Proc. Acad. nat. Sci. Philad.* 1857: 204.

Leknaes, C. A. T., (1903): Wandernde *Oestrus*-Larven bei einen 3jährigen Hengstfohlen. *Jber. Leist. Geb. VetMed.* 22 (1902): 109.

Léniez, A., (1875): L'oestre et sa larve. *Bull. Soc. linn. N. Fr.* 2 (1874-75) :353-355, 363-368.

León, L. A. & M. Andrade, (1955a): Caso de miíasis ocular a *Oestrus ovis* (Diptera, Oestridae) observado en Quito. *Boln inf. cient. nac., Quito* 7(63): 171-184, 2 pls. [Also in *Boln Lab. clín. Luis Razetti* 16: 621-639, figs., 1955].

León, L. A. & M. Andrade, (1955b): Observaciones sobre la miíasis ocular producida por el *Oestrus ovis* en el Ecuador (Diptera – Oestridae). *Revta ecuat. Ent. Parasit.* 2(3-4) (1954-55): 377-388, pls.

Leonardi, G. & A. Lunardoni, (1900): Gli insetti nocivi ai nostri orti, campi, frutteti e boschi, all'uomo ed agli animali domestici, 3 (Imenotteri e Ditteri): xxi + 549 pp., figs. Napoli.

Linnaeus, C., (1737): Flora Lapponica, exhibens plantae per Lapponiam crescentes, secundum systema sexuale, collectas in itinere impensis Soc. Reg. Litter. et Scient. Sveciae A. 1732 instituto. Additis synonymis & locis natalibus omnium, descriptionibus & figuris rariorum, viribus medicatis & oeconomicis plurimarum, xxxvii + 372 pp., 12 pls. Amstelodami.

Linnaeus, C., (1758): Systema naturae per regna tria naturae, secundum classes, ordines, genera, species, cum characteribus, differentiis, synonymis, locis (Ed. 10), 824 pp. Holmiae.

Linnaeus, C., (1761): Fauna Suecica, sistens animalia Sueciae regni: Mammalia, Aves, Amphibia, Pisces, Insecta, Vermes, distributa per classes et ordines, genera et species. Cum differentiis specierum, synonymis, auctorum, nominibus incolarum, locis natalium, descriptionibus insectorum. Editio altera, aucta, 578 pp. Stockholmiae.

LIU, Y. C., C. M. KAO & M. TS'UI, (1964): Conjunctival myiasis due to *Oestrus ovis*. Report of 3 cases. *Chin. med. J.* 83(3): 190-193, figs.

LIZCANO HERRERA, J., (1950): Contribución al estudio del *Oestrus ovis*, Linné, 1761 [sic]. *Boln Zootecn., Córdoba* 6: 92-94.

LOCHHEAD, W., (1916): Some notes regarding nose and other bot flies. 46th A. Rep. ent. Soc. Ontario: 102-108.

LOEW, H., (1863): Enumeratio dipterorum quae C. Tollin ex Africa Meridionali misit. *Wien. ent. Mschr.* 7(1): 9-13.

LÖWE, W., (1854): Die falsche Drehkrankheit der Schafe, hervorgerufen durch die Schafbremse (*Oestrus ovis*). Gründliche Anleitung diesen Krankheitszustand zu erkennen, zu behandeln und radical zu halten. Für Schäfereibesitzer, Thierärzte, Schäfer, etc. (4th Ed.), 8 + 31 pp., 1 pl. Braune.

LORA D., C., M. VÁSQUEZ D. & C. MARCHINARES A., (1966): Tratamiento de la míasis cavitaria por *Oestrus ovis* con Neguvon. *Revta Centro nac. Pat. anim., Lima* 5(8-9): 69-74.

LOTIN, A. V., (1946): O zabolevaniyakh glaz, vysyvaemykh lichinkami mukh i ovodov. [On the eye maladies produced by flies and bot-flies larvae]. Sb. nauch. Rab., posv. 70-letiyu akad. V. P. Filatova, Odessa: 314-317.

LUCAS, H., (1838): Oestre, *Oestrus*. *Dictionnaire Pittoresque d'Histoire Naturelle*, 6: 225-232.

LUCAS, P. H., (1876): [*Cephalemyia ovis*]. *Bull. Soc. ent. Fr.* (5) 6: xcv-xcvi.

LUGGER, O., (1897): Parasites of man and the domesticated animals. 2nd Rep. St. Ent. of Minnesota 1896.

LÜHE, (-), (1906): Die thierischen Parasiten des Elchs. *Königsberg. Schr. phys. Ges.* 46 (1905): 177-180.

LUMBRERAS CRUZ, H. & F. POLACK, (1955): Primer caso peruano de oculomíasis producida por larvas de *Oestrus ovis* Linneo, 1758. *Revta méd. peru.* 26: 95-99, figs.

LUTZ, A., (1917): A occurrencia [sic] do *Oestrus ovis* no Rio de Janeiro e nos estados visinhos, p. 111, in his Contribuições ao conhecimento dos oestrideos brazileiros. *Mems Inst. Oswaldo Cruz* 9: 94-112, pls. 27-29.

LYON, M. W., (1934): Conjunctival ophthalmomyiasis caused by *Oestrus ovis*. *J. Parasit.* 20(6): 337.

LYONS, E. T., J. H. DRUDGE & F. W. KNAPP, (1967): Controlled tests of anthelminthic activity of trichlorphon and thiabendazole in lambs, with observations on *Oestrus ovis*. *Am. J. vet. Res.* 28(125): 1111-1116.

LYUBIMOV, M. & S. AL'F, (1932): Hadwen's zametki ob istorii zhizni ovod severnogo olenya *Oedemagena tarandi* L. i *Cephenomyia trompe* Modeer. [Hadwen's observations on the life history of the reindeer flies *O. tarandi* and *C. trompe*]. Sb. Olenev., tundr. Vet. Zootekhn. 1932: 296-298.

McCARTHY, P. H., (1961): Two parasites of marsupials in Central Queensland. *Aust. vet. J.* 37: 405.

MACHIDA, M., (1961): [On the nostril fly of sheep (*Oestrus ovis*) in Hokkaido]. *Meguro Kiseichu Kan Geppo* 30: 2. [In Japanese].

MACLEAY, W. S., (1825a): On the insect called Oistros by the ancient Greeks and Asilus by the Romans. *Trans. Linn. Soc. Lond.* 14(2): 353-359.

MACLEAY, W. S., (1825b): Ueber den Oistros der Griechen und Asilus der Römer [Abstr. of 1825a]. *Isis* (Oken's) 12: 1341-1342.

MACLEAY, W. S., (1830): On the *Oestrus* of M. B. CLARK. *Zool. J.* 5 (1829-30): 18-25, 276.

MACQUART, J., (1834): Insectes diptères du nord de la France. Athéricères: Créophiles, Oestrides, Myopaires, Conopsaires, Scénopiniens, Céphalopsides. *Mém. Soc. r. Sci., Agric. Arts, Lille* 1833: 137-368, 6 pls. [Also sep. publ., (Vol. 5), 233 pp., 6 pls., Lille, 1834].

MACQUART, J., (1835): Histoire naturelle des insectes. Diptères, 2: 703 pp., 12 pls. [In N. E. RORET (ed.), Collection des suites à Buffon]. Paris.

MACQUART, J., (1843): Diptères exotiques nouveaux ou peu connus. Mém. Soc. r. Sci., Agric. Arts, Lille 1842: 162-460, 36 pls. [Also sep. publ., 2(3): 5-304, 36 pls., Paris, 1843].

MAGRI, (-), (1868): [Title unknown; cf. 1869]. Medico vet., Torino (3)3: ?

MAGRI, (-), (1869): Kolik von Bremselarven. Reprium Thierheilk. 30: 78-79 [Abstr. of 1868].

MAIR, W. G., (1880): [Note on O. ovis]. Trans. N.Z. Inst. 12: 446.

MAKEVMIN, S. G., (1959): Osobennosti techeniya estroza ovets v 1958 godu. [Special trends on ovine oestrosis in the year 1958]. V. pomoshch'-sel'skokhozyaistven-nomu Proisvodstvu, Stalingrad: 26-27.

MALAGUTI, G. B., (1842): Due larve del genere Oestrus cavate dall'orecchio di un contadino. Raccoglitore, Fano 10(1): 1-9, pl.

MALLOCH, J. R., (1919): The Diptera collected by the Canadian Expedition 1913, 18 (3,C):34c-90c.

MALLOCH, J. R., (1934): Chironomidae, Sciaridae, Phoridae, Syrphidae, Piophilidae, Helomyzidae, Calliphoridae, Oestridae and Tachinidae (with a note on Oedemagena tarandi (Linné) by W. J. HOLLAND), in The exploration of Southampton Island, Hudson Bay, 1929-30. Mem. Carnegie Mus. 12(2,4): 13-32, figs.

MANINE, A., (1941): Un cas de myiase oculaire à Oestrus ovis Linné dans le Sahara Central (Fort Flatters, Sahara Constantinois). Archs. Inst. Pasteur Algér. 19(2): 287-289.

MARION DE PROCÉ, (-), (1827): Observation relative à la présence de plusieurs larves d'oestres (?) dans le canal digestif d'un individu de l'espèce humaine. J. Sect. Méd. Soc. acad. Loire-inf. 3: 89-95, figs. 1-4.

MASSONAT, E. & C. VANEY, (1913): Ethologie et pupation chez les diptères pupipares et les oestrides. C. r. Séanc. Soc. Biol. 75: 49-51.

MAZIDIS, S. P. & C. XIROCOSTAS, (1958): [Un cas de myiase oculaire due aux larves du diptère Oestrus ovis]. Delt. 'ell. mikrobiol. 'ug. 'Etair. 3(2-3): 89-99. [In Greek].

MAZINA, R. O., (1948): Sluchai orbital'nogo miaza vyavannogo lichinkoi ovech'ego ovoda Oesterus [sic] ovis. [A case of orbital myiasis caused by the larva of the sheep bot-fly, O. ovis]. Vest. Oftal. 27(6): 33-34.

MAZZOLENI, R., (1965): La miasis ovina da Oestrus ovis. Veterinaria, Milano 14(4): 355-360.

MÉGNIN, J. P., (1868): Diptères parasites des animaux. Insectol. agric. 2: 170-172, 1 pl.

MÉGNIN, J. P., (1878a): Note sur les oestres indigènes. Bull. Soc. ent. Fr. 1878: xl-xli.

MÉGNIN, J. P., (1878b): Oestride très rare en France. Ibid. 1878: lxxiv.

MÉGNIN, J. P., (1878c): Sur des mouches provenant de larves trouvées dans le pharynx d'un cerf. Recl Méd. vét. Éc. Alfort 55 [= 6(5)]: 601-602.

MÉGNIN, J. P., (1878d): Sur une larve d'oestre trouvé dans l'un des pédoncules cérébraux chez un cheval mort d'apoplexie. Ibid. 55: 602-603.

MÉGNIN, J. P., (1879): Note sur l'acclimatation d'une nouvelle espèce d'oestrides en France. Ibid. 56 [= 6(6)]: 787-789.

MÉGNIN, J. P., (1880): Les parasites et les maladies parasitaires chez l'homme, les animaux parasites et les animaux sauvages avec lesquels ils peuvent être en contact. Insectes, arachnides, crustacés, 484 pp., 63 figs. & Atlas: 26 pls. Masson, Paris.

MÉGNIN, J. P. & (-) BOUTHERY, (1880): [O. ovis]. Bull. Soc. ent. Fr. 1880: cxxxiv-clii.

MEIGEN, J. W., (1824): Systematische Beschreibung der bekannten europäischen zweiflügeligen Insekten, 4: xii + 428 pp., pls. 33-41. Hamm.

MEIRA, M. T. V. DE, J. RUFFIÉ & H. T. DE SOUZA, (1957): La myiase humaine à Oestrus ovis dans l'Archipel du Cap Vert. Anais Inst. Med. trop., Lisb. 14(3-4): 407-423, 2 pls.

MELENEY, W. P. & S. A. APODACA, (1969): Regeneration of a population of Oestrus ovis in sheep on an isolated range. J. Am. vet. med. Ass. 155(2): 136-138.

MELENEY, W. P., N. G. COBBETT & H. O. PETERSON, (1962): The natural occurrence of *Oestrus ovis* in sheep from the southwestern United States. *Am. J. vet. Res.* 23(97): 1246-1251, 4 figs.

MELENEY, W. P., N. G. COBBETT & H. O. PETERSON, (1963): Control of *Oestrus ovis* in sheep on an isolated range. *J. Am. vet. med. Ass.* 143(9): 986-989; 144(7): 756-758.

MELLO, M. J., (1941): A respeito do *Oestrus ovis*, parasito das cavidades dos ossos da cabeça de ovinos. *Biológico* 71(10): 290.

MERDIVENCI, A., (1957): Türkiye'de *Oestrus ovis* (Linné, 1761 [sic]) in insalarda sebeniyet verdiği bir myiasis ocularis olayi. [A case of human myiasis ocularis by *O. ovis* in Turkey]. *Sağl. Derg.* 31(4): 274-283, illus.

MERK-BUCHBERG, (-), (1918): Zur Fortpflanzung der Oestriden. *Wild u. Hund* 24: 231.

MIDDLETON, I. E., (1907): The camel bot fly. *Pharm. J.* (4) 25 [= vol. 79]: 212.

MIDDLETON, I. E., (1912): [Note on *O. ovis*]. *Proc. ent. Soc. Lond.* 1911: lvii.

MILLER, D., (1922): Bot-flies in New Zealand. *N.Z. J. Agric.* 25(3): 144-148, figs. 1-9.

MILLER, J. H., H. E. JOHNSON & A. L. STOUT, (1961): Control of nasal bot fly in sheep. *J. Am. vet. med. Ass.* 138(8): 431-433.

MILOSHEV, B., (1955): Sluchai na spontanna ruptura na belodroben ekhinokok, pridruzhen s faringonazomiaza na *Oestrus ovis*. [Case of spontaneous rupture of pulmonary echinococcal cyst associated with pharyngonasal myiasis caused by *O. ovis*]. *Sovrem. Med., Sof.* 6(10): 106-108. [In Bulgarian].

MIRÉ, P. DE, J. A. RIOUX & D. JARRY, (1960): Oestres et oestroses sur le massif de l'Émi-Koussi. Mission épidem. au N.-Tchad par J. A. RIOUX: 112-114, 2 figs.

MISHIN, I. P., (1954): Zarazhennost' severnykh olenei o Sakhalii nosovym ovodom v zavisimisti ot razvitiya vibriss. [Infection of Sakhalin reindeer by nasal bot flies in relation to the development of their vibrissae]. *Zool. Zh.* 33(1): 162-165.

MISHIN, I. P., (1957): Idem. *Trudy nauchno-issled. Inst. sel'.-khoz. krain. Sev.* 3: 82-88, figs.

MITCHELL, W. C. & N. G. COBBETT, (1933a): Notes on the life cycle of *Oestrus ovis*. *J. Am. vet. med. Ass.* 35(5) [= vol. 82]: 780-781.

MITCHELL, W. C. & N. G. COBBETT, (1933b): Field investigations relative to control of *Oestrus ovis*. *Ibid.* 26(2) [= vol. 83]: 247-254, 2 figs.

MODEER, A., (1786): Styng-Flug-Slägtet. *K. svenska VetenskAkad. Nya Handl.* 7: 125-158, 180-185.

MÖSCHLER, A., (1912): Entomologische Beobachtungen von der Kurischen Nehrung. *Königsberg. Schr. phys. Ges.* 52: 273-277.

MÖSCHLER, A., (1928): Die Larven der Elchrachenbremse *Cephenomyia ulrichi* (Brauer). *Schr. phys.-ökon. Ges. Königsberg* 65: 253-254.

MÖSCHLER, A., (1935): Beobachtungen über die Lebensweise und die Schädlichkeit der Elchrachenbremse, *Cephenomyia ulrichi* Brauer, auf der Kurischen Nehrung. *Z. ParasitKde* 7: 572-578.

MORTON, A., (1884): *Oestrus ovis*, or gadfly of the sheep. *Pap. Proc. R. Soc. Tasm.* 1883: 258-259.

MÜLLER, O. A., (1903): Die Schafbremse (*Oestrus ovis*). *Natur Haus* 11: 167-168, 1 fig.

MURIE, J., (1870): On a larval *Oestrus* found in the hippopotamus. *Proc. zool. Soc. Lond.* 1870: 77-80, figs. A-C.

MYKYTOWYCZ, R., (1964a): Occurrence of bot-fly larvae *Tracheomyia macropi* Froggatt (Diptera: Oestridae) in wild red kangaroos, *Megaleia rufa* (Desmarest). *Proc. Linn. Soc. N.S.W.* 88[3(= n° 403)](1963): 307-312, 2 figs., 2 tables.

MYKYTOWYCZ, R., (1964b): [Note on *T. macropi*], in A survey of the endopaparasites of the red kangaroo *Megaleia rufa* (Desmarest). *Parasitology* 54(4): 677-693, 10 figs.

'v. N.', (1918): Fortpflanzung der Oestriden. *Wild u. Hund* 24: 314.

NATVIG, L. R., (1918): Beitrag zur Biologie der Dasselfliegen des Renntieres. *Tromsø Mus. Årsh.* 38-39: 112-132, pl.

NATVIG, L. R., (1933): Anoplura pinnipediorum, Culicidae und Oestridae, mit Berücksichtigung ihrer generellen geographischen Verbreitung. *Scient. Ser. norweg. N. Pole Exp.* 1918-25, Bergen 5 (16a): 1-20. 3 figs. [Also sep. pub., Bergen, 20 pp., 3 figs., 1933].

NEGRU, D. & I. MAY, (1964): Date biologice, epizootologice si experimentări, terapeutice în estroza ovină. *Revtă Zooteh. Med. vet.* 14(10): 71-77, figs.

NEGRU, D., I. MAY, D. MARICA & A. BEJAN, (1962a): Experimentări privind acţiunea larvicidă a hexaclorciclohexanului (aerosoli) şi a preparatelor organofosforice in estroza ovină. *Lucr. Inst. Cerc. vet. Bioprep. Pasteur* 1: 623-634.

NEGRU, D., I. MAY, D. MARICA & A. BEJAN, (1962b): Cercetări privind epizootologia estrozei ovine in R.P.R. *Ibid.* 1: 535-547, map.

NEGRU, D., I. MAY & E. SIRBU, (1964): [Recherches sur l'application de la préparation organophosphorée 'Trichlorphon Wolfen' dans l'oestrose ovine]. *Ibid.* 3(2): 305-312, 1 graph [In Romanian].

NEIVA, A., (1930): [Note on *O. ovis*]. *Revta Ind. anim.* 1: 628.

NEPOKLËNOV, A. A. & V. I. BUKSHTYNOV, (1972): ['Estrosol' (an aerosol preparation of dichlorfos) against sheep nose fly larva]. *Veterinariya* 11: 72-73. [In Russian].

NEPOKLËNOV, A. A., V. I. BUKSHTYNOV & K. F. ZABALOTNY, (1972): [Efficacy of Ftalofos (phosmet) against *O. ovis* and its excretion from the body of the sheep]. *Problemy vet. Sanit.* 41: 296-301. [In Russian].

NEVILL, E. M. & P. A. BASSON, (1966): A description of the first stage larva of *Oestrus aureoargentatus* Rodhain and Bequaert (1912) obtained by artificial mating (Diptera: Oestridae). *Onderstepoort J. vet. Res.* 33(2): 287-296, 3 figs.

NICOLI, R. M., J. LANFRANCHI & D. JARRY, (1962): Note d'entomologie médicale; les myiases humaines à Oestridae (oestrosis) (Diptera Brachycera). *Bull. Soc. ent. Fr.* 67 (5-6): 108-115, pls.

NIKITIN, I. N., (1962): Lechenie ovets pri polostnym ovode. [Sheep treatment in the case of bot fly occurrence]. *Veterinariya* 39(10): 47-48.

NIKOLAEV, V. A., (1959a): Opyt primeneniya dymovykh geksakhloranovykh shashek dlya zashchity olenstad ot ovodov i gnusa v tundrakh. Nenetskogo natsional'nogo okruga. [Experiments in the use of hexachlorane smoke draught for the protection of reindeer herds against botflies and tabanids in the tundras of the Nenets national district]. *Trudy vses. nauchno-issled. Inst. vet. Sanit. Ektoparazit.* 14: 13-15.

NIKOLAEV, V. A., (1959b): Zashchita stad ot ovodov i letayushchikh krovososushchikh nasekomykh kak metod profilaktiki zabolevaniya severnykh olenei nekrobatsillëzom. [Protection of the herd against botflies and flying blood-sucking Diptera as a method of prophylaxis in necrobacillosis of reindeer]. *Ibid.* 14: 11-13.

NIKOLIĆ, N., (1952): Ophthalmomyiasis externa (conjunctivitis parasitaria) professionalna bolest ovčara. *Arh. Hig. Rada* 3(3): 315-318, pl.

NITZULESCU, V., M. NITZULESCU & T. CHISSIM, (1966): Considérations à propos d'un cas d'ophthalmomyiase externe à larves d'*Oestrus ovis*. *Annls Parasit. hum. comp.* 41(4): 379-386, 3 figs.

NORDKVIST, M., (1968): Treatment experiments with systemic insecticides against the larvae of the reindeer grub fly (*Oedemagena tarandi* L.) and the reindeer nostril fly (*Cephenomyia trompe* L. [sic]). *Meddn St. vet.-med. Anst.* 38: 281-293.

NOSIK, A. F. & A. P. GONCHAROV, (1954): [Abwehrmassnahmen gegen die Rachenbremsen]. *Sots Tvarynn.* 8: 54-56. [In Ukrainian].

NOSIK, A. F. & A. P. GONCHAROV, (1958): K estrozu ovets. *Sb. Trudy khar'kov. vet. Inst.* 23: 287-293.

NOSIK, A. F. & A. P. GONCHAROV, (1960): [*O. ovis* en Ukraine]. [Problèmes Parasit., 3e Conf. scient. Parasitologistes d'Ukraine]: 329. [In Ukrainian].

NUMAN, A., (1851): Bijdrage tot de kennis van de Schaapneushorzel, *Oestrus nasalis ovinus. Tijdschr. wis. natuurk. Wet.* 4: 132-152, 2 pls.

OLIVIER, G. A., (1811-12): Insectes [i.e., Arthropoda, Pt. 5], in Société de Gens de Lettres, de Savans et d'Artistes, Encyclopédie Méthodique, 8: 1-360 (= livr. ?), 1811; 361-722 (= livr. 77), 1812. Paris.

OPPERMANN, T., (1919): Reinvasion eines Jährlingsschafes mit Oestruslarven. *Dt. tierärztl. Wschr.* 27(38): 420-421, 2 figs.

ORMEROD, E. A., (1884): Some observations on the Oestridae, commonly known as bot-flies, especially on the ox warble fly, 24 pp., figs. London.

ORMEROD, E. A., (1885): Report of observations of injurious insects and common farm pests during the year 1884, with methods of prevention and remedy. Eighth Report, vii + 122 pp., figs. London.

ORMEROD, E. A., (1888): Report of observations of injurious insects and common farm pests during the year 1887, with methods of prevention and remedy. Eleventh Report, 130 pp. London.

ORTECHO, C. L. & D. MARBLE, (1966): Efectividad del O, O dimethyl O-P-sulfamoyl phenil prosphorothioate contra la larva del *Oestrus ovis. Revta Centr. nac. Pat. anim.* 6(10): 47-54.

ORTEGA REYES, M., (1887): Las larvas de las moscas en las fosas nasales, o la enfermedad llamada myiasis. *Gac. med., Méx.* 22(1): 3-16, 1 pl.

OSBORN, H., (1896): Insects affecting domestic animals: An account of the species of importance in North America, with mention of related forms occurring on other animals. *Bull. U.S. Dep. Agric.* (n.s.) 5: 1-302, 170 figs.

OSOMANI, J. J. & R. SALSAMENDI, (1945): *Oestrus ovis* L. Su frecuencia en el Uruguay. Algunas consideraciones sobre su biologia. *An. Fac. Med. Univ. Montevideo* 30(1-4): 381-385, figs. [Also in *Revta Med. vet.; B. Aires* 27(3-4): 131-134, 1945].

OTTEN, E., (1944): Massenbefall bei·Schafen mit der Rachenbremse *Oestrus ovis* L. *Z. Fleisch- u. Milchhyg.* 54(13): 121.

PAINTER, R. H., (1930): Notes on Kansas bot-flies (Oestridae, Diptera). *J. Kans. ent. Soc.* 3: 32-35.

PAMPIGLIONE, S., (1957): Sulla miasi dei seni frontali da *Oestrus ovis* nell'uomo. *Ig. Sanità pubbl.* 13(11-12): 731-732.

PAMPIGLIONE, S., (1958a): Indagine epidemiologica sulla miasi congiuntivale umana da *Oestrus ovis* in Italia. Nota I. Inchiesta tra i medici italiani. *Nuovi Annali Ig. Microbiol.* 9: 242-263.

PAMPIGLIONE, S., (1958b): Indagine epidemiologica sulla miasi umana da *Oestrus ovis* in Italia. Nota II. Inchiesta tra i pastori. *Ibid.* 9: 494-517.

PAMPIGLIONE, S., (1958c): La miasi da *Oestrus ovis* nell'uomo in Italia, malattia dei pastori. *Attual. med., Roma* 23(5): 5-6, 8-9.

PAMPIGLIONE, S., (1958d): Sulla miasi dei pastori dell'Etna descritta da G. A. Galvani del 1838–primo valido contributo scientifico alla conoscenza della miasi umana da '*Oestrus ovis*'. *Ig. Sanità pubbl.* 14(7-8): 389-395.

PAMPIGLIONE, S., (1958e): K voprosu o rasprostranenii miaza, vyzyvaemogo *Oestrus ovis* sredi ital'yanskikh pastukhov. *Usloviya Zhisni i Zdorov'e*, 1(4): 259-261.

PANICH, P., (1953): [Bot flies]. *J. vet. Ass. Thailand* 5(3): 28-31. [Thai text].

PANZER, G. W. F., (1809): Faunae insectorum germanicae initiae oder Deutschlands Insecten [H.] 104: 24 pp., 24 pls.; [H.] 105: 24 pp., 24 pls.; [H.] 108: 24 pp., 24 pls. Nürnberg.

PARADOKSOV, L. F., (1931): Oftalmomiaz ot lichinok polostnykh ovodov i vol' fartovoi mukhi. [Ophthalmomyiasis by larvae of nose bot flies and *Wolfahrtia*]. *Russk. oftal. Zh.* 14(3): 221-224.

PARAMONOV, S. J., (1944): Ueber die Begriffsverwirrung um die Gasterophiliden und Oestriden. *Dt. tierärztl. Wschr.* 52(1-2): 16.

132

PARAMONOV, S. J., (1953): Notes on Australian Diptera, IX. The description of the adult *Tracheomyia macropi* Frog., first endemic Australian oestrid. *Ann. Mag. nat. Hist.* (12) 6: 195-199, 3 figs.

PASĂRE, G. & I. VICEA, (1968): Miaze la om cu *Oestrus ovis. Microbiologia, Parazit., Epidem.* 13(3): 265-265, illus.

PASĂRE, G. & I. VICEA, (1969): Sur un cas de myiase causé par *Oestrus ovis* dans une tannerie. *Annls Parasit. hum. comp.* 44(1): 101-105.

PASCHEFF, C., (1937): Ophthalmomyiasis externa, *Oestrus ovis* (Beitrag zu den Myiasis in Bulgarien). *Klin. Mber. Augenheilk.* 98: 721-727, figs.

PATTON, W. S., (1920): Some notes on the arthropods of medical importance and veterinary importance in Mesopotamia and on their relation to disease. *Indian J. med. Res.* 8: 1-16.

PATTON, W. S., (1937): Studies on the higher Diptera of medical and veterinary importance. The bot flies of the subfamily Oestrinae. *Ann. trop. Med. Parasit.* 31(1): 113-125, figs.

PAULL., (-), (1844): Die Bildung der *Oestrus*-Larven im Schlunde, Magen, etc., der Pferde, die Nachtheile welche sie hervorbringen, und die Mittel ihre Entwicklung zu verhüten. *Mag. Beob. Erfahr. Geb., Zücht., etc, Hausthiere* 2(3): ? [Abstr. in *Reprium Thierheilk.* 6(2): 170, 1845].

PAVLOVSKII, E. N., (1929): Lichinki polostnykh ovodov, kak parazit glaza cheloveka. Zhivotnye parazyti i nekotorye parazitarnye bolezii cheloveka v Tadzhikistane [Bot fly larvae as parasites of the human eye. Animal parasites and some parasitary diseases of men in Tadzhikistan]: 60-69. Leningrad.

PAVLOVSKII, M. M., (1909): Ovodovaya bolezh' severnykh olenei (kratkoe soobshchenie). [The bot fly disease of reindeer (short communication)]. *Vest. obshch. Vet.* 6: 288-291.

PAVLOWSKA, M., (1911): Sur les myiases produites chez l'homme par les oestrides. Thèse. Paris.

PEARSE, A. S., (1916): Oestridae; *Oestrus. Taf. Handb. Med. Sci.* (3rd ed.), 1: 844.

PEREIRA, (-) S., (1957): Luta contra o catarro nasal dos ovinos – *Oestrus ovis. Gazeta agríc. Angola* 2(3): 598-600, 605, figs.

PERKINS, H. F., (1914): The fly, *Oestrus ovis*, parasite in man. *Science, N.Y.* (n.s.) 39(1004): 476.

PETERSON, H. O., N. G. COBBETT & W. P. MELENEY, (1959): Treatment of *Oestrus ovis* with dimethoate. *Vet. Med.* 54(7): 377-383, 4 figs.

PETERSON, H. O., E. M. JONES & N. G. COBBETT, (1958): Effectiveness of Dow ET-57 (Trolene) against the nasal bot fly of sheep. *Am. J. vet. Res.* 19(70): 129-131, figs.

PETROV, D. & V. BRATANOV, (1963): Prinos k'm prouchvane na estrozata po ovtsete u nas [Contribution towards the study of the sheep nasal botfly], ? pp. [In Bulgarian].

PETROV, D., G. GECHEVA & S. MESHKOV, (1966): [The biology of *Oestrus ovis*]. *Vet.-Med. Nauki, Sof.* 3(10): 1053-1058. [In Bulgarian].

PETTIT, R. H., (1933): Sheep botfly (*Oestrus ovis* L.). *Insect Pest Surv. Bull. U.S.* 13(3): 95.

PFADT, R. E., (1964): Sheep bot fly control tests. *J. econ. Ent.* 57(6): 928-931.

PFADT, R. E. & J. CAMPBELL, (1963): Sheep bot fly control tests with DDVP. *Ibid.* 56(4): 530-531.

PINKERTON, A. W., (1971): Conjunctival myiasis [*O. ovis, Rh. purpureus*]. *J. Am. med. Ass.* 215(5): 797.

PITTENGER, B. N., (1958): Ocular myiasis caused by *Oestrus ovis. Archs Ophthal., N.Y.* 60 (6): 1107-1108.

PLESKE, F. D., (1926): Obzor palearkticheskikh vidov Oestridae i ob'yasnitel'nyi katalog sostava kolletsii etikh dvukrylykh v Zoologicheskom Musee Akademii Nauk. [Revision of the palaearctic species of Oestridae and an annotated catalogue of the extant collections of these flies in the Zoological Museum of the Academy of Sciences]. *Erzheg. zool. Muz.* 26(3-4): 215-230.

PLESKE, F. D., (1930): Résultats scientifiques des expéditions entomologiques du Musée Zoologique dans la région de l'Oussouri. II. Diptera: Les Stratiomyiidae, Erinnidae, Coenomyiidae et Oestridae. *Ibid.* 31: 181-206.

POKIDOV, I. I., (1973): [Measures for controlling *Oestrus ovis* infection in sheep]. *Veterinariya* 4: 68-71.

POLYAKOV, V. A., (1965): O meste vstrechi samtsov i samok nosoglotochnogo ovoda severnogo olenya. [On the meeting place of the male and female of Oestridae of the reindeer]. *Trudy vses. nauchno-issled. Inst. vet. Sanit. Ektoparazit.* 26: 212-213.

PORTIER, P., (1909): Physiologie de l'appareil respiratoire des larves d'oestres. *C. r. Séanc. Soc. Biol.* 67: 568-571.

PORTSCHINSKY, J., (1884): Diptera europaea et asiatica nova aut minus cognita. *Horae Soc. ent. Ross.* 18 (1883-84): 122-134.

PORTSCHINSKY, J., (1906-15): Russkii ovod (*Rhinoestrus purpureus* Br.) parazit loshadi, vypryskivayushchii lichinok v glaza lyudei. [*Rh. purpureus*, a parasite of the horse, injecting its larva in the eyes of men]. *Trudy Byuro Ent.* 6(1): 1-44, 1906; (2): 1-41, 1908; (3): 1-47, 1915.

PORTSCHINSKY, J., (1913): Ovechi' ovod' (*Oestrus ovis* L.), ego zhizn', svoistva, sopoby bor'by i otnosheni'e ego k' chelov'ku. [*O. ovis*, its biology and relation with man]. *Ibid.* 10(3): 1-63.

PROKŮPEK, K. & J. WILLOMITZER, (1960): Boj proti střeckům. [The fight against Oestridae]. *Veterinářství* 10(3): 87-89.

PUSTOVOI, I. F., (1958): O merakh bor'by s polostnym ovodom ovets. [Control measures against the sheep nasal bot fly]. *Sel'. Khoz. Tadzhik.* 4: 23-28.

RABELLO, E. X. & D. DE M. MALHEIRO, (1954): Presença de larvas de *Oestrus ovis* L. 1761 [sic] (Diptera)Muscoidea [sic]–Oestridae) em *Capra hircus* L. 1761 [sic] no estado de São Paulo, Brasil. *Revta Fac. Med. vet., S Paulo* 5: 41-47, 5 pls.

RAILLET, A., (1918): Sur la nomenclature de deux oestrides du cheval. *Bull. Soc. Zool. Fr.* 43: 102-104.

RAKUSIN, W., (1970): Ocular myiasis interna caused by the sheep nasal bot fly (*Oestrus ovis* L.). *S. Afr. med. J.* 44(40): 1155-1157, 5 figs.

RANATUNGA, P. & P. RAJAMAHENDRAN, (1972): Observations on the occurrence of *Oestrus ovis* L. (Dipt. Oestridae) and pleuropneumonia in goats on a dry-zone farm in Ceylon. *Bull. ent. Res.* 61(4): 657-659, 4 figs.

RANATUNGA, P. & D. J. WEILGAMA, (1972): Efficacy of trichlorphon and ruelene pour-on on *Oestrus ovis* in sheep and goats in Ceylon. *Ceylon vet. J.* 20(1-2): 11-13.

RATZEBURG, J. T. C., (1844): Die Ichneumon der Forstinsecten in forstlicher und entomologischer Beziehung, als Anhang zur Abbildung und Beschreibung der Forstinsecten, 1 (Cent. 1-4): 8-224, Berlin.

RAY, J., (1710): Historia Insectorum: Opus posthumum cui subjungitus appendix De Scarabaeis Britannicis auctore M. Lister, 15 + 400 pp. Churchill, London.

RÉAUMUR, R. A. F., (1734): Mémoires pour servir à l'histoire des insectes, 1: 4 + 654 pp., 50 pls. Paris.

RÉAUMUR, R. A. F., (1738): Idem, 4: 36 + 636 pp., 44 pls. Paris.

RÉAUMUR, R. A. F., (1740): Idem, 5: 44 + 728 pp., 38 pls. Paris.

REDI, F., (1668): Esperienze intorno alla generazione degl'insetti fatte da Fr. Redi e da lui scritte in una lettera all'illustrissimo Sgr. Carlo Datti, 228 pp., 29 pls. Firenze.

REHBINDER, C., (1970): Observations of 1st instar larvae of nostril fly *Cephenomyia trompe* L. [sic] in the eyes of reindeer and their relation to keratitis in this animal. *Acta vet. scand.* 11(2): 338-339.

REYNON, M. P., (1953): Oculomyiase à *Oestrus ovis. Bull. mens. Soc. méd. Mil. Fr.* 47(10):

227-228.

RILEY, C. V., (1869): The sheep bot fly, or horse maggot. 1st An. Rep. on noxious, benef. and other Insects of Missouri: 161-165, figs.

ROBERTS, I. H. & H. P. COLBENSON, (1963): Larvae of *Oestrus ovis* in the ears of a sheep. *Am. J. vet. Res.* 24: 628-630, figs.

ROBINEAU-DESVOIDY, J. B., (1863): Histoire naturelle des diptères des environs de Paris, 1: XVI + 1143 pp. Paris.

RODHAIN, J., (1915): *Herpetomonas* parasites de larves d'oestrides cavicoles. *Bull. Soc. Path. exot.* 8: 369-372, pl. 2, figs. 1-8.

RODHAIN, J., (1926a): *Herpetomonas rhinoestri*, sp.n., parasite des larves de *Rhinoestrus nivarleti* Rodh. et Beq. *C. r. Séanc. Soc. Biol.* 95: 1124-1127.

RODHAIN, J., (1926b): Le mode de transmission de *Herpetomonas rhinoestri* Rd. *Ibid.* 95: 1128-1130.

RODHAIN, J., (1926c): Larves d'oestrides cavicoles chez *Okapia johnstoni* Scl. *Rev. Zool. afr.* 14: 137-139.

RODHAIN, J., (1927): Contribution à la faune des oestrides du Congo Belge. *Annls Parasit. hum. comp.* 5(3): 193-213, figs. 1-7, pl. 1, figs. 1-6.

RODHAIN, J. & J. BEQUAERT, (1912): Sur deux oestrides nouveaux parasites du potamo-chère et de l'antilope chevaline au Congo Belge. *Rev. Zool. afr.* 1(3): 365-383, 7 figs.

RODHAIN, J. & J. BEQUAERT, (1913): *Gedoelstia cristata* nov. gen. nov. sp. d'oestride para-site de *Bubalis Lichtensteini* au Katanga. *Ibid.* 2(2): 171-186, 4 figs.

RODHAIN, J. & J. BEQUAERT, (1915): Sur quelques oestrides du Congo. *Bull. Soc. Path. exot.* 8(7): 452-458; 8(9): 687-695; 8(10): 765-778.

RODHAIN, J. & J. BEQUAERT, (1916): Matériaux pour une étude monographique des diptères parasites de l'Afrique. Deuxième partie. Révision des Oestrinae du Conti-nent Africain. *Bull. scient. Fr. Belg.* 50(1-2): 53-165, figs. 1-30, pl. 2, figs. 1-4.

RODHAIN, J. & J. BEQUAERT, (1919): Matériaux pour une étude monographique des diptères parasites de l'Afrique. Troisième partie. Diptères parasites de l'éléphant et du rhinocéros. *Ibid.* 52: 379-465, pl. 3.

RODHAIN, J. & J. BEQUAERT, (1920): Oestrides d'antilopes et de zèbres recueillis en Afrique Orientale avec un conspectus du genre *Gasterophilus*. *Rev. Zool. afr.* 8(2): 169-228, figs. 1-3.

RODRÍGUEZ NOVOA, L. J., (1935): 'La estridia de los ovinos' o falsa locura de los lanares. *Boln Dir. Agric. Ganad., Lima* 5: 226, 228-232, 234.

RODRÍGUEZ NOVOA, L. J., (1942): Gusanos de la cabeza de los carneros, 'estridia de los ovinos'. *Vida agríc., Lima* 19: 59-61.

RÖSE, A., (1865): Ueber die Oestriden (Dasselfliegen) und die Beobachtung derselben in zoologischen Gärten. *Zool. Gart., Frankf.* 6(7): 225-266.

RÖSE, A., (1866): Weitere Beobachtungen über die Oestriden (Pferd- oder Biesflie-gen). *Ibid.* 7(11): 416-420.

ROGERS, C. E., F. W. KNAPP, D. COOK & M. V. CROWE, (1968): A temperature study of the overwintering site of the sheep bot fly *Oestrus ovis. J. Parasit.* 54(1): 164-165.

RONDANI, C., (1854): Sulla pretesa identità specifica degl'estridi del cavallo. *Nuovi Annali Sci. nat. Bologna* (3) 9: 67-71, fig.

RONDANI, C., (1857): Dipterologiae italicae prodromus. Vol. 2: Species italicae ordines dipterorum in genera characteribus definita, ordinatim collectae, methodo analitica distinctae, et novis vel minus cognitis descriptis, Pars prima: Oestridae, Syrpfhidae [sic], Conopidae, 264 pp., 1 fig. Parmae.

RONDANI, C., (1865): Diptera italica non vel minus cognita descripta vel annotata ob-servationibus nonnullis additis. Fasc. I. Oestridae-Syrphidae-Conopidae; Fasc. II. Muscidae. *Atti Soc. ital. Sci. nat.* 8: 127-146, 193-231.

ROTH, H. H., (1964): Misplacement of throat bots (*Pharyngobolus africanus* Brauer) in the African elephant. *Jl S. Afr. vet. med. Ass.* 35(3): 392.

Roubaud, E., (1914a): Oestrides gastricoles et cavicoles de l'Afrique Occidentale Française. *Bull. Soc. Pathol. exot.* 7(3): 212-215.

Roubaud, E., (1914b): Études sur la faune parasitaire de l'Afrique Occidentale Française, 1: 251 pp., 4 pls. Paris.

Rubtsov, I. A., (1940): Geograficheskoe rasprostrannenie i evolyutsiya ovodov v svyazi s istoriei ikh khozyaev. [Geographical distribution and evolution of bot flies in relation to the history of their hosts]. *Priroda, Mosk.* 6: 48-60.

Rubtsov, I. A., (1948): [A new palaearctic species of the cavicole bot-fly (Diptera, Oestridae)]. *Ent. Obozr.* 30: 138-142, 6 figs.

Ruppert, F., (1913): Untersuchungen über die Entwicklung der Oestruslarven und die Bekämpfung der Oestruslarvenkrankheit. *Z. InfektKrankh., parasit. Krankh. Hyg. Haustiere* 13(7): 469-474, figs. 1-3. [Also in *Dt. Fleishbesch.-Ztg* 11(3): 30-31, 1 fig., 1914].

Ruser, (-), (1895): Ueber das Vorkommen von *Oestrus*-Larven in Rückenmarkskanal des Rindes. *Z. Fleisch- u. Milchhyg.* 6(7): 127-129, figs. 1-3.

Ruser, (-), (1896): Zur Entwicklungsgeschichte der *Oestrus*-Larven. (Nachweiss der Larven in Schlunde). *Ibid.* 6(7): ?

Ruser, (-), (1899): [*Oestrus*-Larven; Abstr. of 1896]. *Mschr. prakt. Thierheilk.* 11(2): 87. [Also in *Thierarzt* 38(12): 268, 1899].

Sabbagh, M., (1934): Un premier cas de myiase oculaire due à *Oestrus ovis* en Syrie. *Annls Oculist.* 171(12): 1016-1018.

Sabrosky, C. W., (1950): Date of publication of James' 'The flies that cause myiasis in man'. *Proc. ent. Soc. Wash.* 52: 315.

Saccà, G., L. Gabrielli & E. Stella, (1965): Note su *Oestrus ovis* L. (Diptera, Oestridae) e descrizione di alcuni casi di miiasi nell'uomo. *Annali Inst. sup. Sanità* 1(1-6): 73-94, 2 pls.

Sachs, R., (1970): Ueber den Befall ostafrikanischer Wildtiere mit parasitischen Fliegenlarven (Diptera, Oestridae). *Acta trop.* 27(4): 281-290, figs.

Saitta, S., (1903): Myiasis da *Cephalomyia ovis*. *Gazz. Osp., Milano* 24 (128): 1357-1359.

Salamatin, V. N., (1930): O lichinkakh loshadinogo ovoda. [On the larva of the horse bot flies]. *Prakt. Vet., Mosk.* 7(7): 624-626.

Salcés Fermín, (-) & S. P. Calvo, (1945): Ensayo experimental con D.D.T. sobre *Oestrus ovis*. *Revta Med. vet., B. Aires* 27(3-4): 137-142, figs.

Salm, H. A. von, (1818): Ueber die Erscheinung des *Oestrus ovis* (Schaf-Bremse) in den abgesägten Schaf-Widder Hörnern. *Verh. Arb. ökon. patriot. Soc. Fürstenth. Schweidnitz u. Jauer* 1818: 320.

Salvador Yépez, M. & F. Gallardo Z., (1971): Presencia de *Oestrus ovis* L. (Diptera, Oestridae) en ovinos y caprinos del Estado Lara. *Revta vet. venez.* 31(183): 234-237.

Salvador Yépez, M., F. Gallardo Z. & R. Torres Artigas, (1968): Contribución al estudio de la parasitología venezolana. *Acta cient. venez.* 19(1): 46 [*O. ovis*].

Samuel, W. M., D. O. Trainer & W. C. Glazener, (1971): Pharyngeal botfly larvae in white-tailed deer. *J. Wildl. Dis.* 7(3): 142-146.

Sapogov, A. G., (1947): Preismushestva bor'by s polostnym ovodom ovets metodom ruchnogo shora. [The advantage of controlling the sheep bot fly by the method of hand collecting]. *Veterinariya* 24(1): 41. [Abstr. in *Rev. appl. Ent.* 36(9): 154-155, 1958].

Saxesen, (-), (1850): [Title unknown]. *Stettin. ent Ztg* 11: ?

Sayin, F., (1967): [*Rhinoestrus purpureus* infesting donkeys in Turkey]. *Vet. Fak. Derg. Ankara Üniv.* 14(4): 535-540, 2 figs.

Sayin, F., I. Meric, H. Köseoglu, N. Sincer & S. Ayabakan, (1972): [The use of Neguvon for control of grubs in Angora goats]. *Ibid.* 19(3): 338-348.

SCHEBEN, L., (1910): [Title unknown]. *Zentbl. Bakt. ParasitKde* 56: 50-54.

SCHEIBER, S. H., (1860-62): Vergleichende Anatomie und Physiologie der Oestriden-Larven. *Sber. Akad. Wiss. Wien* (math.-naturw. Cl.) 41 (15): 409-496, pls. 1-2, figs. 1-33, 1860; [2. Teil]45(1): 7-68, pls. 1-3, figs. 34-62, 1862.

SCHERF, H., (1960): Ein neuer Fall von Ophthalmomyiasis, hervorgerufen durch *Oestrus ovis* L. *Beitr. Ent.* 10(3-4): 402-404, figs.

SCHINER, J. R., (1862): Vorläufiger Commentar zum dipterologischen Theile der 'Fauna Austriaca'. III. *Wien. ent. Mschr.* 5(5): 137-144.

SCHINER, J. R., (1861): H. 6/7, pp. 441-656, H. 8 (part), pp. 657-674, + i-lxxx, pl. 2. In: (L. REDTENBACHER & I. R. SCHINER), Fauna Austriaca. Die Fliegen (Diptera) [by SCHINER], 1: lxxx + 674 pp., 2 pls. Wien.

SCHIPP, (-) VON, (1927): Zur Biologie der Rachenbremsenlarven. *St Hubertus* 45(32): 514. [Abstr. in *Jb. Jagdk.* 9 (1927): 84, 1928].

SCHIRRE, L., (1968): Conjunctival myiasis due to *Oestrus ovis*. *S. Afr. med. J.* 42(30): 765-766, 1 fig.

SCHMIDT, (-), (1915): Ueber Rachenbremsen bei Damwild und das Zusammenleben der Rachenbremsenlarven. *Wild u. Hund* 21(31): 485-486, figs.

SCHMIDT, (-): (1918): Rachenbremsen beim Damwild. *Ibid.* 24: 314.

SCHNEIDER, F. L., (1953a): Sheep nose grub treatment pays off. *New Mexico Stockman* 17(12): 79, figs.

SCHNEIDER, F. L., (1953b): New Mexico Indians treat sheep for nose grubs. *Ibid.* 19(7): 20.

SCHNEIDER, F. L., (1954): Pueblo Indians continue nose grub fight. *Ibid.* 19(7): 20.

SCHOLTZ, H., (1850): Ueber den Aufenthalt der Dipteren während ihrer ersten Stände. *Ent. Z.* 4: 25-34.

SCHOUTENDEN, H., (1912): Note sur l'hôte de l'*Oestrus macdonaldi* Gedoelst. *Rev. Zool. afr.* 2(1): 142.

SCHRANCK, F. VON P., (1781): Enumeratio insectorum Austriae indigenorum, 9-548 pp., 4 pls. Klett.

SCHRANCK, F. VON P., (1803): Fauna Boica. Durchgedachte Geschichte der in Baiern einheimischen und zahmen Thiere, 3(1): 1-272. Landshut.

SCHREBER, J. C. D. VON, (1759): Novae species insectorum, 16 pp., 1 col. pl. Halae.

SCHROETER, J. S., (1776): Von der Bissel-Mücke der Thüringer, einer besonderer Gattung der Fliegen [*Oestrus*]. *Schroeter's Abhandl.* 1: 316-322.

SCHUURMANS STEKHOVEN, J. H., (1935): Oestridae (Dipt.). *Wiss. Ergebn. niederl. Exp. Karakorum* (1922-30), 1: 414.

SCHWAB, K. L., (1840): Die Oestraciden–Bremsen der Pferde, Rinder u. Schafe. Eine naturgeschichtliche thierärztliche Abhandlung, 83 pp. München. [2nd ed., 1858, 10 + 93 pp.].

SCOTT, H. G., (1964): Human myiasis in North America (1952-1962 inclusive). *Fla. Ent.* 47(4): 255-261.

SCOTT, J. W., (1942): *Oestris* [sic] *ovis* in the Rocky Mountain bighorn, *Ovis canadensis*. *J. Mammal.* 23(3): 345-346.

SDOBNIKOV, V. M., (1933): Nekotorye dannye po biologii olenya i olenevodstvu v severo-vostochnoi chasti Malozemel'skoi tundry. [Some data on reindeer biology and reindeer breeding in the northeastern part of the Malozemelskaya tundra]. *Sb. Olen. Past. sev. Kraya*, [Inst. Olenev., Leningrad] 2: 185-229, graph.

SDOBNIKOV, V. M., (1935): [Information on the biology of the reindeer and on reindeer husbandry in the northeastern section of the Malaya Zemlya tundra]. *Exp. Stn. Rec.* 72(5): 651.

SÉGUY, E., (1923): Sur le *Cephalopsis titillator* Clark. *Bull. Mus. natn. Hist. nat., Paris* 29: 387-390.

Séguy, E., (1924): Les insectes parasites de l'homme et des animaux domestiques, 422 pp., 463 figs. Paris.

Séguy, E., (1925): Sur les caractères communs aux oestrides et aux calliphorides. *C.r. hebd. Séanc. Acad. Sci., Paris* 181(20): 735-736.

Séguy, E., (1928): Études sur les mouches parasites. Tome I. Conopides, Oestrides et Calliphorides de l'Europe occidentale. *Encycl. Ent.* (A) 9: 1-251. Paris.

Séguy, E., (1948): Introduction à l'étude des myiases. *Revta bras. Biol.* 8(1): 93-111.

Sells, W., (1842): Observations on the Oestridae. *Trans. ent. Soc. Lond.* 3(1-2) (1841-43): 72-78.

Selyaev, V. A., (1933): Nosovoi ovod verblyuda (*Cephalopsis titillator* Cl.). [The nasal bot fly of the camel]. *Konevodstvo* 6: 39-42.

Semenov, P. V., (1962): Primenenie khlorofoza dlya bor'by polostnym ovodom ovets. [Use of chlorofos for control of sheep bot fly]. *Ovtsevodstvo* 8(5): 37-38.

Semenov, P. V., (1964): Novye sredstva i metody profilaktiki i lecheniya ovets, zarazhenikh polostnym ovodom. [New ways and methods of prophylaxis and treatment of sheep infected with bot fly]. *Problemy vet. Sanit.* 1964: 262-268.

Senger, C. M., (1963): Some parasites of Montana deer. *Montana Wildl. Bull.* (Autumn) 1963: 5-13, figs., pl.

Sergent, E., (1952a): Répartition géographique de la 'thimni', myiase oculo-nasale de l'homme, due à l'oestre du mouton. *Bull. Acad. natn. Méd.* 136(27-30): 519-520.

Sergent, E., (1952b): La thimni, myiase oculo-nasale de l'homme causée par l'oestre du mouton. *Archs Inst. Pasteur Algér.* 30(4): 319-361, figs.

Sergent, E. & E. Sergent, (1907a): La 'thim'ni', myiase humaine d'Algérie causée par *Oestrus ovis* L. *Annls Inst. Pasteur, Paris* 21(5): 392-399, 1 map. [Abstr. in: *Bull. Inst. Pasteur, Paris* 5(16): 708-709, 1907; *J. trop. vet. Sci.* 2(4): 436-437, 1907; and *Revue gén. Méd. vét.* 10: 563, 1907].

Sergent, E. & E. Sergent, (1907b): Myiasis in Algeria. *Lancet* 2(6)[= vol. 173]: 399-400.

Sergent, E. & E. Sergent, (1908a): Human myiasis due to *Oestrus ovis. Jl R. microsc. Soc.* 1: 44.

Sergent, E. & E. Sergent, (1908b): 'Thim'ni', eine beim Menschen in Algier vorkommende, durch *Oestrus ovis* bedingte Erkrankung. *Z. Fleisch u. Milchhyg.* 18(11): 369.

Sergent, E. & E. Sergent, (1910): La 'thim'ni', myiase humaine d'Algérie causée par '*Oestrus ovis* L.'. In: E. Sergent, Recherches expérimentales sur la pathologie algérienne (microbiologie – parasitologie) 1902-1909, Alger: 226-233, 1 map.

Sergent, E. & E. Sergent, (1913): La 'tamné', myiase humaine des montagnes sahariennes touaregs, identique à la thimni des Kabyles, due à *Oestrus ovis. Bull. Soc. Path. exot.* 6(7): 487-488.

Serres, J. R., (1913): Estridia de los ovinos. *Gac. rural* 6(69): 759-761. [Abstr. in *Rev. appl. Ent.* (B) 1(8): 148, 1913].

Shannon, R. C., (1922): The bot-flies of domestic animals. *Cornell Vet.* 1922(july): 240-262.

Shannon, R. C. & E. del Ponte, (1926): Sinopsis parcial de los muscoideos argentinos. *Rev. Inst. bact., B. Aires* 4(5): 3-44, 4 pls.

Sharma, J. N. & B. B. Verma, (1962): An unusual case of 'false gid' in a goat. *Indian vet. J.* 39(10): 568-570.

Shaw, G., (1806): General zoology, or systematic natural history, 6 (Insecta): 197 pp.

Shcherban, N. F., (1973): [Prevention of *Oestrus ovis* infestation (trichlorfos aerosol)]. *Veterinariya* 2: 71-72.

Shcherban, N. F. & L. A. Chukhleb, (1968): [Sheep bot-fly control]. *Ibid.* 9: 44-45.

Shumakovich, E. E., (1934): Pervye itogi rabot po veterinarnoi parazitologii v Mongol'skoi respublike. [Erste Ergebnisse der Arbeiten über veterinäre Parasitologie in der Mongolischen Republik]. *Sov. Vet.* 6: 58-60.

SIBONI, D., (1952): Un cas de myiase oculaire à *Rhinoestrus*. *Ann. Oculist.* 185: 967-969.

SICART, M., J. RUFFIÉ & M. MEIRA, (1958): L'évolution larvaire de *Oestrus ovis* (Linné, 1761 [sic]). *Annls Parasit. hum. comp.* 33(3): 295-302, 2 figs.

SIMISON, E. V., (1956): Conjunctival myiasis with *Oestrus ovis* larvae. *Archs Ophthal., N.Y.* 55(3): 418-419, figs.

SIMONDS, J. B., (1856): The sheep gad fly. *Wool Grow. Stock Reg.* 10(1): 10.

SIMULA, R., (1954): Ulteriore contributo alla conoscenza della difusione della duodenite nodo-fistolosa da *Oestrus meridionalis* in provincia di Sassari. *Atti Soc. ital. Sci. vet.* 7 (1953): 621-622.

SJÖSTEDT, Y., (1908): 2. Oestridae, in Wiss. Ergebn. schwed. zool. Exp. Kilimandjaro-Meru, Deutsch Ostafrikas 1905-06, 10(Diptera)(2): 11-24, pls. 1-2.

SLAVIN, A. P., (1934): Lichinki ovoda kak simulyanti sapa. [Bot fly larvae simulating glanders]. *Sov. Vet.* 2: 30-34.

SMITH, F. F., (1972): A device to aid in examining the brain and meninges of deer for helminth parasites. *J. Wildl. Dis.* 8(2): 109-111, figs.

SMITH, S. M. & G. F. BENNETT, (1966): First record of a wild-caught female of *Cephenemyia phobifera* (Clark) (Diptera: Oestridae). *Can. J. Zool.* 44(2): 346-347.

SNIDERMAN, H. R., (1939): Larval conjunctivitis. Report of a case due to *Oestrus ovis* *Am. J. Ophthal.* 22(11): 1253-1255, figs.

SNIJDERS, A. J. & I. G. HORAK, (1972): Trials with rafoxanide. 4. Efficacy against the larvae of the oestrid fly *Gedoelstia hässleri* in the bluebock (*Damaliscus dorcas philipsi* Harper, 1939). *J. S. Afr. vet. med. Ass.* 43(3): 295-297.

SOFRONOV, N. V. & M. V. LYSOV, (1952): Bor'ba s lichinkami polostnogo ovoda u Karakul'skikh ovets. [Control of the larvae of the cavicole bot fly on Karakul sheep]. *Karakulev. Zverov.* 5(3): 56-58.

SOLIMAN, K. N. & H. EL-HINAIDY, (1966): Myiasis due to species of the family Oestridae in man and animals with special reference to the specific camel myiasis, *Cephalopina titillator* Clark, in Egypt, U.A.R. *Proc. 1st int. Congr. Parasit.*, Roma: 946-947.

SPEISER, P. (1907): Die Dipterenfamilie der Oestriden. *Schr. phys.-ökon. Ges. Königsb.* 47 (1906): 295-303.

SPENCER, G. J., (1958): On the reproductive potential of the sheep nostril fly *Oestrus ovis* L. (Diptera: Oestridae). *Proc. ent. Soc. Br. Columb.* 55: 25-26.

SPIRYUKHOV, I. A. & S. M. MACHUL'SKII, (1958a): K voprosu estroza ovets (Predvaritel'-noe soobshchenie). [On the question of treatment of ovine oestrosis]. *Trudy buryat-mongol'. zoovet. Inst.* 13: 281-289.

SPIRYUKHOV, I. A. & S. M. MACHUL'SKII, (1958b): Lechenie ovets prozhenykh estrozom, geksakhloretanom. [Treatment of sheep infested with oestrosis, hexachlorane and hexachloretan]. *Veterinariya* 35(5): 76-78, figs.

SPIRYUKHOV, I. A. & S. M. MACHUL'SKII, (1959): K biologii polostnogo ovoda ovets Buryatskoi ASSR. [On the biology of the cavicole bot of sheep in the Buryatsk ASSR]. *Trudy buryat-mongol'. zoovet. Inst.* 14: 253-258.

STAMPA, S., (1959): The control of internal parasites of sheep with neguvon and asuntol. A preliminary report. *Jl S. Afr. vet. med. Ass.* 30(1): 19-26.

STAMPA, S. & J. W. POLS, (1961): Field tests to establish the efficacy of the preparation Neguvon A against the larvae of sheep nostril fly (*Oestrus ovis* L.) under South African conditions. *Vet.-med. Nachr.* 4: 239-243, 1 fig.

STARK, H. H., (1923): Ophthalmic myiasis externa due to larvae of *Oestrus ovis*. *J. Am. med. Ass.* 81(20): 1684-1685, 2 figs.

STARTSEV, I. S., (1937): Ovodnaya invaziya severnogo olenya i organizatsiya mer bor'by s nei. [Invasion of bot flies in the reindeer and organization of control methods for it]. *Sov. Vet.* 6: 15-19.

STEEL, J. H., (1887): On bots (larval Oestridae) on the horse and camel. *J. Bombay nat. Hist. Soc.* 2: 27-30.

STEFANI, T. DE, (1915): Note di myiasis negli animali e nell'uomo. *Rinnov. econ.-agr.* 9: 89-92, 110-113.

STEIN, J. P. E. F., (1849): Dipterologisches. (Inflatae, Oestracides, Coriaceae). *Stettin. ent. Ztg* 10: 117-120.

STEVENSON, L., (1934): Bot flies and their control. *Bull. Ont. Dep. Agric.* 378: 1-13, illus.

STEWART, J. R., (1932): Treatment for *Oestrus ovis. J. Am. vet. med. Ass.* 33(1): 108.

STEWART, N. H., (1920): *Cephenemyia*, a probable cause of death among deer. *Proc. Pa Acad. Sci.* 3: 97, 1 fig.

STEWART, N. H., (1930): Preliminary report on the occurrence of the nose fly (*Cepheno-myia*) in the deer of Pennsylvania. *Bull. Bd Game Com. Pa.* 12: 61-65, 2 figs.

STOUDER, K. W., (1930): Nose fly survey. Floyd, Wright, country results satisfactory; extension service offers cooperation; early organization necessary. *Iowa Vet.* 1(3): 32, 34.

STOUDER, K. W., (1931): Summary of nose fly control work. *Ibid.* 2(11): 22-25.

STOUDER, K. W., (1932): Bot eggs viable for long period. *Ibid.* 3(1): 34.

STOUDER, K. W., (1940): Sheep gad fly – grub in the head. *Iowa Wool News* 1(7): 3.

STRAND, E., (1928): Miscellanea nomenclatorica zoologica et palaeontologica. *Arch. Naturgesch.* (A) 92(8): 30-75.

SULTANOV, Z. N., (1941): Sluchai nakhozhdeniya lichinok ovech'ego ovoda v konyunk-tive veka i glaznogo ybloka y cheloveka. [Case of presence of sheep bot fly larva in the human eye]. *Azerb. med. Zh.* 2: 57-59.

SULTANOV, M. A. & A. BEKUZIN, (1949): Materialy k biologii ovech'ego ovoda u slo-viyakh gornykh pastbishch. [Materials for the biology of the sheep bot fly under mountainous pasture conditions]. *Dokl. Akad. Nauk uzbek. SSR* 7: 26-28.

SURCOUF, J. & L. GEDOELST, (1909): Description d'un oestride nouveau, parasite de l'hippopotame. *Bull. Soc. Path. exot.* 2(10): 615-619, pl. 7.

SURCOUF, J. & L. GUYON, (1925): Recherches pcéliminaires sur la morphologie et la biologie des larves d'oestrides. (1re. note). *Bull. Soc. ent. Fr.* 1925: 68-72, illus.

SUVORTSEV, I. S., (1889): Svedeniya o lashadninom ovode (*Rhinoestrus purpureus* Br.) na padayushcheon na cheloveka. [Notice on the horse bot fly attacking man]. *Trudy russk. ent. Obshch.* 23: xvii-xix.

SVYABOVETS, G., (1956): Bor'ba s estrozom ovets v plemovtsesovkhoze Aidar. [Treat-ment of ovine oestrosis on the state sheep breeding farm 'Aidar']. Trudy 2. nauch. Konf. Parazit. SSSR: 185.

SYMES, C. B. & J. I. ROBERTS, (1932): A list of the Muscidae and Oestridae causing myiasis in man and animals in Kenya, recorded at the Medical Research Laboratory, Nairobi. *E. Afr. med. J.* 9(1): 18-20.

SZIDAT, L. & E. HEINEMANN, (1937): Neue Beobachtungen über die Morphologie und Biologie des ersten Larvenstadiums der Elchrachenbremse (*Cephenomyia ulrichi* Brauer). *Dt. Jagd* (A) 6(52): 995-996, figs.; 7(1): 20.

SZILÁDY, Z., (1935): [Die ungarischen Dasselfliegen]. *Állat. Közl.* 32: 136-139. [In Hungarian, with German abstr.].

TARNANI, I. K., (1903): Noviya nablyudeniya nad ovodami (Diptera, Oestridae). [New observations on Oestridae]. *Ent. Obozr.* 3(2): 101-102.

TARNANI, I. K., (1905): Biologie der *Oestrus*-Larve der Rinder. *Tierarzt* 44(4): 77.

TEMPLE, (-), H. HARANT & H. VIALLEFONT, (1939): Deux cas de myiase oculaire par *Oestrus ovis. Archs Soc. Sci. méd. biol. Montpellier* 15(2): 26-28.

TEPPER, J. G. O., (1900): Botflies, gadflies and breeze-flies. *J. Agric. Ind. S. Aust.* 3(7) (1899-1900): 564-566.

TERENT'EV, F. A., (1930): [The botfly of reindeer]. *Zap. Okh. nauchno-issled. Prom.* 4: 55.

TERENT'EV, F. A., (1933a): K voprosu bor'by i invaziei ovoda severnykh olenei. [The question of the struggle against bot fly infestation of reindeer]. *Trudy vses. Inst. eks-per. Vet.* 9: 121-126, 3 figs.

TERENT'EV, F. A., (1933b): Bor'ba s invaziei ovoda severnykh olenei. [Control of infestation of the bot fly of reindeer]. *Sev. Olenev.* 20-21: 9-11.

TERENT'EV, F. A. & N. D. TERENT'EV, (1933): Kozhnyi i nosovoi ovod severnogo olenya i mery bor'by s nim [Warble and bot flies of the reindeer and methods of control], 31 pp. Ed. KOIZ.

TERNOVOI, V. I., (1973): [On flight-distance of flies of *O. ovis*]. *Parazitologiya* 7(2): 123-127.

THEOBALD, F. V., (1903): The sheep nasal fly (*Oestrus ovis* Linnaeus). *Jl S -east. agric. Coll. Wye* 12: 68-74.

THEOBALD, F. V., (1910): The sheep nasal fly (*Oestrus ovis* Linn.), in his Report on economic zoology for the year ending September 31st, 1910. *Ibid.* 19: 87, pl. 1, figs. 1-4.

THIERRY, E., (1900): L'oestre du mouton. *J. Agric. prat., Paris* 2: 224-226, fig. 28.

THIERRY, E., (1905): Idem. *Ibid.* 10: 344-345, fig. 56.

THOMANN, H., (1947a): Ueber ein Massenschwärm von *Cephenomyia stimulator* Clark (Dipt.). *Mitt. schweiz. ent. Ges.* 20(4): 304-305.

THOMANN, H., (1947b): *Cephenomyia stimulator* Clark. *Ibid.* 20(5): 540.

THOMPSON, W., (1889): Larvae of *Cephenomyia* in a man's head. *Insect Life* 2(4): 116.

THON, D., (1832): *Oestrus. Allg. Encycl. Wiss. Künste* (3) 2: 248-249.

TONKOZHENKO, A. P., V. I. BUKSHTYNOV & V. M. REPIN, (1972): [Use of aerosols against *O. ovis*]. *Veterinariya* 6: 82-83.

TOULANT, P. & F. MÉDINGER, (1935a): Myiase oculaire à *Oestrus ovis*. *Presse méd.* 43(46): 921.

TOULANT, P. & F. MÉDINGER, (1935b): Nouveaux cas de myiase oculaire à *Oestrus ovis*. *Algérie méd.* 95: 828-830.

TOWNSEND, C. H. T., (1912): A readjustment of muscoid names. *Proc. ent. Soc. Wash.* 14: 45-53.

TOWNSEND, C. H. T., (1915): New genera and species of Australian Muscoidea. *Can. Ent.* 47: 151-160.

TOWNSEND, C. H. T., (1916a): Designation of muscoid genotypes, with new genera and species. *Insecutor Inscit. menstr.* 4: 4-12.

TOWNSEND, C. H. T., (1916b): New genera and species of Australian Muscoidea. *Can. Ent.* 48(5): 151-160.

TOWNSEND, C. H. T., (1917): The head and throat bots of American game animals. *J. N.Y. ent. Soc.* 25: 98-105.

TOWNSEND, C. H. T., (1918): New muscoid genera, species and synonymy. *Insecutor Inscit. menstr.* 6: 151-182.

TOWNSEND, C. H. T., (1921): Some new muscoid genera ancient and recent. *Ibid.* 9: 132-134.

TOWNSEND, C. H. T., (1927): On the *Cephenemyia* mechanism and the day-night-day circuit of the earth by flight. *J. N.Y. ent. Soc.* 35: 245-252.

TOWNSEND, C. H. T., (1932): New genera and species of Old World oestromuscoid flies. *Ibid.* 40: 439-479.

TOWNSEND, C. H. T., (1934): New neotropical oestromuscoid flies. *Revta Ent., Rio de J.* 4: 201-212, 390-406.

TOWNSEND, C. H. T., (1935): *Manual of Myiology*, 2: 289 pp., 9 pls. Itaquaquecetuba.

TOWNSEND, C. H. T., (1936): Idem, 3: 249 pp. S. Paulo.

TOWNSEND, C. H. T., (1939): Speed of *Cephenemyia*. *J. N.Y. ent. Soc.* 47: 43-46.

TOWNSEND, C. H. T., (1941): An undescribed American *Cephenemyia*. *Ibid.* 49: 161-163.

TRABUT, G., (1931): Un cas de myiase oculaire à *Oestrus ovis* à Alger. *Archs Inst. Pasteur Algér.* 9(4): 638.

TRABUT, G., (1933): Un nouveau cas de myiase oculaire à *Oestrus ovis* à Alger. *Ibid.* 11(4): 598.

TROSCHELL, (-), (1881): [Title unknown]. *Sber. Ver. Rheinl.* 38: 119-121.

TROSCHELL, (-), (1882): [Title unknown]. *Ver. preuss. Rheinl. Westphal.* 38: ?

TSAI, H. C., (1961): Chetyrekhkhloristyi ugledov dlya lecheniya estroza ovets. [Carbon tetrachloride in the treatment of ovine oestrosis]. *Trudy vses. nauchno-issled. Inst. vet. Sanit. Ektoparazit.* 19: 37-38.

TSELISHCHEVA, L. M. & A. M. KRIVKO, (1958): Zheludochnye i polostnye ovody sel'-skokhozyaistvennykh zhivotnykh Kazakhstana. [Gastric and cavicole bot flies of agricultural animals in Kazakhstan]. IV mezhdunarodn. reg. Konf. Stran. Azii po parazit. Bolezn. Zhivotn., Alma-Ata: 1-2.

U., H. (= Ullrich?), (1926): Einige Anregungen zur Bekämpfung der Rachenbremsen und Dasselfliegen. *Dt. Jäger-Ztg* 87(15): 259. [Abstr. in *Jb. Jagdk.* 8 (1926): 28, 1927].

ULLRICH, H., (1935): Ueber das Vorkommen der Rachenbremse beim Damwild (*Cephenomyia multispinosa* spec. nov.). *Zool. Anz.* 111(1-2): 43-45, figs. 1-6.

ULLRICH, H., (1936): Untersuchungen über die Biologie der Rachenbremse (Genus *Cephenomyia* Latreille), über die pathogenen Einflüsse der Rachenbremsenlarven auf ihre Wirtsthiere und über Bekämpfungsmöglichkeit der Rachenbremsenplage, 69 pp., pls. Neudamm.

ULLRICH, H., (1937a): Die Parasitenfauna unseres Elches. *Dt. Jagd* (A) 9 (40): 771-772, figs.

ULLRICH, H., (1937b): Wo sitzt das erste Larvenstadium der Elchrachenbremse, *Cephenomyia ulrichi* Brauer. *Ibid.* (A) 7(17): 312.

ULLRICH, H., (1939): Zur Biologie der Rachenbremsen unseres einheimischen Wildes, Genus *Cephenomyia* Latreille und Genus *Pharyngomyia* Schiner. Insekten als Parasiten unseres einheimischen Wildes. *Verh. VII. int. Kongr. Ent.,* Weimar 3: 2149-2162, 2163-2171, pls.

UNSWORTH, K., (1943): Observations on the occurrence of larvae of *Oestrus ovis* in the nasal cavities and frontal sinuses of goats in Nigeria. *Ann. trop. Med. Parasit.* 42(2): 249-250.

UNSWORTH, K., (1949): Observations on the seasonal incidence of Oestrus infection among goats in Nigeria. *Ibid.* 43(3-4): 337-340.

VAINSHTEIN [WEINSTEIN], A. A., (1913): K voprosu o znachenii lichinok ovech'ego ovoda (*Oestrum* [sic] *ovis*) v patologii glaza. [On the question of the importance of the sheep bot fly larva in eye pathology]. *Vest. Oftal.* 30(1): 91-92.

VALENTINE, C. J., (1900): Gadfly and botfly. *J. Agric. Ind. S. Aust.* 3(6): 516-517.

VALLISNIERI, A., (1713): Esperienze ed osservazioni intorno all'origine, sviluppi e costumi di varii insetti, 232 pp., 12 pls. Padova.

VALLISNIERI, A., (1733): Opere fisico-chimiche continenti un gran numero di trattati, osservazioni, ragionamenti e dissertazioni sopra la fisica, la medicina e la storia naturale, 3 vols. Venezia.

VARTIC, N., E. ŞUTEU & Z. TRICĂ, (1965): Cercetari experimentale asupra eficacitatii preparatului 'Bubulin' in estroza la ovine. *Lucr. stiint. Inst. agron. Cluj* (Ser. Med. vet. zooteh.) 21: 169-173.

VARTIC, N., E. ŞUTEU & Z. TRICĂ, (1967): Experimentelle Untersuchungen über die therapeutische Wirkung des 'Bubulins' bei der Oestrose der Schafe. *Mschr. vet.-Med.* 22(3): 93-94.

VERMEIL, C., (1954): Deux cas tunisiens de myiase oculo-nasale due à *Oestrus ovis. Annls Parasit. hum. comp.* 29(3): 324-325.

VICEA, I., G. PASĂRE & G. PETRESCU, (1968): [Considérations portant sur deux cas de myiase oculaire externe à *O. ovis*]. *Oftalmologia, Buc.* 3: 263-266, 3 figs. [In Romanian].

VILENBERG, G., A. PERDRIX & P. DUBOIS, (1971): Traitement de l'oestrose ovine par injection d'un insecticide organophosphoré, le diméthoate. *Revue Élev. Méd. vét. Pays trop.* 24(1): 43-46.

VILLENEUVE, J., (1925): L'oestridomorphisme. *Encycl. ent.* (B) 2(Diptera)(2): 1-4.

VILLERS, C. J. DE, (1789): Caroli Linnaei entomologia, fauna suecicae descriptionious aucta; DD. Scopoli, Geoffroy, De Geer, Fabricius, Schranck, etc. speciebus vel in systemate non enumeratis, vet nuperrime detectis, vel speciebus Galliae australis locupletata, generum specierumque rariorum iconibus ornata, curante et augente Carolo de Villers, 3: 656 pp., 4 pls. Lugduni.

VIMMER, A., (1925): Larvy a kukly dvolkridleho hmyzu stědo evropskeho se svlastnim zretelem na skudce rostlin kulturnich. Praha.

VOBLIKOVA, N. V., (1960): Khlorofos dlya unichtozheniya lichinok 1. stadii nosovogo ovoda severnogo olenya. [Chlorofos for the destruction of 1st stage larvae of reindeer bot fly]. *Veterinariya* 37(4): 79-80.

VOBLIKOVA, N. V., (1961): [Phosphamid in the control of reindeer bot fly]. *Ibid.* 39(7): 56-58.

VOBLIKOVA, N. V., (1961): [Eigentümlichkeiten der Entwicklung der Larve des 1. Stadiums der Nasenbremse des Renntieres und die Methode ihrer Vertilgung]. *Trudy nauchno-issled. Inst. sel'.-khoz. krain. Sev.* 11: 109-112.

VOSTRYAKOV, P. N. & D. V. SAVEL'EV, (1957): O vozmozhnosti snizheniya kontsentratsii DDT i geksakhlorana pri protivoovodovykh obrabotkakh severnykh olenei. [DDT and hexachlorane against reindeer bot flies]. *Byull. nauchno-issled. Inst. sel'.-khoz. krain. Sev.* 2: 24-25.

VRBA, Č., (1960): Dosavadní prostředky proti střečkovitosti a výrobni perspektivy našeho chemického průmuslu. [The present means against Oestridae and productive perspective of our chemical industry]. *Veterinářství* 10(3): 89-91.

WACHTL, F. A., (1886): Ueber ein aussergewöhnliches Vorkommen der Larven von *Cephenomyia stimulator* Clk. *Wien. ent. Ztg* 5: 305-6.

WALKER, F., (1849): List of the specimens of dipterous insects in the collection of the British Museum, 2: 231-484; 3: 485-687; 4: 688-1172. London.

WALKER, F., (1853): Vol. 2, 297 pp., pls. 11-20. In: (F. WALKER, H. T. STAINTON & S. J. WILKINSON), Insecta Britannica. London.

WALTON, C. L., (1930): The occurrence of males of the horse botflies. *NWest Nat.* 5: 224-226.

WARE, G. W., L. W. DEWHIRST & R. ECHEVERRIA, (1969): Sheep bot fly in Arizona. *Progve Agric. Ariz.* 21(5): 20-21.

WEINBERG, M., (1909): Substances hémotoxiques sécrétées par les larves d'oestres. *C. r. Séanc. Soc. Biol.* 65: 75-77.

WESTWOOD, J. O., (1840): Order XIII. Diptera Aristotle (Antliata Fabricius. Halteriptera Clairv.), pp. 125-128 (= signature I, part), 129-144 (= signature K), 145-158 (= signature L), in his An introduction to the modern classification of insects. Synopsis of the genera of British Insects, 158 pp. London.

WESTWOOD, J. O., (1843): Notice on some oestrideous insects (*Oe. tarandi* and *trompe*, spec. differ.). *Proc. Linn. Soc. Lond.* 1: 179.

WETZEL, H., (1969): Okulovaskuläre Myiasis bei Haustieren in Süd- und Südwest-Afrika. *Berl. Münch. tierärztl. Wschr.* 82(17): 330-332.

WETZEL, H., (1970a): Die Fliegen der Unterfamilie Oestrinae (Diptera: Oestridae) in der Aethiopischen Region und deren veterinär-medizinische Bedeutung. *Z. angew. Ent.* 66(3): 322-336.

WETZEL, H., (1970b): Beschreibung des zweiten und dritten Larvenstadiums von *Rhinoestrus giraffae* Zumpt (1965) (Diptera, Oestridae). *Acta trop.* 27(4): 291-300, figs.

WETZEL, H., (1971a): Beschreibung des ersten Larvenstadiums von *Rhinoestrus vanzyli* Zumpt und Bauristhene (1962) (Diptera: Oestridae). *Zool. Anz.* 186(1-2): 103-105, figs.

WETZEL, H., (1971b): Anatomie des Verdauungskanals und der inneren Geschlechtsorgane der Hoeckerfliegen *Gedoelstia* Rodhain und Bequaert. *Z. angew. Ent.* 68(3): 289-296, figs.

WETZEL, H., (1973): Zur Kenntnis der Rachenfliege des Flusspferdes, *Rhinoestrus hippopotami* Gruenberg (Diptera: Oestridae). *Ibid.* 60(1): 5-14, figs.

WETZEL, H. & E. BAURISTHENE, (1970): The identity of the third larval stage of *Oestrus ovis* Linnaeus (1758) (Diptera: Oestridae). *Zool. Anz.* 184(1-2): 87-94, figs.

WHITWORTH, C., (1880): The *Oestrus ovis. Vet. J. Ann. comp. Path.* 10: 249.

WIEDEMANN, C. R. W., (1830): Aussereuropäische zweiflügelige Insekten, 2: xii + 684 pp., pls. 7-10b. Hamm.

WILKINSON, F. C., (1958): The sheep nasal bot. *J. Dep. Agric. West. Aust.* 7(5): 513-515, illus.

WINCHESTER, J. F., (1882): *Oestrus ovis – Strongylus filaria. Am. vet. Rev.* 6(8): 347-348.

WINGE, E., (1872): Oestruslarver. *Norsk Mag. Laegevidensk.* (3) 2: 89-90.

WRIGHT, W. H., (1933): Treatment for bots. *Success. Fmg* 31(12): 30.

YANOVICH, G. I., (1959): Primenenie aerosolei dlya bor'by s estrozom ovets. [The use of aerosol for control of ovine oestrosis]. *Trudy vses. nauchno-issled. Inst. vet. Sanit. Ektoparazit.* 14: 5-6.

YERBURY, J. W., (1901): An appeal for assistance in collecting gad-flies, bot-flies and warble-flies. *J. Bombay nat. Hist. Soc.* 13: 683-686.

YERBURY, J. W., (1904): [Title unknown]. *Proc. ent. Soc. Lond. 1904*: lxvii-lxviii.

YOSHIKAWA, M., (1927): [A note on the larvae of *Cephenomiya* [sic] *trompe* Modeer of the reindeer in Karafuto]. *Dobuts. Zashi, Tokyo* 39: 82-87, pl. [Abstr. in *Jap. J. Zool.* 2: 88]. [In Japanese].

ZAIDENOV, A. M., (1958): Sluchai oftalmiaza v g. Chite, vyzyvannyi lichinkoi ovech'ego ovoda. [Case of ophthalmyiasis in Chita, caused by the larva of the sheep bot fly]. *Nauch. Zap. Chit. nauchno-issled. Inst. Epidem., Mikrobiol. Gigieny* 4: 192.

ZELLER, P. C., (1841): Nachricht über die Seefelder bei Reinerz in entomologischer Beziehung. *Stettin. ent. Ztg* 2: 171-176, 178-182.

ZETTERSTEDT, J. W., (1822): Resa fenom Sveriges och Norriges Lappmarken, 2 vols. Lund.

ZETTERSTEDT, J. W., (1838): Dipterologis Scandinaviae. Sect 3: Diptera, pp. 477-868, in his Insecta Lapponica, vi + 1140 pp. Lipsiae.

ZETTERSTEDT, J. W., (1844): Diptera Scandinaviae disposita et descripta, 3: 895-1280. Lund.

ZÜRN, F. A., (1872): Die Schmarotzer auf und in den Körpern unserer Hausthiere, 1 (Die thierischen Parasiten): 236 pp. Weimar.

ZUMPT, F., (1957): Some remarks on the classification of the Oestridae s. lat. (Diptera). *J. ent. Soc. Sth Afr.* 20(1): 154-161, pl.

ZUMPT, F., (1958a): On *Rhinoestrus steyni* n. sp. and *Gasterophilus zebrae* Rodhain & Bequaert (Diptera), parasites of Burchell's zebra (*Equus burchelli* Gray). *Ibid.* 21(1): 56-65, 7 figs.

ZUMPT, F., (1958b): Remarks on the systematic position of myiasis producing flies (Diptera) of the African elephant, *Loxodonta africana* (Blumenbach). *Proc. R. ent. Soc. Lond.* (B) 27(1-2): 8-14, 3 figs.

ZUMPT, F., (1959): The *Rhinoestrus* species of equids in Africa south of the Sahara (Diptera: Oestridae). *Novos Taxa ent.* 14: 1-10, pls.

ZUMPT, F., (1961a): *Oestrus bassoni*, nov. spec., a new nasal fly from South Africa (Diptera: Oestridae). *Ibid.* 24: 3-11, 3 figs.

ZUMPT, F., (1961b): Nasenfliegen beim afrikanischen Wild. *Natur Volk* 91(7): 251-256, illus., pl.

ZUMPT, F., (1962a): Die oestroiden Fliegen des Wildes in der äthiopischen Region. *Verh. 11. int. Kongr. Ent., Wien* 2(7-14): 454-457.

Zumpt, F., (1962b): The oestroid flies of wild and domestic animals in the Ethiopian region, with a discussion of their medical and veterinary importance (Diptera: Oestrinae & Gasterophilinae). *Z. angew. Ent.* 49(3): 393-419.

Zumpt, F., (1963): Fruitful scientific work on parasites infesting wild animals in the Transvaal. *Fauna Flora, Pretoria* 14: 15-33.

Zumpt, F., (1965a): Myiasis in man and animals in the Old World, xv + 267 pp., illus. London.

Zumpt, F., (1965b): XXV. The oestroid flies of Africa south of the Sahara. Their classification, biology and practical importance, pp. 341-352, 2 figs. In: D. H. S. Davis, (ed.), Ecological studies in Southern Africa. Monografiae biol. 14. Junk, Den Haag.

Zumpt, F., (1968): The enigma of *Oestrus macdonaldi* Gedoelst solved. *Jl S. Afr. vet. med. Ass.* 39(1): 99-100.

Zumpt, F., (1970): Arthropod parasites of the African elephant (*Loxodonta africana* Blumenbach) and the Indian elephant (*Elephas maximus* Linnaeus) – Taxonomic and biological aspects. *J. Parasit.* 56(4): 380.

Zumpt, F. & E. Bauristhene, (1962): Two new *Rhinoestrus* from the Springbuck (*Antidorcas marsupialis* (Zimmermann)) in South Africa (Diptera: Oestridae). *Novos Taxa ent.* 28: 3-33, 13 figs.

Zumpt, F. & H. Wetzel, (1970): Fly parasites (Diptera: Oestridae) of the African elephant *Loxodonta africana* (Blumenbach) and their problems. *Koedoe* 13: 104-121, 9 figs.

PART C. CHARACTERISTICS OF THE HOSTS OF OESTRIDAE

Once a tentative phylogeny for the Oestridae has been established, it now becomes necessary to understand why only a few mammals serve as their hosts. This is the problem discussed in this section.

For the history of mammals and their classification I have used as bases the works of SIMPSON (1945), ROMER (1967), the treaty of palae-ontology of PIVETEAU (1958, 1961), and especially WALKER et al. (1964); more specific papers are cited in the text.

It is a well known fact that the classification and zoogeography of mammals is very tentative and hampered by three factors in a special way: 1. The lack of a good phylogenetic analysis and therefore of a sounder basis for the establishment of a plausible zoogeography. As remarked by CRACRAFT (1973: 489): 'The study of historical zoo-geography includes the construction of hypotheses about (1) phylo-genetic (cladistic) relationships, (2) the probable distribution of an-cestral species, and (3) the probable pathways of dispersal and those paleogeographic and paleoclimatological factors that influenced this dispersal. I consider the inclusion of knowledge of phylogenetic rela-tionships within the realm of historical biogeography because "a bio-geographic analysis implies, logically follows from, and at best can be no more reliable than, a prior phyletic analysis" (NELSON, 1969: 246). Once relationships have been postulated, then hypotheses about centers of origin and dispersal routes take on maximum precision.' 2. The lack of attempts at explaining the distribution of certain groups by continental drift (following, of course, a phylogenetic analysis of these groups). Matthewsianism was responsible for a widely accepted idea – that almost all groups of mammals spread from the northern continents to the southern ones; alternative hypotheses were not even attempted, or entirely discarded. COOKE (1972: 89-90) commented: 'The presence of fossil mammals of Ethiopian type in Eurasia and their subsequent extinction there have also led to the widespread belief that Africa possesses its present character primarily because it has served as a "refuge" for the survival of archaic forms of life. W. D. MATTHEWS' classic study, "Climate and Evolution" (1915), did much to foster a vision of Asia as the major center from which the various groups dispersed, and the views of such workers as PILGRIM (1941) in favor of Africa as an important evolutionary center and a source for diffusion *into* the Palearctic region have received scant approval or acceptance. Further paleontological discoveries [and a phylogenetic analysis] are needed before the extent of the interchanges between the Palearctic and Ethiopian regions can be fully evaluated, but in the last few years new evidence has accumulated which emphasizes the essentially indigenous nature of much of the living and extinct African mammalian fauna.' 3. The exaggerated stress laid upon 'centres of origin'. For a discussion of this problem, see CROIZAT, NELSON & ROSEN (1974).

Therefore, some of the conclusions reached here will necessarily be somewhat speculative, as a reformulation of the classification and zoogeography of mammals remains a task for the future.

Tables showing dimensions and weights of mammals were compiled from the data furnished by WALKER *et al.* (1964). Time-ranges for the several groups of mammals were extracted from ROMER (1967), but some genera were not included in the subfamilies and tribes, when not cited in SIMPSON (1945) and PIVETEAU (1958, 1961).

Abbreviations used in the illustrations (and sometimes in the text) are the following:

Af	Africa	Mad	Madagascar
As	Asia	Med	Mediterranean
Aus	Australia	NA	North America
CA	Central America	NAf	North Africa
CAs	Central Asia	NAs	North Asia
EAf	East Africa	NZ	New Zealand
EAs	East Asia	SA	South America
EEu	East Europe	SAf	South Africa
EInd	East Indies	SWAs	Southwest Asia
Eu	Europe	WAf	West Africa
Gr	Greenland	WInd	West Indies
Ind	India	WNA	Western North America

The name of a genus followed by an asterisk (*) indicates that members of that genus are parasitized; the generic name in parentheses and with an asterisk indicates that parasitism is accidental, and that the larva does not complete development in that host.

The somewhat lengthy discussion of the classification and palaeontology of some groups of mammals, apparently of little significance for this section, seems desirable for two reasons: 1. To provide general information about mammals for dipterists not conversant with this class; 2. To provide bases for the more detailed discussion about the history of mammalian groups and their relations with Oestridae, which will be presented in Part D.

I. LIST OF PRESENT HOSTS

1. Marsupialia

Macropus (Megaleia) rufus (Desmarest) and *Macropus (Macropus) robustus* Gould (Perameloidea, Macropodidae)–parasitized by *Tracheomyia macropi* (Froggatt) (Tracheomyiinae).

2. Proboscidea

Loxodonta africana (Blumenbach) (Elephantoidea, Elephantidae)–parasitized by *Pharyngobolus africanus* Brauer (Pharyngobolinae).

3. Artiodactyla

3.1. Suina

Potamochoerus porcus (Linnaeus) (Suoidea, Suidae)–parasitized by *Suinoestrus nivarleti* (Rodhain & Bequaert) (Oestrinae, Rhinoestrini).

Phacochoerus aethiopicus (Pallas) (Suoidea, Suidae)–parasitized by *Rhinoestrus phacochoeri* Rodhain & Bequaert (Oestrinae, Rhinoestrini).

Hippopotamus amphibius Linnaeus (Hippopotamoidea, Hippopotamidae)–parasitized by *Rhinoestrus hippopotami* Grünberg (Oestrinae, Rhinoestrini).

3.2. Tylopoda

Camelus bactrianus Linnaeus and *Camelus dromedarius* Linnaeus (Cameloidea, Camelidae)–parasitized by *Cephalopina titillator* (Clark) (Oestrinae, Cephalopinini).

3.3. Pecora

3.3.1. Family Cervidae
Dama dama (Linnaeus) (Cervinae)–parasitized by *Pharyngomyia picta* (Meigen) (Cephenemyiinae, Pharyngomyiini), *Acrocomyia auribarbis*

(Meigen) and *Cephenemyia multispinosa* Ullrich (Cephenemyiinae, Cephenemyiini).

Cervus (Cervus) elaphus Linnaeus (Cervinae)–parasitized by *Pharyngomyia picta* (Meigen) (Cephenemyiinae, Pharyngomyiini) and *Acrocomyia auribarbis* (Meigen) (Cephenemyiinae, Cephenemyiini).

Cervus (Cervus) canadensis Erxleben (Cervinae)–parasitized by *Cephenemyia jellisoni* Townsend (Cephenemyiinae, Cephenemyiini).

Cervus (Sika) nippon hortulorum Swinhoe (Cervinae)–parasitized by *Pharyngomyia picta* (Meigen) (Cephenemyiinae, Pharyngomyiini).

Odocoileus (Eucervus) hemionus (Rafinesque) (Odocoileinae, Odocoileini)–parasitized by *Cephenemyia jellisoni* Townsend, *C. apicata* Bennett & Sabrosky, and *C. pratti* Hunter (Cephenemyiinae, Cephenemyiini); secondarily parasitized by *C. trompe* (Modeer).

Odocoileus (Odocoileus) virginianus (Zimmermann) (Odocoileinae, Odocoileini)–parasitized by *Cephenemyia phobifera* (Clark) and *C. pratti* Hunter (Cephenemyiinae, Cephenemyiini).

Alces alces (Linnaeus) (Odocoileinae, Alcini)–parasitized by *Pharyngomyia picta* (Meigen) (Cephenemyiinae, Pharyngomyiini) and *Cephenemyia ulrichi* Brauer, *C. jellisoni* Townsend, and *C. phobifera* (Clark) (Cephenemyiinae, Cephenemyiini).

Rangifer tarandus (Linnaeus) (Odocoileinae, Rangiferini)–parasitized by *Cephenemyia trompe* (Modeer) (Cephenemyiinae, Cephenemyiini).

Rangifer arcticus (Richardson) (Odocoileinae, Rangiferini)–parasitized by *Cephenemyia trompe* (Modeer) (Cephenemyiinae, Cephenemyiini).

Rangifer caribou (Gmelin) (Odocoileinae, Rangiferini)–parasitized by *Cephenemyia trompe* (Modeer) (Cephenemyiinae, Cephenemyiini).

Capreolus capreolus (Linnaeus) (Odocoileinae, Capreolini)–parasitized by *Pharyngomyia picta* (Meigen) (Cephenemyiinae, Pharyngomyiini) and *Procephenemyia stimulatrix* (Clark) (Cephenemyiinae, Cephenemyiini).

3.3.2. Family Giraffidae
Giraffa camelopardalis Linnaeus (Giraffinae)–parasitized by *Rhinoestrus giraffae* Zumpt (Oestrinae, Rhinoestrini).

Okapia johnstoni (Sclater) (Okapiinae)–parasitized by an undescribed species of Oestrinae (cf. RODHAIN, 1926; ZUMPT, 1965).

3.3.3. Family Bovidae
Bos taurus Linnaeus and *Bos indicus* Linnaeus (cattle) (Bovinae, Bovini)–accidentally infested by *Gedoelstia cristata* Rodhain & Bequaert and *G. haessleri* Gedoelst (Oestrinae, Gedoelstiini) and *Oestrus ovis* Linnaeus (Oestrinae, Oestrini).

Hippotragus equinus (Desmarest) (Hippotraginae, Hippotragini)–parasitized by *Oestrus aureoargentatus* Rodhain & Bequaert (Oestrinae, Oestrini).

Hippotragus niger (Harris) (Hippotraginae, Hippotragini)–parasitized by *Oestrus aureoargentatus* Rodhain & Bequaert and *Loewioestrus variolosus* (Loew) (Oestrinae, Oestrini).

Oryx gazella (Linnaeus) (Hippotraginae, Hippotragini)–accidentally infested by *Oestrus sp., Rhinoestrus sp.* and *Loewioestrus variolosus* (Loew) (Oestrinae, Oestrini and Rhinoestrini).

Damaliscus dorcas (Pallas) (Hippotraginae, Alcelaphini)–parasitized by *Oestroides macdonaldi* (Gedoelst) and *Loewioestrus variolosus* (Loew) (Oestrinae, Oestrini), and *Gedoelstia haessleri* Gedoelst (Oestrinae, Gedoelstiini).

Damaliscus korrigum (Ogilby) (Hippotraginae, Alcelaphini)–parasitized by *Kirkioestrus blanchardi* (Gedoelst) and *K. minutus* (Rodhain & Bequaert) (Oestrinae, Kirkioestrini); *Oestrus aureoargentatus* Rodhain & Bequaert, *Oestroides macdonaldi* (Gedoelst) and *Loewioestrus variolosus* (Loew) (Oestrinae, Oestrini); and *Gedoelstia cristata* Rodhain & Bequaert and *G. haessleri* Gedoelst (Oestrinae, Gedoelstiini).

Damaliscus korrigum jimella Matsch. (Hippotraginae, Alcelaphini)–parasitized by *Loewioestrus variolosus* (Loew) (Oestrinae, Oestrini).

Alcelaphus buselaphus (Pallas) (Hippotraginae, Alcelaphini)–parasitized by *Kirkioestrus blanchardi* (Gedoelst) and *K. minutus* (Rodhain & Bequaert) (Oestrinae, Kirkioestrini); *Gedoelstia cristata* Rodhain & Bequaert and *G. haessleri* Gedoelst (Oestrinae, Gedoelstiini); and *Oestrus aureoargentatus* Rodhain & Bequaert and *Loewioestrus variolosus* (Loew) (Oestrinae, Oestrini).

Alcelaphus caama (Cuvier) (Hippotraginae, Alcelaphini)–parasitized by *Oestrus aureoargentatus* Rodhain & Bequaert and *Loewioestrus variolosus* (Loew) (Oestrinae, Oestrini); and *Gedoelstia cristata* Rodhain & Bequaert and *G. haessleri* Gedoelst (Oestrinae, Gedoelstiini).

Alcelaphus lelwel jacksoni Thom. (Hippotraginae, Alcelaphini)–parasitized by *Kirkioestrus blanchardi* (Gedoelst) and *K. minutus* (Rodhain & Bequaert) (Oestrinae, Kirkioestrini).

Alcelaphus lichtensteinii (Peters) (Hippotraginae, Alcelaphini)–parasitized by *Kirkioestrus blanchardi* (Gedoelst) and *K. minutus* (Rodhain & Bequaert) (Oestrinae, Kirkioestrini); *Gedoelstia cristata* Rodhain & Bequaert and *G. haessleri* Gedoelst (Oestrinae, Gedoelstiini); and *Oestrus aureoargentatus* Rodhain & Bequaert, *Oestroides macdonaldi* (Gedoelst), and *Loewioestrus variolosus* (Loew) (Oestrinae, Oestrini).

Connochaetes gnou (Zimmermann) (Hippotraginae, Alcelaphini)–parasitized by *Gedoelstia cristata* Rodhain & Bequaert and *G. haessleri* Gedoelst (Oestrinae, Gedoelstiini).

Connochaetes taurinus (Burchell) (Hippotraginae, Alcelaphini) – parasitized by *Kirkioestrus blanchardi* (Gedoelst) and *K. minutus* (Rodhain & Bequaert) (Oestrinae, Kirkioestrini); *Gedoelstia cristata* Rodhain & Bequaert and *G. haessleri* Gedoelst (Oestrinae, Gedoelstiini); and by '*Oestrus' dubitatus* Basson & Zumpt, *Oestrus aureoargentatus* Rodhain & Bequaert, and *Loewioestrus variolosus* (Loew) (Oestrinae, Oestrini).

Antidorcas marsupialis (Zimmermann) (Antilopinae, Antilopini) – parasitized by *Rhinoestrus antidorcitis* Zumpt & Bauristhene, and *R. vanzyli* Zumpt & Bauristhene (Oestrinae, Rhinoestrini); also by *Oestrus aureoargentatus* Rodhain & Bequaert and *Loewioestrus variolosus* (Loew) (Oestrinae, Oestrini).

Procapra gutturosa (Pallas) (Antilopinae, Antilopini) – parasitized by '*Pharyngomyia' dzerenae* Grunin (Cephenemyiinae, Pharyngomyiini).

Capra aegagrus Erxleben (Caprinae, Caprini) – parasitized by *Oestrus caucasicus* Grunin (Oestrinae, Oestrini).

Capra cylindricornis (Blyth) (Caprinae, Caprini) – parasitized by *Oestrus caucasicus* Grunin (Oestrinae, Oestrini).

Capra hircus Linnaeus (Caprinae, Caprini) – parasitized by *Oestrus ovis* Linnaeus (Oestrinae, Oestrini).

Capra ibex Linnaeus (Caprinae, Caprini) – parasitized by *Oestrus caucasicus* Grunin and *Oestrus ovis* Linnaeus (Oestrinae, Oestrini).

Capra sibirica Meyer (Caprinae, Caprini) – parasitized by *Oestrus caucasicus* Grunin and *Oestrus ovis* Linnaeus (Oestrinae, Oestrini).

Capra sp. (Caprinae, Caprini) – parasitized by Gedoelstia *cristata* Rodhain & Bequaert and *G. haessleri* Gedoelst (Oestrinae, Gedoelstiini); and *Oestrus caucasicus* Grunin (Oestrinae, Oestrini).

Ovis ammon (Linnaeus) (Caprinae, Caprini) – parasitized by *Oestrus ovis* Linnaeus and *Gruninia tshernyshevi* (Grunin) (Oestrinae: Oestrini and Rhinoestrini).

Ovis aries Linnaeus (Caprinae, Caprini) – parasitized by *Oestrus ovis* Linnaeus (Oestrinae, Oestrini).

Ovis canadensis Shaw (Caprinae, Caprini) – parasitized by *Oestrus ovis* Linnaeus (introduced) (Oestrinae, Oestrini).

Ovis orientalis Gmelin (Caprinae, Caprini) – parasitized by *Gruninia tshernyshevi* (Grunin) (Oestrinae, Rhinoestrini).

Ovis vignei bochariensis Nasonov (Caprinae, Caprini) – parasitized by *Gruninia tshernyshevi* (Grunin) (Oestrinae, Rhinoestrini).

4. Perissodactyla

Equus asinus Linnaeus (Equidae) – parasitized by *Rhinoestrus latifrons* Gan, *R. purpureus* (Brauer) and *R. usbekistanicus* Gan (Oestrinae, Rhinoestrini); accidentally infested by *Gedoelstia* (Oestrinae, Gedoelstiini).

Equus burchelli (Gray) (Equidae) – parasitized by *Rhinoestrus usbekistanicus* Gan and *R. steyni* Zumpt (Oestrinae, Rhinoestrini).

Equus caballus Linnaeus (Equidae) – same parasites of *asinus*.

Equus zebra Linnaeus (Equidae) – parasitized by *Rhinoestrus steyni* Zumpt (Oestrinae, Rhinoestrini).

5. Common names of the hosts

This list includes those common names most frequently met with in the literature on oestrids.

1. African bush pig – *Potamochoerus porcus* (Linnaeus)
2. African elephant – *Loxodonta africana* (Blumenbach)
3. African river hog – *Potamochoerus porcus* (Linnaeus)
4. African water hog – *Potamochoerus porcus* (Linnaeus)
5. American elk – *Cervus canadensis* Erxleben
6. Ass – *Equus asinus* Linnaeus
7. Barren ground caribou – *Ranginer arcticus* (Richardson)
8. Black-tailed deer – *Odocoileus* (*Eucervus*) *hemionus* (Rafinesque)
9. Black wildebeest – *Connochaetes gnou* (Zimmermann)
10. Blagorodnyi olen – *Cervus elaphus elaphus* Linnaeus
11. Blesbock – *Damaliscus dorcas* (Linnaeus)
12. Blue wildebeest – *Connochaetes taurinus* (Burchell)
13. Bucharskii olen – *Cervus elaphus bactrianus* Lydekker
14. Buibol – *Bubalus bubalis* (Linnaeus)
15. Buffalo (Indian or Water) – *Bubalus bubalis* (Linnaeus)
16. Buffalo (African) – *Syncerus caffer* (Sparrman)
17. Caribou – *Rangifer caribou caribou* (Gmelin)
18. Chevreuil – *Capreolus capreolus* (Linnaeus)
19. Common hartebeest – *Alcelaphus buselaphus* (Pallas)
20. Dvugorbyi verblyud – *Camelus bactrianus* Linnaeus
21. Dzerena – *Procapra gutturosa* (Gmelin)
22. Donkey – *Equus asinus* Linnaeus
23. Elan – *Alces alces* (Linnaeus)
24. Eland – *Alces alces* (Linnaeus)
25. Elch – *Alces alces* (Linnaeus)
26. Elk (European) – *Alces alces* (Linnaeus)
27. Elk (American) – *Cervus canadensis* Erxleben
28. Fallow deer – *Dama dama* (Linnaeus)
29. Gemsbock – *Oryx gazella* (Linnaeus)
30. Giraffe – *Giraffa camelopardalis* Linnaeus
31. Gornyi karibu – *Rangifer tarandus* (Linnaeus)
32. Great grey kangaroo – *Macropus* (*Macropus*) *major* Shaw
33. Great red kangaroo – *Macropus* (*Megaleia*) *rufus* (Desmarest)
34. Hartebeest – genus *Alcelaphus*
35. Hippopotamus – *Hippopotamus amphibius* Linnaeus
36. Horse – *Equus caballus* Linnaeus
37. Ibex – *Capra ibex* Linnaeus

38. Izyubr' – *Cervus elaphus xanthopygus* Milne-Edwards
39. Kangaroo – genus *Macropus*
40. Karibu – *Rangifer tarandus* Linnaeus
41. Kenguru – genus *Macropus*
42. Khangule olen' – *Cervus elaphus bactrianus* Lydekker
43. Korrigum – *Damaliscus korrigum* (Ogilby)
44. Kosul' – *Capreolus capreolus* (Linnaeus)
45. Koza – *Capra hircus* Linnaeus
46. Lan' – *Dama dama* (Linnaeus)
47. Lapland reindeer – *Rangifer tarandus* (Linnaeus)
48. Lichtenstein's hartebeest – *Alcelaphus lichtensteinii* (Peters)
49. Los' – *Alces alces* (Linnaeus)
50. Los' (Ussuriski) – *Alces alces americanus bedfordi* Lydekker
51. Loshad – *Equus caballus* Linnaeus
52. Maral' – *Cervus canadensis* Erxleben
53. Mongolian gazelle – see Dzerena
54. Moose – *Alces alces* (Linnaeus)
55. Mule deer – *Odocoileus (Eucervus) hemionus* (Rafinesque)
56. Odnogorbyi verblyud – *Camelus dromedarius* Linnaeus
57. Olen' – genus *Cervus*
58. Orignal – *Alces alces* (Linnaeus)
59. Osël – *Equus asinus* Linnaeus
60. Ovtsa – *Ovis aries* Linnaeus
61. Piatnistyi olen' – *Cervus (Sika) nippon hortulorum* Swinhoe
62. Red deer – *Cervus elaphus* Linnaeus
63. Red hartebeest – *Alcelaphus buselaphus* (Pallas)
64. Red river hog – *Potamochoerus porcus* (Linnaeus)
65. Ree – *Capreolus capreolus* (Linnaeus)
66. Rehe – *Capreolus capreolus* (Linnaeus)
67. Reindeer – *Rangifer tarandus* (Linnaeus)
68. Ren – *Rangifer tarandus* (Linnaeus)
69. Renntiere – *Rangifer tarandus* (Linnaeus)
70. Roan antelope – *Hippotragus niger* (Harris)
71. Roe deer – *Capreolus capreolus* (Linnaeus)
72. Sable antelope – *Hippotragus niger* (Harris)
73. Schafe – *Ovis aries* Linnaeus
74. Severnyi olen' – *Rangifer tarandus* (Linnaeus)
75. Sheep – *Ovis aries* Linnaeus
76. Sika deer – *Cervus nippon* Temminck
77. Springbuck – *Antidorcas marsupialis* (Zimmerman)
78. Tsesseby – *Damaliscus lunatus* (Burchell)
79. Tugainoi olen' – *Cervus elaphus bactrianus* Lydekker
80. Tur – *Capra cylindricornis* (Blyth)
81. Tur (Dagestanski) – *Capra cylindricornis* (Blyth)
82. Verblyud – genus *Camelus*
83. Wallaroo – *Macropus (Osphranter) robustus* (Gould)
84. Wapiti – *Cervus canadensis* Erxleben
85. Wart hog – *Phacochoerus aethiopicus* (Pallas)
86. White-tailed deer – *Odocoileus (Odocoileus) virginianus* (Zimmerman)
87. Wildebeest – genus *Connochaetes*
88. Woodland caribou – *Rangifer caribou caribou* (Gmelin)
89. Zebra – *Equus burchelli* (Gray) and *E. zebra* Linnaeus

II. GENERAL CONSIDERATIONS ABOUT THE HOSTS

The above list of hosts and examination of Table I show how limited is the parasitism of mammals by oestrids – among the 998 recent genera of mammals (according to WALKER *et al.*, 1964), only 25 were listed as hosts, or, only 2.5% of the recent genera of mammals. These 25 genera belong to only 4 out of 18 orders. These four orders (Marsupialia, Proboscidea, Perissodactyla and Artiodactyla) make up a total of 170 genera; only 25 of these are parasitized, or a percentage of 14.7. Separately we have – Marsupialia: 1 genus in 80 (1.25%); Proboscidea: 1 in 2; Perissodactyla: 1 in 6; Artiodactyla: 22 in 82 (26.8%).

These figures emphasize the questions already posed:

1. Why, among all recent orders of mammals, do oestrid hosts belong to only four?

2. Why, in these four orders, is only a relatively small fraction parasitized?

3. What characteristics do oestrid hosts have in common?

As we have virtually no information on the physiological (biochemical) interactions between parasites and hosts, answers to these questions have to be sought in the morphological, geographical and palaeontological characteristics of the mammalian orders parasitized.

1. Marsupialia

1.1. Generalities

The marsupials were classified by Romer (1967: 379-380) into 4 suborders: (i) Polyprotodonta, with the families Didelphidae (SA, NA), †Caroloameghinidae (SA), †Borhyaenidae (SA), †Necrolestidae (SA), Dasyuridae (Aus) and Notoryctidae (Aus); (ii) Peramelida, with the single family Peramelidae (Aus); (iii) Caenolestidia, with 4 families: Caenolestidae (SA), †Groeberiidae (SA), †Polydolopidae (SA) and †Argyrolagidae (= Microtragulidae; SA); (iv) Diprotodonta, with the families Phalangeridae (Aus), †Thylacoleonidae (Aus), Phascolomidae (= Phascolomyidae, Vombatidae; Aus), Macropodidae (Aus) and †Diprotodontidae (Aus).

In 1964 RIDE proposed another system for the Australian marsu-

Table 1. Incidence of parasitism by Oestridae on the families and genera of mammals.

Orders	No. of fams.	No. of fams. parasitized	No. of genera	No. of genera parasitized
1. Monotremata	2	–	3	–
2. Marsupialia*	8	1	80	1
3. Insectivora	8	–	63	–
4. Dermoptera	1	–	1	–
5. Chiroptera	17	–	178	–
6. Primates	11	–	60	–
7. Edentata	3	–	14	–
8. Pholidota	1	–	1	–
9. Lagomorpha	2	–	10	–
10. Rodentia	35	–	352	–
11. Cetacea	8	–	38	–
12. Carnivora	7	–	101	–
13. Tubulidentata	1	–	1	–
14. Proboscidea*	1	1	2	1
15. Hyracoidea	1	–	3	–
16. Sirenia	2	–	3	–
17. Perissodactyla*	3	1	6	1
18. Artiodactyla*	9	6	82	22
Totals	120	9	998	25

pials, dividing them into 3 'orders': (i) Marsupicarnivora, with the Superfamily Dasyuroidea (fams. Dasyuridae and Thylacinidae, Aus); (this order would also include the South American Didelphoidea and †Borhyaenoidea); (ii) Peramelina, with the single family Peramelidae; (iii) Diprotodonta, with the families †Wyniardiidae, Phalangeridae, Macropodidae, †Diprotodontidae and Vombatidae; the Notoryctidae were left 'incertae sedis'.

Neither system agrees with the available serological data. KIRSCH (1968), studying the comparative serology of Australian marsupials, concluded that (a) the South American didelphoids and the Australian marsupials represent two different lineages; (b) the honey possum, *Tarsipes*, is distant from the other Phalangeridae; (c) the koala, *Phascolarctos*, is more related to the wombats and is relatively distant from the Phalangeridae; (d) the bandicoots (Peramelidae) are related to the Dasyuroidea and not to the Phalangeroidea.

Later studies by HAYMAN, KIRSCH, MARTIN & WALLER (1971) have suggested that there are three basic stocks of contemporary marsupials: (a) the South American caenolestoids; (b) the South American didelphoids; (c) the Australian marsupials as a unit. The Caenolestoidea were probably the first to segregate.

Recent studies of albumin of both South American and Australian marsupials by MAXSON, SARICH & WILSON (1975) have also furnished very much the same results, i.e., that South American and Australian marsupials represent two different branches that radiated independently, although monophyletic. Moreover, these authors concluded that: 'The rate constant for albumin evolution in placental mammals and iguanid lizards is about 1.7 units per Myr per pair of lineages compared. If this holds for marsupials and hylines then for the separation of the Australian and New World faunas, one obtains a figure of 73 Myr for marsupials and 68 Myr for hylines. These figures are consistent with current geological dating for South America-Antarctica separation as occurring in the late Cretaceous 70.Myr ago.'

The distinction between Australian and South American marsupials is also demonstrated by the morphology and development of spermatozoa (HUGHES, 1965; BIGGERS & LAMATER, 1965), as well as by the difference in the Mallophaga (Phthiraptera) associated with them. At this point a small digression about the Mallophaga may be introduced.

HARRISON (1916: 257), dealing with the Boopidae of Australian marsupials, had already predicted that South American marsupials should also be parasitized by Mallophaga. In 1922 he had the satisfaction of describing the Trimenoponidae, considering them as related to the Boopidae, thus implying a phylogenetic relationship between Mallophaga and marsupials from South America and Australia. HOPKINS (1949: 536-537) also considered as probable a relationship between the two families of ectoparasites:

'No sucking lice [Anoplura] are known from the marsupials, and their Mallophaga all belong to the more primitive suborder, Amblycera. The distribution on marsupials of the latter group of lice is peculiar and interesting, the Boopidae occurring only on Australian marsupials, whereas the Trimenoponidae occur on South American marsupials and on certain South American rodents; ...my own belief is that the two groups [of Mallophaga] are quite closely related, that the differences between them can be accounted for by evolution during the very great length of time which has elapsed since the Australian and South American marsupials separated off from their common stock, and that their infestation of marsupials is probably primary.'

However, more recent specialists are not in accordance with HOPKINS considering these two families as monophyletic. Thus, MURRAY & CALABY (1971: 82) declare:

'CLAY (1970) casts doubt on the commonly held view that the Boopidae are primary parasites of Australian marsupials, and suggests that infestation of these animals was comparatively late and arose from an avian menoponid ancestor. This might have become established on an ancestral phalangeroid stock (= Order Diprotodonta of KIRSCH, 1968) and diversified on that group, the parasites now found on the other groups of marsupials being due to secondary infestations. A consideration of the known

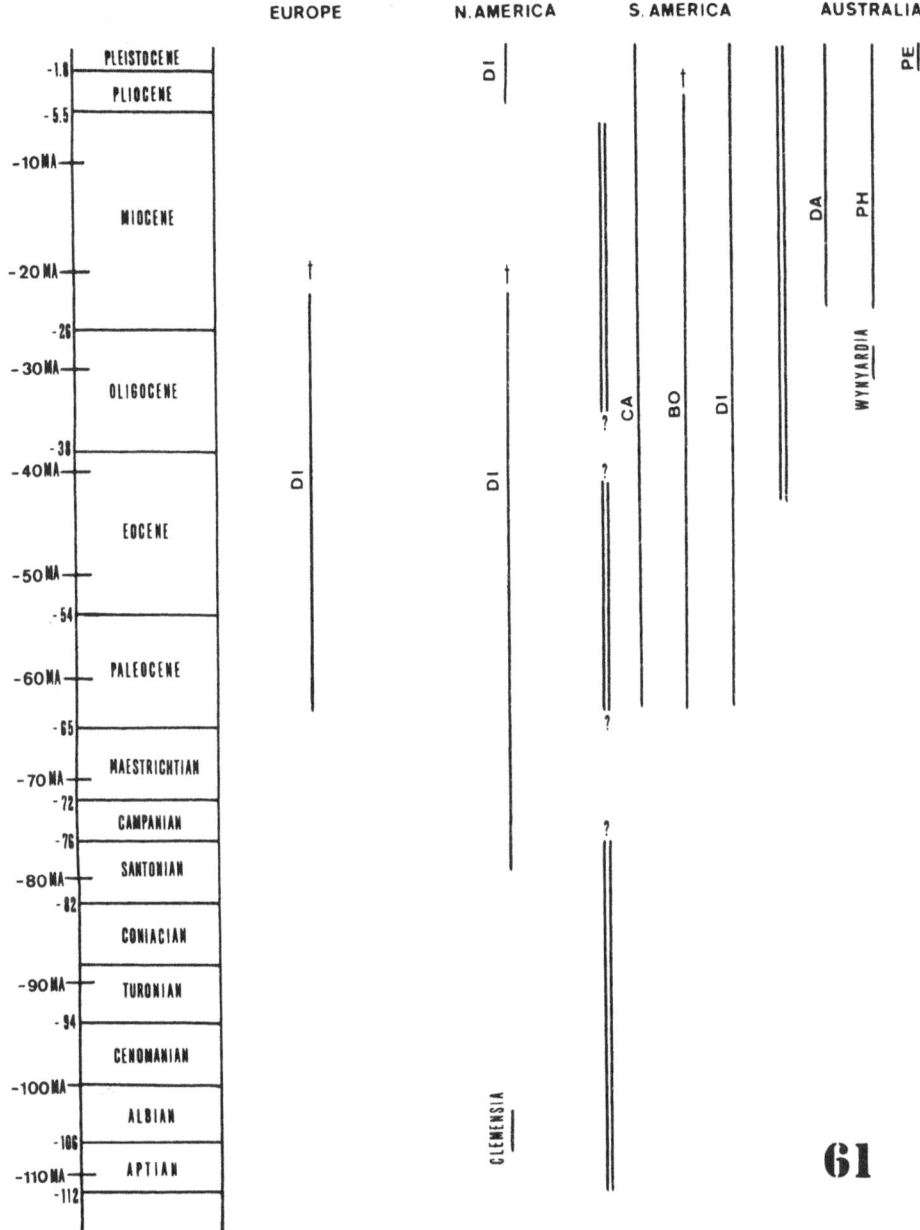

Fig. 61. Time-range of world marsupials (Double bars: barriers between zoogeographical regions); DI: Didelphoidea; CA: Caenolestoidea; BO: Borhyaenoidea; DA: Dasyuroidea; PH: Phalangeroidea: PE: Perameloidea. (Adapted from JARDINE & McKENZIE, 1972; *Clemensia* should be renamed *Holoclemensia*, cf. SLAUGHTER, 1968b).

distribution of members of the Boopidae from the possum families leads us to go further and suggest that the ancestral boopids became established in the first place on the family Macropodidae. Typical members of this family were in existence in the Miocene and probably Oligocene, and by the Pliocene the family had radiated widely (STIRTON, TEDFORD & WOODBURNE, 1968). This seems to be early enough to allow sufficient time to elapse for the Boopidae to reach its present diversity, which, as pointed out by CLAY, is not very great. There may of course be something peculiar about the arboreal possum-like marsupials that has discouraged an association of lice with them, but in the absence of any knowledge of the biology of these lice we can make no suggestions on this point. We agree also with CLAY that the Boopidae and Trimenoponidae are not closely related and almost certainly had different origins.'

1.2. Palaeontological history (Fig. 61)

JARDINE & McKENZIE (1972) have summarized the fossil history of marsupials, as follows:

'The earliest known marsupials are didelphoids from the Cretaceous of North America. SLAUGHTER (1968a) recognized a marsupial from the mid-Cretaceous (Albian, about -100 Ma) on the basis of molar teeth. Undoubted marsupials have been described from the Upper Cretaceous (Santonian, about -80 Ma; RUSSELL, 1952). In the late Cretaceous didelphoid marsupials (which ranged from the mouse-like *Pediomys* to the dog-sized *Didelphodon*) underwent considerable differentiation in North America (CLEMENS, 1966; LILLEGRAVEN, 1969). At the end of the Cretaceous the North American marsupials were reduced in number and variety, probably because of competition with more specialized carnivorous placental mammals. Failure to find marsupials in the late Cretaceous of Europe or Asia [see below] is not evidence that marsupials originated in North America. Only one late Cretaceous deposit of the Old World is known to have an extensive mammal fauna (CLEMENS, 1968) and, because North American and Eurasia were effectively a single continent in the Cretaceous, it would be surprising if Cretaceous marsupials are not eventually discovered in the Old World.'

'Didelphoid marsupials have been found in the Paleocene (about -60 Ma) of North America, Europe and South America. One of the North American late Cretaceous forms, *Alphadon*, had dental characteristics which suggest that it may have been ancestral to all the known Paleocene didelphoids [sic!] (CLEMENS, 1968). Didelphoid marsupials became extinct in North America and Europe in the early Miocene. In the Paleocene of South America appear the first known representatives of the caenolestoids and the carnivorous borhyaenoids (COUTO, 1952).'

'SIMPSON (1950) divided the Tertiary mammals of South America into three groups according to the presumed times at which their ancestors entered it from North America. First, there are mammals derived from late Cretaceous and earliest Paleocene immigrants. These include the marsupials; a largely herbivorous group of placental mammals (litopterns, notoungulates and so on) derived from relatively unspecialized North American placentals (ferungulates); and a group of "exotic" placentals (xenarthrans – including sloths, ant-eaters and armadillos) derived from another unspecialized North American stock (palaeonodonts). Second, there are New World monkeys and caviomorph rodents, probably derived from late Eocene or early Oligocene immigrants from North America. Third, there are the numerous and diverse specialized placental mammals which have invaded from north America since the late Miocene and which have largely displaced the older elements of the mammal fauna.'

'The hypothesis that the South American marsupials are derived from marsupial immigrants from North America is indirectly supported by evidence for the derivation of the other old elements of the South American fauna from North American immigrants [sic]. Borhyaenoids, the dominant carnivores of South America from the Paleocene to the Pliocene, probably evolved from didelphoids in South America (SIMPSON, 1941) but the place of origin of the caenolestoids is uncertain. Their Paleocene representatives were already very distinct from didelphoids.'

'The hypothesis that all Australian marsupials are derived from immigrants from South America by way of Antarctica is based on more conclusive evidence. From the Albian/Aptian (about -106 Ma) when Africa separated from South America and Antarctica to -43 Ma (when Australia separated from Antarctica) this was the only available overland source for Australian marsupials. Cox's hypothesis (1970) that marsupials reached Australia from Africa through Antarctica is ruled out by the early separation of Africa. Colonization of Antarctica by mammals during the late Cretaceous or early Tertiary is ecologically plausible. Antarctica was probably not glaciated during this period and the only known early Tertiary flora of Antarctica (Seymour Island, 64°S) indicates a cool temperate flora comparable to that of parts of New Zealand today (DUSÉN, 1908; CRANWELL, 1959).'

'The earliest known Australian marsupial, *Wynyardia*, is from the upper Oligocene (about -30 Ma) (GILL, 1957). RIDE has shown that it has several primitive "didelphoid" features not found in later Australian marsupials (1964). By the Miocene Australian marsupials had undergone considerable differentiation, and fossils referable to three living families and one extinct family are known (STIRTON, TEDFORD & MILLER, 1961).'

A possible Cretaceous representative of Palaearctic marsupials may be *Deltatheridium*, formerly included among the Insectivora. The finding of more complete material in the Djadokhta (?Coniacian or Santonian) and Barun Goyot (middle Campanian) formations of Mongolia brought new doubts about the position of *Deltatheridium*, which may be a very primitive marsupial (cf. BUTLER & KIELAN-JAWOROWSKA, 1973).

It is too early, therefore, to produce a conclusive explanation of marsupial zoogeography.

1. 3. Hypotheses on the origin of Australian marsupials

This is one of the most fertile fields of speculation. Several papers, some even absurd, have been published. Several alternative and often conflicting explanations have been proposed by Cox (1970, 1973), MARTIN (1970), HOFFSTETTER (1970), DALZIEL (1973), FOODEN (1972), JARDINE & MCKENZIE (1972), McGOWRAN (1973) and COUTO (1975), to cite only a few.

The greatest difficulty is to explain the alleged absence of Eutheria in Australia. As K. A. JOYSEY (in JARDINE & MCKENZIE, 1972) pointed out, 'it remains possible that placentals and marsupials both reached Australia and that placentals subsequently became extinct.'

That some kind of mammal must have been present in Australia during the Lower Cretaceous seems to be indicated by the presence of fossil fleas found in siltstones at Koonwarra, southern Gippsland. RIEK (1970) described his findings thus:

'One flea is of normal pulicid form and size. In many respects it is ancestral to the modern stickfast fleas (*Echidnophaga*), some species of which are endemic on Australian marsupials. This flea has very elongated mouthparts similar to those of the stick-fast fleas.'

'The second flea is most interesting, for it combines a typical flea abdomen and male genitalia with a very primitive head structure. The head is less compressed and the antennae are not markedly shortened and recessed in grooves along the side of the head as in modern fleas; they are prominent although the segments are short and cone-in-cone form. The legs are unusually long and the tibiae bear pseudocombs of short stout bristles: the hind femur is not enlarged for jumping. The structure of the legs suggests that this insect lived on a sparsely-haired (furred) animal and that it clung to the outer portions of the hairs rather than burrowed between the hairs. The specimen, with a body length of 7 mm, is large as compared with most modern fleas, especially as it is a male; male fleas are distinctly smaller than females. Females of some modern fleas, however, are of this order of size. The nematocerous-type antennae tend to support the more usually accepted conclusion that the fleas evolved from a nematocerous type ancestor.'

'Deductions concerning the ecological association of at least the more primitive of the two species with a furred animal, and not a bird, indicates that marsupials [or some other mammal] must have been present in Australia at a very much earlier period than has hitherto been conceded, and this species thus sheds new light on the probable centre of origin [sic], and on the early dispersal of the marsupials, as it affects the zoogeography of the southern continents.'

1.4. Characteristics of Australian marsupials

Among marsupials parasitism by oestrids has only been reported in Australia. Our discussion will therefore be limited to the Australian forms. The modern representatives of Australian marsupials are classified in three superfamilies. The classification adopted here is a compromise between the one employed by WALKER et al. (1964) and that adopted by KEAST (1968, 1972). For the Macropodidae the classification presented by RAVEN & GREGORY (1946) was accepted.

1.4.1. Superfamily Dasyuroidea (= Order Marsupicarnivora, part) (Table 2)

The Dasyuroidea include the families Dasyuridae (with subfamilies Dasyurinae and Myrmecobiinae) and Thylacinidae.

(i) The Dasyuridae are represented by 19 genera and about 45 species, with a distribution in Australia, Tasmania, Normanby and Aru Islands, from sea-level to altitudes of almost 3400 m. They are predominantly nocturnal, preying on other animals which they are able to kill. The genera are mainly terrestrial, although some 'rats' and 'cats' in this family are primarily arboricolous. During the day they take shelter inside hollow trees, holes on the ground or in caves. According to their ecological characteristics, they may be classified as:

(a) Marsupial 'rats' – insectivorous forms, similar to rats; this group includes the smallest living marsupials, with a length of less than 95 mm: *Phascogale, Antechinus, Planigale, Murexia, Neophascogale, Parantechinus, Phascolosorex, Pseudantechinus, Myoictis, Dasyuroides, Dasycercus, Sminthopsis* and *Antechinomys*. Their weight varies between 5 and 175 grammes;

(b) Marsupial 'cats' – forms slightly similar to cats and weasels, principally carnivorous: *Dasyurus, Dasyurinus, Satanellus* and *Dasyurops*; their weight is from 0.5-3 kg;

(c) Tasmanian devils (*Sarcophilus*), similar to small dogs; nocturnal, carnivorous or sarcophagous, attacking even snakes;

(d) Marsupial 'anteaters' – small forms, primarily insectivorous,

Table 2. Superfamily Dasyuroidea.

1. Family Dasyuridae		
Genera	Body Length (mm)	Weight (g)
Phascogale	160-220	–
Antechinus	90-170	24-49
Planigale	50-92	5
Murexia	105-200	–
Neophascogale	170-230	–
Parantechinus	111-120	–
Phascolosorex	117-226	–
Pseudantechinus	75-105	–
Myoictis	170-250	–
Dasyuroides	165-182	–
Dasycercus	125-220	122-175
Sminthopsis	70-120	10-32
Antechinomys	80-100	–
Dasyurus	350-450	1130
Dasyurinus	290-650	550
Satanellus	240-350	–
Dasyurops	400-750	2000-3000
Sarcophilus	525-800	4530-9070
Myrmecobius	175-275	275-450
2. Fam. Thylacinidae		
Thylacinus	1000-1300	–

Table 3. Family Peramelidae.

Genera	Body Length (mm)	Weight (g)
Perameles	200–425	550
Echymipera	200–500	up to 2000
Thylacomys	200–550	300–1600
Chaeropus	230–250	–
Isoodon (= *Thylacis*)	240–290	1100–1400
Rhynchomeles	320	–
Perodyctes	190–500	–
Microperodyctes	174	–

feeding principally on termites. The marsupial 'anteaters' (genus *Myrmecobius*, the only representative of Subfamily Myrmecobiinae) differ from the other Dasyuridae in being diurnal;

(e) Tasmanian wolves – the Thylacinidae include only the genus *Thylacinus*, confined to forested areas of that island. These are large forms, exclusively carnivorous, hunting principally kangaroos and wallabies but not despising birds and other small mammals. They hunt alone, in pairs or in family groups. They are essentially nocturnal.

1.4.2. Superfamily Perameloidea (= Order Peramelina) (Table 3)
The Perameloidea contain a single family. Eight genera and about 19 species are known, found in Australia, Tasmania, New Guinea (at

Table 4. Family Phalangeridae.

Genera	Body Length (mm)	Weight (g)
Phalanger	325–650	–
Trichosurus	320–580	1300–5000
Acrobates	60–80	12–14
Distoechurus	100–120	–
Cercartetus	75–100	15–25
Eudromicia	70–120	–
Gymnobelideus	152–168	–
Petaurus	120–320	90–130
Dactylopsila	170–320	–
Dactylonax	195–280	–
Wyulda	362	–
Tarsipes	88–100	13–17
Phascolarctos	600–850	4000–15000
Pseudocheirus	180–450	up to 1500
Hemibelideus	300–400	–
Petropseudes	325–450	–
Schoinobates	300–480	1360

altitudes up to 4200 m), in the islands of Kei and Aru, the archipelagos of Entrecasteaux and Bismarck and in Ceram. They inhabit plains, banks of rivers and swamps with tall grass, scrub and forest. Members of this family are terrestrial and nocturnal; only representatives of the genus *Thylacomys* excavate burrows for protection. Most species have mixed feeding habits but are essentially insectivorous.

1.4.3. Superfamily Phalangeroidea (= Order Diprotodonta)
The Phalangeroidea include the following families: (i) †Wynyardii-dae, based on a unique fossil from the Upper Oligocene or Lower Miocene (RIDE, 1964; QUILTY, 1966); (ii) Phalangeridae (Table 4), with subfamilies Phalangerinae, Tarsipedinae, Phascolarctinae and †Thyla-coleoninae (a marsupial 'lion' from the Upper Oligocene or Lower Miocene to the Lower Recent; extinct). The Phalangeridae comprise 17 genera and about 42 species, occurring in Australia, Tasmania, New Guinea, west to Ceram, Timor, Celebes and adjacent islands, and east to the Bismarck Archipelago and the Solomon Islands. Some have been introduced in New Zealand. The Phalangeridae have the widest distribution among the Australian marsupials. They live in forested areas, from sea level to elevations above 4000 m. Ecologically they may be considered as occupying the niches of primates (lemurs), rats, squirrels, Dermoptera and sloths (genus *Phascolarctos*, the koala). They are predominantly herbivorous but among the more primitive recent genera, such as *Distoechurus* and *Cercartetus*, insectivorous forms still exist. Some of the smaller phalangers feed on nectar.

(iii) †Diprotodontidae–giant quadrupeds, such as †*Diprotodon*, †*Nototherium* and †*Euryzygoma*, known from the Lower Miocene to the Lower Recent; extinct.

(iv) Phascolomidae (also known as Phascolomyidae and Vombati-dae) (Table 5), with only two recent genera which occur in Australia, Tasmania and Flinders Islands (Bass Strait). They are inhabitants of forests, savannas and prairies, in valleys or mountainous areas. They look like small bears, with a robust body and short legs, being considered nocturnal. They live in burrows which they excavate, feeding on roots and grass. As noted by KEAST (1968, 1972) they represent an interesting parallel with the genus *Marmota*.

Table 5. Family Phascolomidae.

Genera	Body Length (mm)	Weight (kg)
Phascolomis	700-1200	15-35
Lasiorhinus	870-1030	25.45-27.27

Table 6. Family Macropodidae.

Genera	Body Length (mm)	Weight (kg)
Hypsiprymnodon	235–335	0.50
Bettongia	280–450	1.59
Aepyprymnus	380–520	2.50
Caloprymnus	270–440	0.85–1.06
Potorous	300–400	1.36–1.81
Setonix	475–600	2–5
Dendrolagus	520–810	–
Dorcopsis	490–800	3.9–6.8
Dorcopsulus	340–550	–
Lagorchestes	325–500	1.70–3.00
Lagostrophus	400–460	–
Petrogale	500–800	3–9
Peradorcas	310–400	–
Onychogale	450–670	–
Thylogale	530–770	–
Wallabia	450–1050	4–24
Macropus[*1]	800–1600	23–70

1. Including *Megaleia* and *Osphranter*.

(v) Macropodidae (Table 6). This family is currently represented by 17 genera and approximately 52 species, distributed throughout Australia, Tasmania, New Guinea, some adjacent islands and the Bismarck Archipelago, and has been introduced in New Zealand. Most forms are nocturnal and during the day remain in nests of grass or shallow burrows. Some species are active during certain intervals of the day. They are herbivorous, eating many kinds of plants. All are predominantly terrestrial, except *Dendrolagus*, which is secondarily arboricolous. The Macropodidae include the following subfamilies: (a) Hypsiprymnodontinae – with only one genus and species (*H. moschatus*), an inhabitant of the rain-forest of the northern coast of Queensland; it occurs also frequently in the dense vegetation bordering rivers and lakes. It is the smallest of the 'rat-kangaroos', feeding on insects and worms, but also the fruit of a palm (*Ptychosperma*) and tubercles. This subfamily seems to represent an annectent group between the primitive Phalangeroidea and the Macropodidae; the dentition is primitive and differs from that of other kangaroos. However, the feet are already developed and adapted for jumping, a constant trait among the Macropodidae (RAVEN & GREGORY, 1946: 4–5). Some observers, however, affirm that *Hypsiprymnodon* runs on all four legs and not by jumps as the other kangaroos; (b) †Sthenurinae, an extinct subfamily, known from the Upper Pliocene to the Lower Recent;

according to RAVEN & GREGORY (1946) some genera inhabited forests and some others grasslands; (c) Potoroinae – including the 'rat-kanga-roos'. As demonstrated by RAVEN & GREGORY (1946) the Potoroinae have had an initial radiation in the forest, passing later to areas of transitional vegetation, until the conquest of grasslands and deserts. According to WALKER *et al.* (1964), *Potorous* 'usually inhabits thick scrub and dense undergrowth, but it also occurs in open woodland and cleared areas'; *Caloprymnus* 'is known at the present time only from the Lake Eyre basin in Central Australia; it probably had a wide range formerly in the Central Desert; … distinctive physical features of the area in which it now occurs are sand ridges, claypans, and rocky plains'; *Aepyprymnus* occurs 'both in open and forested country; it inhabits the open plains and open grass-tree forests'; *Bettongia*, which inhabits 'open plains, grassy plains, ridge country bordering forests', is the only partially carnivorous Macropodidae; (d) Macropodinae – the true kangaroos may be divided into three ecological groups (fol-lowing RAVEN & GREGORY, 1946): inhabitants of forests (*Dendrolagus, Dorcopsulus, Dorcopsis, Thylogale* and *Setonix*); inhabitants of transitional vegetation (*Lagostrophus* and *Lagorchestes*); and inhabitants of open formations – prairies, grasslands, savannas, etc. (*Wallabia, Macropus, Onychogalea, Petrogale* and *Peradorcas*; the last three in rocky areas and mountain slopes with boulders).

The history of Macropodidae evolution in relation to the shape of the skull, form and structure of limbs and dentition was studied by BENSLEY (1903), JONES (1924), TATE (1948) and RIDE (1964). Physiologi-cal studies have demonstrated the existence of 'pseudorumination' among macropodids, permitting digestion of cellulose (MOIR, SOMERS & WARING, 1965), an interesting convergence with the true ruminants (KEAST, 1968, 1972).

The two subfamilies, Potoroinae and Macropodinae, as pointed out by KEAST (l. c.), are excellent parallels of antelopes of the African veld, of the subfamilies Cephalopinae and Hippotraginae, in relative size, habitat and way of life.

1.5. Characteristics of a marsupial host

As seen above, the Australian marsupials seem to be a monophyletic group (as evidenced by serology and morphology of spermatozoa), which has been in Australia since the Cretaceous (-73 million years, MAXSON *et al.*, 1975), has evolved in isolation for a long time with a total absence of competitors, which either never entered Australia or were totally eliminated by the marsupials. Marsupials have therefore occupied a number of ecological niches which are occupied in other continents by several different orders of Eutheria.

This situation, unique in the world, free from interferences, and, so to say, 'pura pro analyse', is an excellent opportunity for the determination of the characteristics of an oestrid host.

As to the origin of the Australian oestrids, three alternatives must be considered: 1. They invaded Australia from Africa or South America, when those continents were more or less united; 2. They invaded Australia as 'island-hoppers', at a much later period, coming from the Oriental Region; 3. They are relics of a past, very broad, Gondwanan distribution.

This question cannot be discussed here at length (see Part D for a discussion). The only alternative that can be immediately discarded is the second, as the Oriental Region does not have (and almost certainly never had) oestrids. Therefore we are left with two alternatives. Whatever is the origin of the Australian oestrids, one would expect that they underwent an adaptive radiation, as occurred with Australian mammals. But now comes the problem: among 64 genera and over 160 species of Australian marsupials, only 1 genus and 2 species are parasitized! How is this apparent failure of the oestrids explained?

Only *Macropus* is parasitized. The taxonomic situation of this genus is chaotic, and there are considerable differences of opinion as to the composition of the genus. Just to cite two extreme ideas – IREDALE & TROUGHTON (1934: 50-55), in their 'Checklist of the mammals recorded from Australia' accept three genera: *Osphranter*, with 6 species; *Megaleia*, with 1 species; and *Macropus*, with 6 species; at the other extreme, WALKER *et al.* (1964: 93) consider *Osphranter* and *Megaleia* as subgenera, and only three species: *Macropus* (*Macropus*) *giganteus* (Shaw) (= *canguru*; = *major*), the Great Grey Kangaroo; *Macropus* (*Osphranter*) *robustus* (Gould) (= *antilopinus*), the Wallaroo or Euro; and *Macropus* (*Megaleia*) *rufus* (Desmarest) (= *laniger*), the Red Kangaroo. I am accepting here, for convenience, the last classification.

As demonstrated by MYKYTOWYCZ (1964), only *M. rufus* and *M. robustus* are parasitized, each with its own probable species of *Tracheomyia* (*macropi* and an undescribed species); parasitism of *M. giganteus* has never been confirmed; previous reports are erroneous.

A possible explanation of this fact is probably found in the ecology of the three species – *M. rufus* inhabits plains and grasslands of the interior of Australia; *M. robustus* is widely distributed in Australia, especially in the coastal mountains and the rocky chains of the interior; whereas *M. giganteus* lives in open forests and in the dense bush of Australia and Tasmania. If ancestors of *Tracheomyia* had adapted to open areas before becoming parasites of *Macropus*, forests would represent a barrier for them.

But the basic problem remains – why only *Macropus* is parasitized? From the examination of the characteristics of this genus, and of the other genera of Marsupialia, it can be noted that *Macropus* presents the following characteristics: (i) Its species are terrestrial and strictly herbivorous; (ii) Its species are the largest marsupials in Australia, measuring from 800-1600 mm in body length and weighing from 23 to 70 kg.

These could be the basic characteristics required by an oestrid in selecting a host.

In fact, as to the first point: female oestrids larviposit directly into the nasal cavities of the host; a carnivorous or insectivorous mammal would eat the female insect; therefore, only herbivores could be successfully infested; on the other hand, adult oestrids have a very short life span; during this period they must discover an aggregation site, copulate and search for a suitable host in which to deposit the larvae; it is plausible to suppose that the larger the host, the more easily can it be located by the female fly. The host also must have an adequate weight to suffer parasitism and its possible consequences (loss of weight, secondary infections, attacks by other parasites, etc.), without fatal consequences which would be disastrous for both host and oestrid; moreover, the nasal cavities of the host must have sufficient room to harbour several (relatively large) larvae.

It is very possible that these two requirements, apparently very simple, are the basic ones for selection of an oestrid host.

Application of these two principles to the series of Australian marsupials reveals several interesting facts: (i) The Dasyuroidea (insectivorous and carnivorous), the Perameloidea (insectivorous), the Phalangeridae (herbivorous but arboricolous, or insectivorous), Phascolomidae (fossorial), and certain Macropodidae (*Hypsiprymnodon*, insectivorous; *Bettongia*, partially carnivorous), etc., are ruled out as hosts; (ii) Other herbivorous marsupials do not have sufficient weight to be parasitized, as can be seen from the examination of Tables 2-6.

Application of these same principles to South American marsupials shows clearly enough why oestrids do not parasitize them. Recent representatives of South American marsupials, belonging to the families Didelphidae and Caenolestidae, include only relatively small forms (Tables 7 and 8) which are insectivorous, carnivorous, carrion-feeders or have mixed feeding habits (WALKER *et al.*, 1964). And what is even more interesting, South American marsupials could never have been parasitized, as, since their first appearance in that continent to

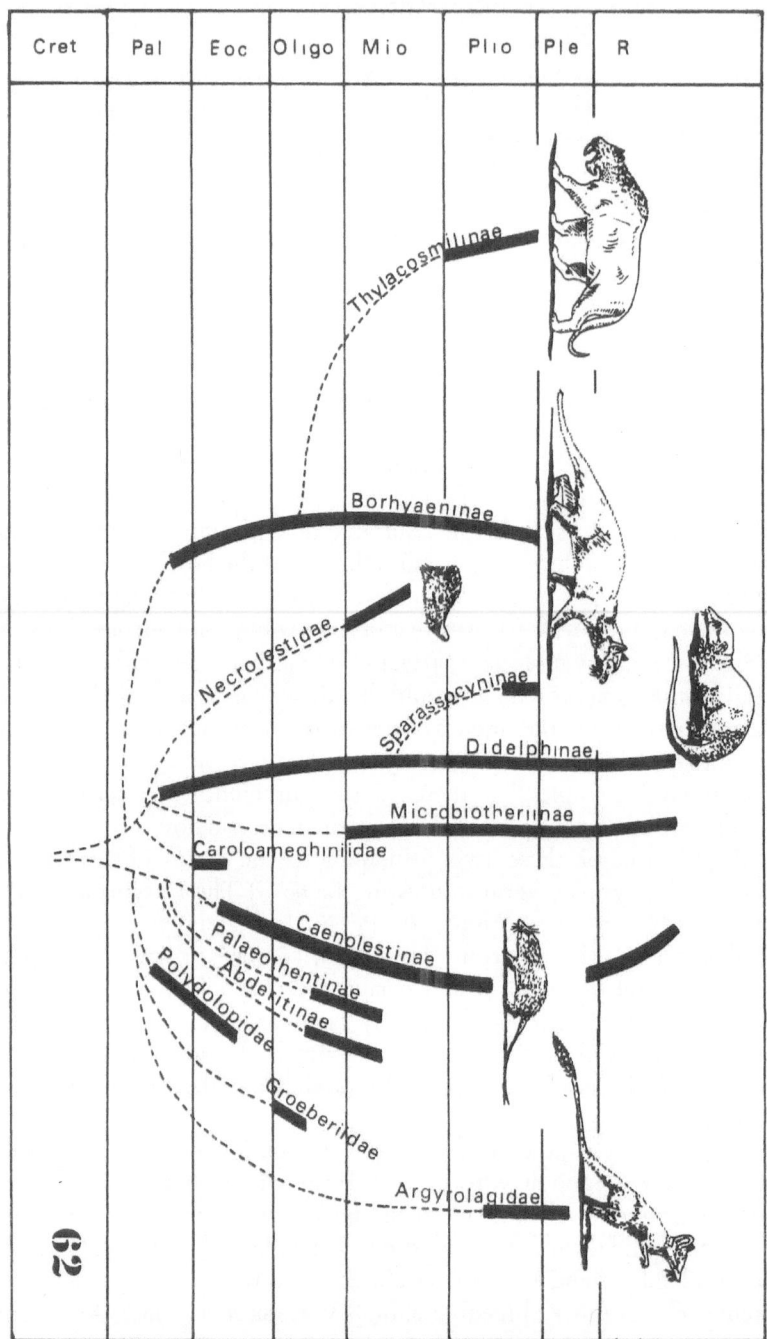

Fig. 62. Phylogeny and time-ranges of South American marsupials (adapted from PAT-
TERSON & PASCUAL, 1968).

Table 7. Family Caenolestidae.

Genera	Body Length (mm)	Weight (g)
Caenolestes	93-135	–
Lestoros	90-120	–
Rhyncholestes	110–128	–

Table 8. Family Didelphidae.

Genera	Body Length (mm)	Weight (g)
Monodelphis	110-140	–
Dromiciops	125	–
Glironia	160-205	–
Lestodelphys	144	–
Marmosa	85-185	950
Philander	250-350	240-400
Metachirus	265	800
Lutreolina	250-400	200-540
Didelphys	325-500	2000-5500
Chironectes	270-325	–

the present, they have never evolved herbivorous forms. These marsupials only gave rise to carnivorous forms (Borhyaenidae), fossorial forms similar to chrysochlorids (Necrolestidae) and forms resembling insectivores, rodents and lagomorphs (Fig. 62).

Therefore, the only genus of marsupials which could ever be parasitized is *Macropus*. This poses an intriguing problem (how did oestrids pass to *Macropus*) which will be discussed in Part D.

Two final words about the characteristics of a host – it is logical to assume that hosts should be diurnal, as oestrids, like other flies, are sun-loving creatures; and gregarious hosts should be favoured by the flies, as chances of being parasitized increase if hosts are easily found and in great numbers. Also in this respect *Macropus* is the 'elect' group; while the majority of species of Australian marsupials are solitary or live in small groups, certain species of *Macropus*, like *rufus*, congregate in herds of some hundreds of individuals.

1.6. Implications in the case of the Eutheria

As said before, Australian marsupials have occupied ecological niches which, in other continents, are occupied by different orders of Euthe-

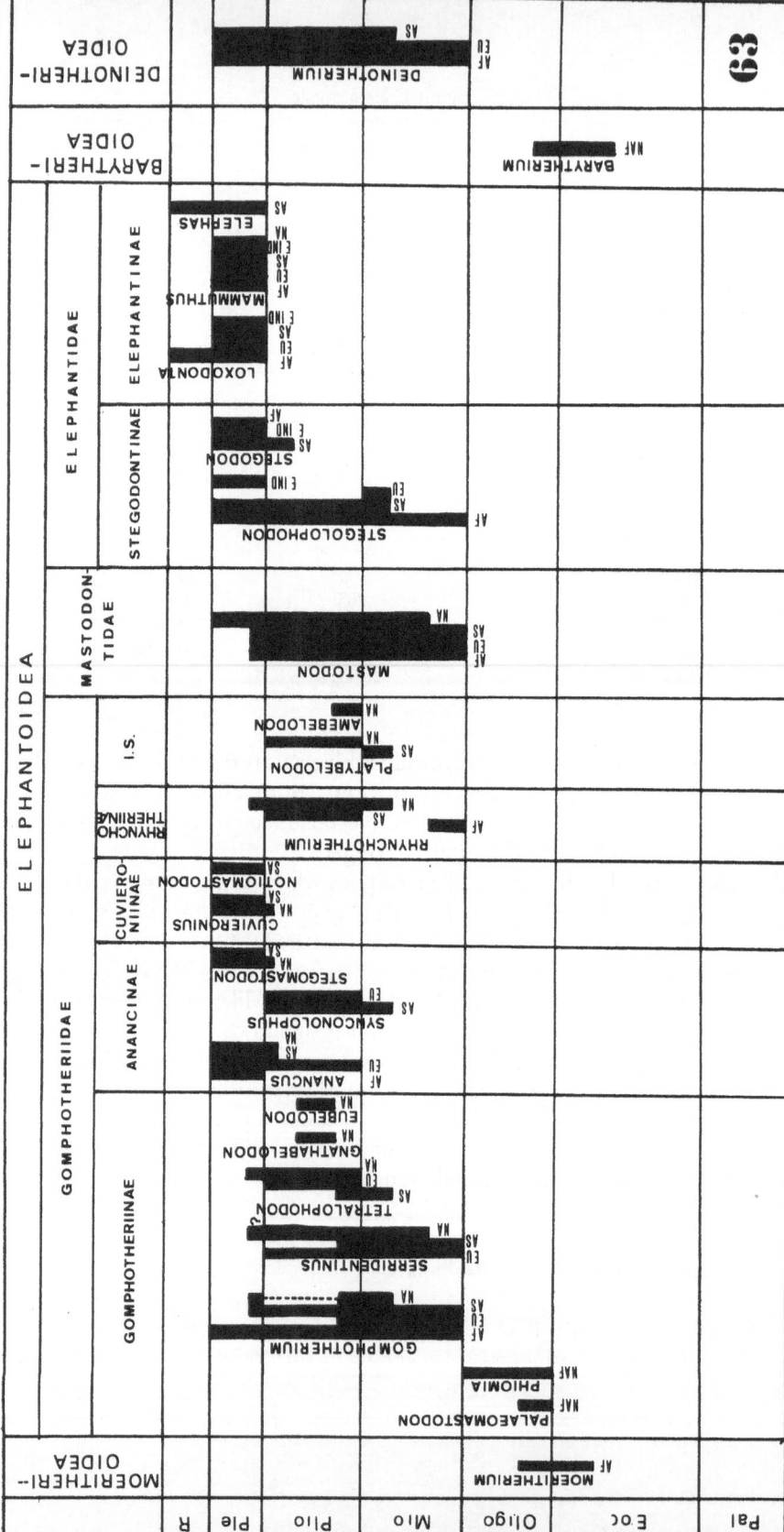

Fig. 63. Classification and time-range of the Proboscidea.

ria. This parallelism provides a good opportunity for testing the two basic principles which characterize oestrid hosts.

Marsupials which have occupied the ecological niches of Primates (Phalangeridae), Dermoptera (Phalangeridae), Rodentia (Dasyuridae, Peramelidae, Phalangeridae), Lagomorpha (Peramelidae), Carnivora (Dasyuridae, Thylacinidae), Edentata (Dasyuridae Myrmecobiinae, Phalangeridae Phascolarctinae), Insectivora (Dasyuridae, Peramelidae, Phalangeridae) and so on, are not parasitized; their Eutherian counterparts are not either. An extension of these two principles immediately excludes the orders Chiroptera, Pholidota, Cetacea, Tubulidentata, Sirenia and Hyracoidea.

On the other hand only *Macropus* and some representatives of the orders Proboscidea, Artiodactyla and Perissodactyla fall within the category of terrestrial herbivores with sufficient weight. And only representatives of these groups have been reported as hosts of Oestridae!

The next step is to apply these two basic principles to these three eutherian orders, to see whether they are also fully applicable here.

2. Proboscidea

The Proboscidea (Fig. 63) are classified into 4 superfamilies and their first known fossils date from the Upper Eocene. As pointed out by DARLINGTON (1963: 351, fig. 40), they must have been evolving and diversifying in Africa for some time before that. From Africa several groups passed to Eurasia and thence to North America, finally reaching South America. This varied group underwent a notable extinction, and now only two species remain: (i) *Elephas maximus* Linnaeus, the Asiatic elephant, an inhabitant of mountainous regions with bamboo forests; this species also frequents alluvial plains with tall grass. It is distributed throughout the Oriental region south of the Himalayas, in India, Ceylon, Burma, Thailand, Vietnam, Malaya and Sumatra (ANDERSON & JONES, 1967: 360-361, fig. 60; WALKER et al., 1964: 1321). (ii) *Loxodonta africana* (Blumenbach), the African elephant, found in many parts of Central, Eastern and Southern Africa. It lives in a great variety of habitats, such as savannahs, river valleys, dense forest, desert scrub and thornbush (WALKER et al., 1964: 1323; ANDERSON & JONES, 1967: 360-361, fig. 60).

Only the African elephant is parasitized, by a specialized subfamily, the Pharyngobolinae (only one species, *Pharyngobolus africanus* Brauer). Cases of parasitism were reported from several localities in the Congo area, and ZUMPT (1965) recorded cases in Uganda and Northern Rhodesia.

The Pharyngobolinae offer no evidence for an explanation of their distribution. They were either always restricted to the Congo area and never adapted to the savannahs, or they did not enter the Oriental region for some ecological reason.

3. Artiodactyla

The first Artiodactyla appear as fossils in the beginning of the Eocene, being very rare. By the end of this epoch they were already diversified. The more ancient types were very primitive in many aspects and had a simple dentition with features similar to that of primitive carnivores, implying a mixed diet. However, as pointed out by ROMER (1967: 274), from whom I am quoting this text, 'such distinctive artiodactyl features as the double-pulleyed astragallus were already present, and it is obvious that the ancestral forms must have been undergoing development in some unknown area during the Paleocene. Possibly the ancestors were primitive condylarthrs' (sic).

This 'unknown area' referred to by ROMER has been much discussed. Asia or Central Africa have been proposed as the 'centre of origin' for the Artiodactyla.

3.1. Classification

The classification of the Artiodactyla has been proposed by SCOTT and followed, with some modifications, up to the present; SIMPSON (1945), VIRET (1961) and ROMER (1967) added some innovations. I follow here, as a matter of convenience, ROMER's classification (subfamilies in accordance with SIMPSON, VIRET and WALKER et al.). The Artiodactyla are divided into 3 suborders: (i) †Palaeodonta, with the superfamily †Dichobunoidea; (ii) Suina, with the superfamilies †Entelodontoidea, Suoidea and Hippopotamoidea; (iii) Ruminantia, with 2 infraorders: (a) Tylopoda (superfamilies †Cainotherioidea, †Anoplotherioidea, †Merycoidodontoidea, and Cameloidea), and (b) Pecora (superfamilies Traguloidea, Cervoidea and Bovoidea).

3.2. Suborder †Palaeodonta

The †Palaeodonta (Fig. 64) include 'the oldest and seemingly the most primitive of known artiodactyls. They are in many respects close to the primitive placental stock; and, were it not for the discovery in

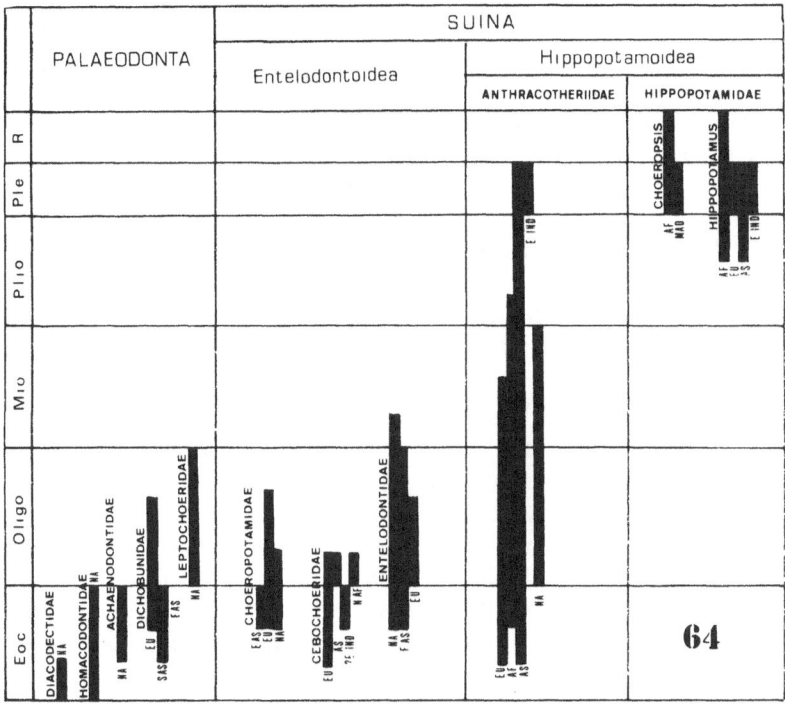

Fig. 64. Classification and time-range of the Palaeodonta and Suina (part) (Artiodactyla).

some instances of a typical artiodactyl astragallus associated with
them, their position as artiodactyls (rather than insectivores or primates) might be questioned.' (ROMER, 1967: 275). They are known
from the Lower Eocene to the Upper Oligocene, principally of North
America (Families †Diacodectidae, Lower Eocene; †Homacodonti-
dae, Lower to Upper Eocene; †Achaenodontidae, Middle to Upper
Eocene; and †Leptochoeridae, Lower to Upper Oligocene); one family, †Dichobunidae (Middle Eocene to Middle Oligocene of Europe;
Middle Eocene of South Asia; Upper Eocene of eastern Asia) is exclusively Palaearctic. They left no trace beyond the Oligocene.

3. 3. Suborder Suina

Includes three superfamilies: (i) †**Entelodontoidea** (Fig. 64). With 3
families: †Choeropotamidae (Upper Eocene of Eastern Asia; Upper
Eocene to Middle Oligocene of Europe; Upper Eocene to Lower
Oligocene of North America); †Cebochoeridae (Middle Eocene to

Lower Oligocene of Europe; Lower Oligocene of Asia; doubtfully recorded from the Lower Oligocene of northern Africa); and †Entelodontidae (Upper Eocene to Lower Miocene of North America; Upper Eocene to Upper Oligocene of Eastern Asia; Lower to Middle Oligocene of Europe). This is another group without recent representatives. VIRET (1961: 910) believes that this group dispersed 'soit de l'Asie Centrale, soit de l'Asie nord orientale.' PILGRIM (1941) believes that the Entelodontoidea, like all other Suina, originated in Central Africa. (ii) **Hippopotamoidea** (Fig. 64). With 2 families: (a) †Anthracotheriidae, with 29 genera (ROMER, 1967: 389), known from Europe (Middle Eocene to Middle Miocene), Asia (Middle Eocene to Pleistocene), Africa (Upper Eocene to Lower Pliocene), North America (Lower Oligocene to Upper Miocene), and the East Indies (Pleistocene). PILGRIM (1941: 145) believes that the Anthracotheriidae arose 'in or near India from some Palaeocene or Lower Eocene dichobunid of a primitive type, which settled in that region in the course of a migration from an area which I can only surmise to have been Central Africa.' (b) Hippopotamidae. The representatives of this family are now distributed all over Africa south of the Sahara, in places with deep water rivers; the present northern limit of the family is Khartoum, the western limit Sierra Leone. In historical times the hippo was known also from Gambia and the mouth of the Nile (ANDERSON & JONES, 1967: 391) and even from India. This family is represented by two genera: *Hippopotamus* and *Choeropsis*.

Hippopotamus has only one species, *H. amphibius*, occurring in most rivers with deep waters in Africa south of the Sahara. The genus is known as fossil since the Upper Pliocene of Africa and Asia, in the Pleistocene of Europe and the East Indies. Hippos are solitary (old males) or live in small groups of 20-30 individuals; they feed on grass, especially at night, and never leave the course of a river and its neighbourhood.

Choeropsis, also with the single species *C. liberiensis*, the pigmy hippo, is known at present only from the Ivory Coast and Liberia, and adjacent parts of Sierra Leone. It is found in rivers and swamps in the interior of damp forests, being less aquatic than *Hippopotamus*; these animals are usually found in pairs, walking through the forest at night, to feed on tender shoots and branches, leaves and fallen fruits. The genus is known as fossil in the Pleistocene of Madagascar and northern Africa.

Only *Hippopotamus* is parasitized, by *Rhinoestrus hippopotami* (Oestrinae, Rhinoestrini). This oestrid occurs only in the Congo and Uganda regions (ZUMPT, 1965: 167), while its host has a much broader distribution.

174

Choeropsis is not parasitized, and this may be explained by (at least) two causes: (i) it seems to be a relatively recent group, and oestrids probably have not had enough time to adapt to it; or (ii) the branch of *Rhinoestrus* parasitizing hippos appeared after *Choeropsis* had become restricted to the Liberian forests; then, the Dahomey Gap (in the territories of Togo and Dahomey, a savannah corridor) would prevent passage of *Rhinoestrus* from the Congo to the Liberian forests. Booth (1954, 1956), studying forest refugia in Equatorial Africa and their role in Quaternary speciation and faunal differentiation of mammals, came to the conclusion that the Dahomey Gap must have fluctuated considerably, but that it always kept 'Liberia' (the forest refugium of Liberia and the high forest areas of the adjacent territories of western Ivory Coast, southeastern Guinea and southern Sierra Leone) in greater isolation than the other two refuge areas ('Gaboon' and Upper Congo). The Dahomey Gap may also have existed for a longer time, thus preventing passage of oestrids from the main Congo forests westwards to the Liberian forests. (iii) **Suoidea** (Fig. 65). The Suoidea comprise two families: (a) Tayassuidae, known from the Lower Eocene to the Recent of North America, whence they passed to South America during the Pliocene, remaining there until the present; in Europe they have been reported from the Middle Oligocene and from the Middle to the Upper Miocene. The Tayassuidae are not parasitized. (b) Suidae, with 5 recent genera and approximately 9 species. They are divided into 6 subfamilies: 1. †Hyotheriinae, with 4 genera, which have existed from the Lower Oligocene to the Upper Pliocene, in Africa, Europe and Asia; 2. †Listriodontinae, with a single genus, from the Lower Miocene to the Upper Pliocene, also in Africa and Eurasia; 3. †Tetraconodontinae, with 2 genera, from the Lower Miocene to the Pleistocene; the group apparently is Palaearctic, being also known from northern Africa; 4. Suinae, with at least 9 genera, 4 of which recent; known since the Lower Miocene (of Africa); 5. †Kubanochoerinae, with only one genus from the middle Miocene of western Asia; 6. Phacochoerinae, with two genera, one from the Pleistocene, the other from the Pleistocene and Recent, both from Africa.

The five recent genera of Suidae belong to the subfamilies Suinae (*Sus, Babyrousa, Hylochoerus* and *Potamochoerus*) and Phacochoerinae (*Phacochoerus*).

Sus, with at least 5 species, is found in Eurasia, Japan, northern Africa and the Malay Archipelago. *Babyrousa*, with a single species, is native to Celebes, Togian Islands, Buru Island (north of the Moluccas) and Sula Island, living in humid forests and margins of rivers and lakes. These two genera from the Palaearctic and Oriental regions are not parasitized, indicating that oestrids parasitizing pigs are probably Ethiopian.

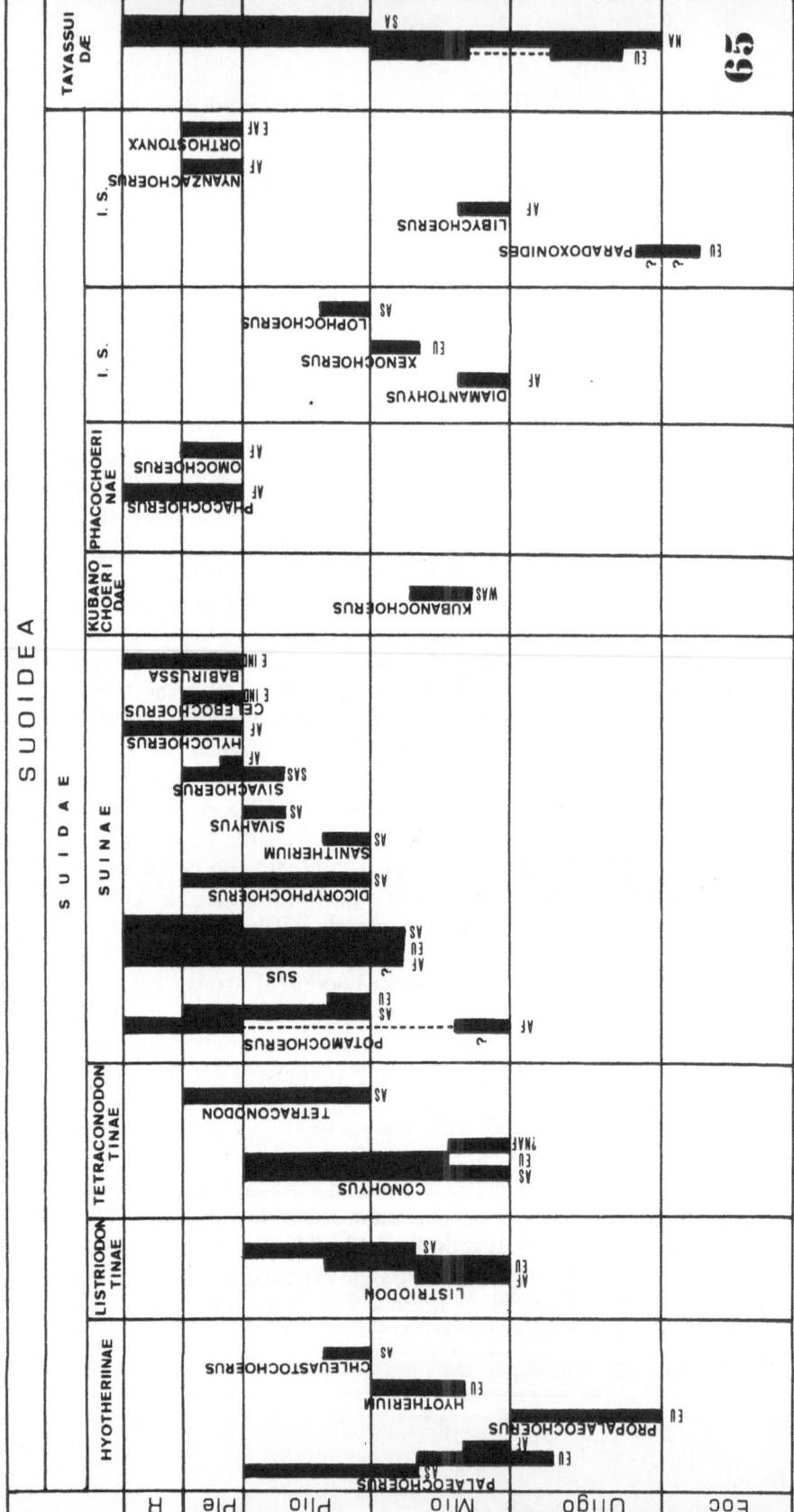

Fig. 65. Classification and time-range of the Suoidea (Artiodactyla, Suina).

The Ethiopian genera of Suidae are characterized thus: (i) *Hylochoerus* has a single species, *meinertzhageni*, an inhabitant of thick forests and bamboo forests of central equatorial Africa, living in groups of 4-20 individuals; these pigs feed on shrubs and grasses; RODHAIN & BEQUAERT (1916: 136) examined heads of these pigs (specimens from Ituri forest, Congo) without finding oestrid larvae; (ii) *Potamochoerus*, also with only one species, *porcus*, occurs in Africa, Madagascar and Impalela Islands; this species is found in a great variety of habitats; these pigs are especially active at night and during the day remain in 'tunnels' opened in the interior of tall grasses and similar places; they are gregarious, forming groups with 4-20 individuals, feeding on roots and fruits; occasionally they will eat eggs, reptiles and small birds; sometimes they cause damage to plantations. *Potamochoerus porcus* is parasitized by *Suinoestrus nivarleti* (Oestrinae, Rhinoestrini); it is to be noted that, in spite of the large distribution of the host, *Suinoestrus* seems to be restricted to the Congo forest area; (iii) *Phacochoerus* has only one species, *aethiopicus*, which occupies a large area in Africa, with greater concentrations in the east and south; they live usually in forests and savannas, being diurnal, generally solitary, feeding on grass, roots, fruits and the bark of trees, and occasionally on carrion; this species is parasitized by *Rhinoestrus phacochoeri*; similarly to the case of *Potamochoerus*, the parasite of *Phacochoerus* also seems to be limited to the Congo forest area, although its host is widely distributed. *R. phacochoeri* occurs in the Congo region, Cameroon and Katanga; RODHAIN & BEQUAERT (1916: 136) also found this species in a region of 'savanne boisée',

It can be assumed that oestrid parasites of pigs are restricted to the forest belt of Equatorial Africa. However, strangely enough, one of the three genera of pigs living in that area is not parasitized, although representatives of all three genera have a large size and sufficient weight (Table 9).

A possible explanation of this fact may reside in the different age of these three groups. *Potamochoerus* is known in Africa since the Miocene; *Hylochoerus* and *Phacochoerus* only from the Pleistocene and Recent;

Table 9. Ethiopian Suidae.

Genera	Body Length (mm)	Shoulder Height (mm)	Weight (kg)
Hylochoerus	1500-1900	762-965	160-275
*Phacochoerus**	1055	635-732	75-100
*Potamochoerus**	1300	585-965	75-130

however, *Hylochoerus* is apparently a representative of a group that came from Asia to Africa, whereas *Phacochoerus* belongs to an exclusively African subfamily, which must have been evolving there for a longer time than is shown by its fossil record (Fig. 65).

Therefore, a reasonable hypothesis is that, in addition to being a terrestrial herbivore with sufficient size and weight, to be a host of an oestrid a mammal must have been in the area of the fly for a certain period of time (which may be tentatively postulated, as a minimum, between the Miocene and the Upper Pliocene). This third requisite seems to be corroborated by the example of the Bovidae (see below).

3.4. Surborder Ruminantia

The Ruminantia are divided into two infraorders: Tylopoda and Pecora. PILGRIM (1941) admits that these two infraorders originated in the Northern Hemisphere – the Tylopoda probably in North America and the Pecora in Asia.

3.4.1. Infraorder Tylopoda (Fig. 66)

This includes 4 superfamilies: (i) †Cainotherioidea, with only one family, exclusively European, known from the Upper Eocene to the Middle Miocene; (ii) †Anoplotherioidea, with the families †Anoplotheriidae, †Xiphodontidae and †Amphimerycidae, also exclusively European, from the Middle Eocene to the Middle Oligocene; (iii) †Merycoidodontoidea, exclusively North American, from the Upper Eocene to the Upper Pliocene; known from Central America in the Middle Miocene; (iv) Cameloidea, with two families: †Oromerycidae, exclusively North American, from the Upper Eocene to the Lower Oligocene; and Camelidae, predominantly American, having invaded Eurasia, northern Africa, and also South America during the Pliocene and Pleistocene. The Camelidae are arranged in 5 subfamilies: (a) †Poebrotheriinae, exclusively North American, known from the Upper Eocene to the Middle Oligocene; (b) †Alticamelinae, also exclusively North American, from the Middle Oligocene to the Upper Pliocene; (c) †Stenomylinae, North American; from the Lower Miocene to the Lower Pliocene; (d) †Pseudolabinae, North American, Upper Oligocene; (e) Camelinae, the only group that has survived to the present, known from the Upper Oligocene to the Recent. The group was predominantly North American, whence it colonized Eurasia and northern Africa in the Pliocene (*Procamelus*) and Pleistocene (*Paracamelus* and *Camelus*). South America was also invaded from North America, and *Lama* and *Vicugna* still survive there.

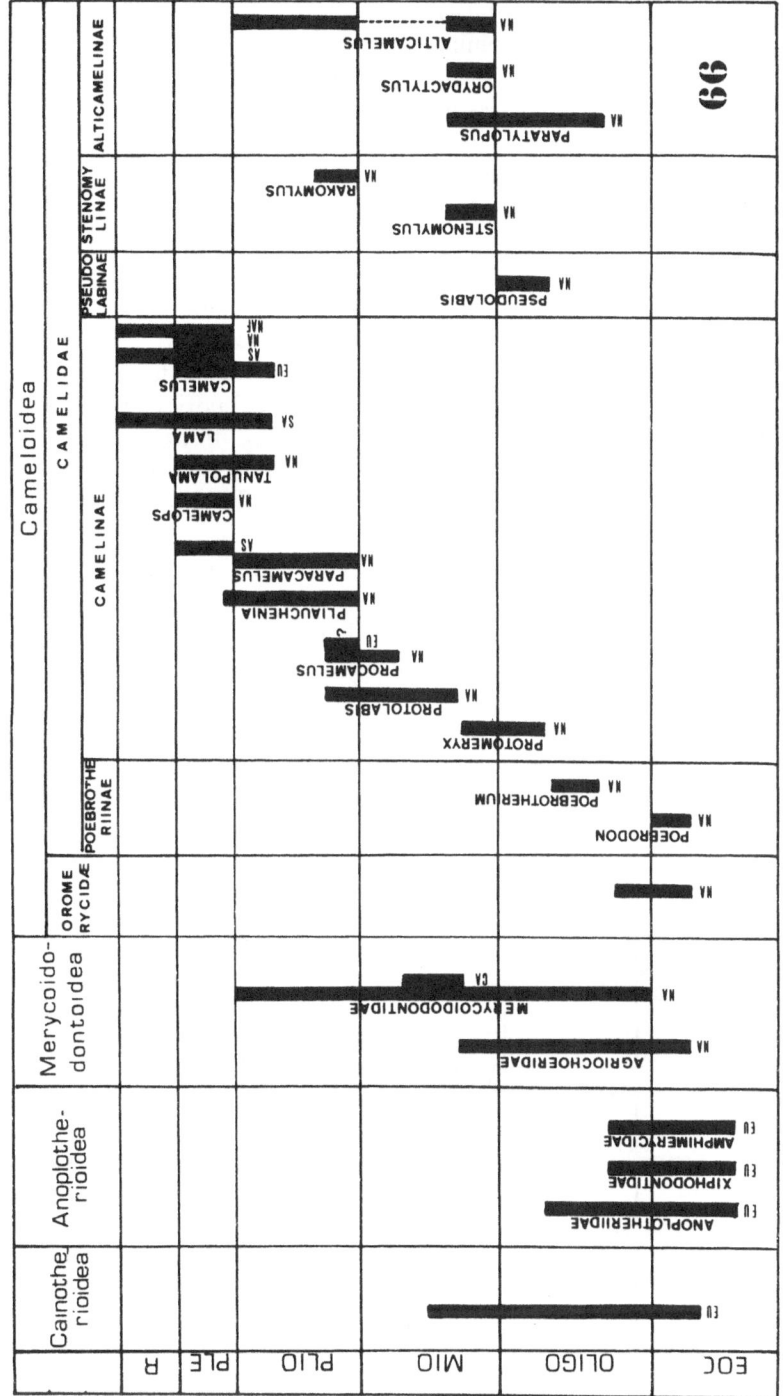

Fig. 66. Classification and time-range of the Tylopoda (Artiodactyla, Ruminantia).

Only the two species of *Camelus* are parasitized, by *Cephalopina titillator* (Oestrinae, Cephalopinini). *Camelus dromedarius* is known only in the domesticated state; *C. bactrianus* is known wild in the Gobi Desert and Mongolia. Both species have been introduced in several parts of the world. Their present area of distribution extends from northern Africa, through Central Asia to Mongolia; wild forms inhabit semi-arid or arid plains, grasslands and deserts.

3.4.2. Infraorder Pecora
This is the richest group of Artiodactyla. Certain authors consider the Tragulina and Pecora as separate infraorders. ROMER (1967) accepts only one suborder, with three superfamilies, Traguloidea, Cervoidea and Bovoidea. Most authors claim that the Pecora originated in Asia, colonizing the other continents from that area. PILGRIM (1941: 162) admits that 'Central Asia was the developmental centre of the Tragulina. Thence they spread over the whole of the Old World. The Pecora branched off from the Tragulina in the Oligocene. From their birth place in Central Asia most of the pecoran families sooner or later invaded other regions.'

A. Traguloidea (Fig. 67). With 4 families: (i) †Hypertragulidae, known from Asia (Upper Eocene to Middle Oligocene), Europe (Lower and Middle Oligocene), and North America (Upper Eocene to Lower Miocene); (ii) †Protoceratidae, exclusively North American, from the Lower Oligocene to the Lower Pliocene; (iii) †Gelocidae, known from Europe (Upper Eocene to Middle Oligocene), Asia

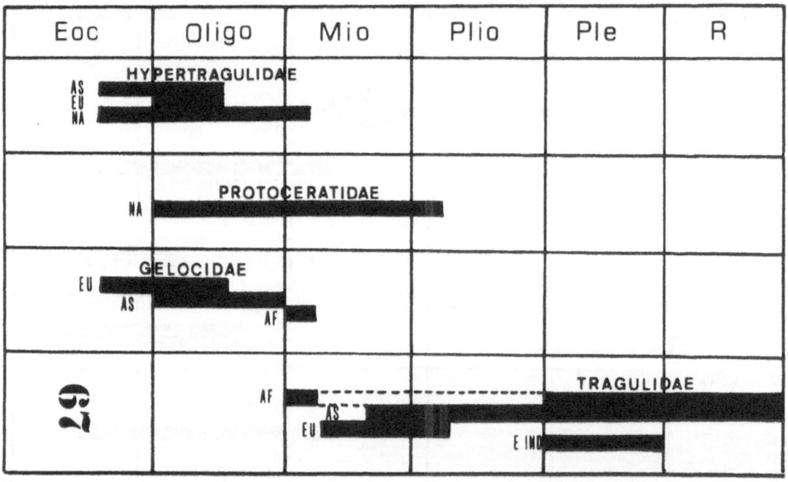

Fig. 67. Time-range of the families of Traguloidea (Artiodactyla, Ruminantia, Pecora).

180

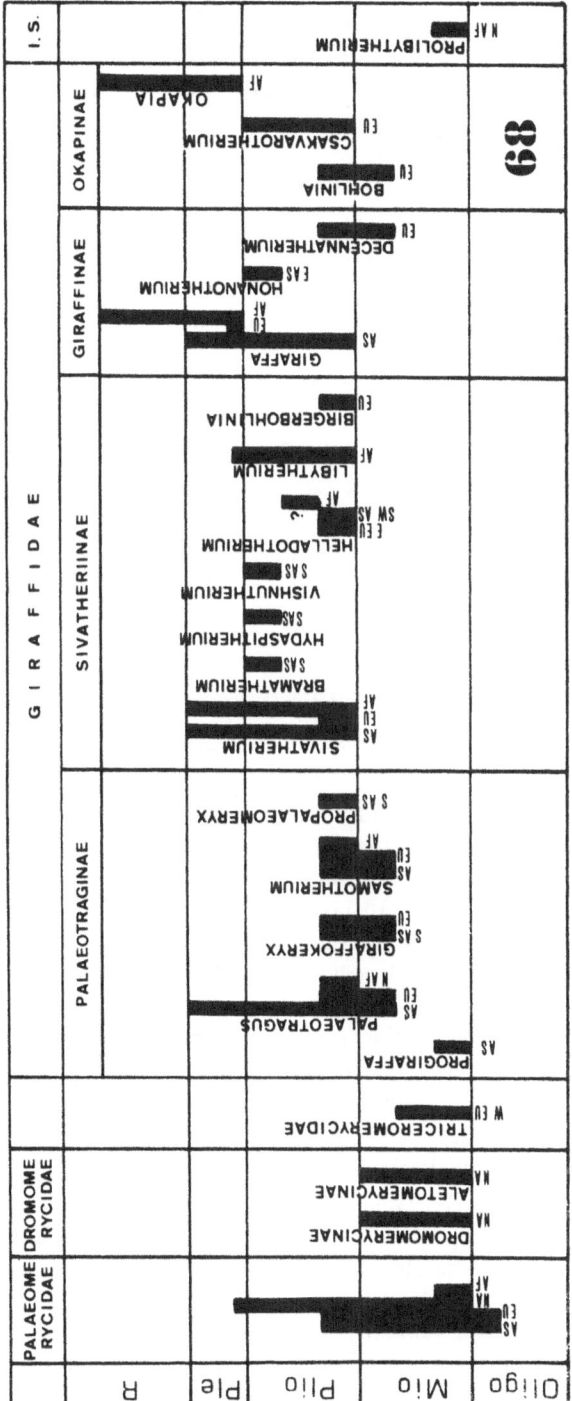

Fig. 68. Time-range of the familie Giraffidae (Artiodactyla, Ruminantia, Pecora, Giraffoidea).

(Oligocene) and Africa (Lower Miocene); (iv) Tragulidae, the only family with recent representatives, known from Asia (Upper Miocene to Recent), the East Indies (Pleistocene), Europe (Middle Miocene to Lower Pliocene), and Africa, where they apparently represent two different invasions (one during the Lower Miocene, the other during the Pleistocene and Recent). The Tragulidae have only two recent genera: *Hyemoschus* and *Tragulus*.

Hyemoschus has a single species, occurring in the dense forests of Gambia and Sierra Leone, and westwards to the Ituri forest on the eastern border of the Congo basin, and then to the south as far as Cameroon. *Tragulus* has about half a dozen species, occurring in India, through southeastern Asia to the Malay Archipelago, including Balabac. The Tragulidae are not parasitized, probably because of their small weight. *Tragulus*, in addition, is Oriental and therefore out of reach.

Table 10. Family Tragulidae.

Genera	Body Length (mm)	Shoulder Height (mm)	Weight (kg)
Hyemoschus	1000	305–355	–
Tragulus	455–560	204–330	2.5–4.5

B. Cervoidea. The Cervoidea include 3 families: (i) †Palaeomerycidae (including †Dromomerycidae and †Triceromerycidae) (Fig. 68). Known from Asia (Upper Oligocene to Lower Pliocene), North America (Lower Miocene to Lower Pleistocene) and Africa (Lower Miocene). (ii) Giraffidae (Fig. 68), with 4 subfamilies: (a) †Palaeotraginae, from the Lower Miocene to the Upper Pleistocene of Asia, Europe and Africa; (b) †Sivatheriinae, also from Asia, Europe and Africa, from the Lower Pliocene to the Upper Pleistocene; (c) Giraffinae, from the Upper Miocene to the Recent, from Asia, Europe and Africa; (d) Okapiinae, known from the Upper Miocene to the Recent, in Europe and Africa.

The palaeontological history of the group was discussed by CRUSAFONT PAIRÓ (1961: 1035–1037).

The two surviving genera, *Giraffa* and *Okapia*, are restricted to Africa. *Giraffa* has a single species, *camelopardalis*, distributed through Africa south of the Sahara; these are gregarious animals, living in herds of up to 70 individuals; they are browzers and cud-chewers, feeding especially on leaves of acacia, mimosas and wild apricots; the genus is known as fossil in Asia (Lower Pliocene to Upper Pleisto-

Table 11. Family Giraffidae.

Genera	Body Length (mm)	Shoulder Height (mm)	Weight (kg)
*Giraffa**	4000	2500-3700	550-1800
*Okapia**	2000	2520	210

cene), in Europe (Lower Pleistocene) and in Africa (since the Lower Pleistocene). *Okapia* is also represented by a single species, *johnstoni*, restricted to the rain forests of the Congo; these diurnal animals prefer dense damp forest, are generally solitary but also are found in pairs, though never in groups; they feed on leaves, fruits and seeds of various plants.

Unfortunately, very little is known about the oestrid parasites of giraffids. ZUMPT (1965: 169) described as *Rhinoestrus giraffae* third stage larvae found in the giraffe; their adult is still unknown. The okapi also is parasitized; RODHAIN (1926) identified the oestrid as *Oestrus ovis*; ZUMPT (1965: 176) believes they represent a new species. (iii) Cervidae. The Cervidae (Fig. 69) are divided into 4 subfamilies, all of them with recent representatives.

(a) Moschinae, known from the Lower Pliocene to the Recent. Only one recent genus exists, *Moschus* (*moschiferus*), found in northeastern and central Asia, from China, Manchuria, Korea and Sakalin Islands to western Mongolia. They live in forests, in elevations of 2600 to 3600 m, being generally solitary and very seldom two or three individuals are found together. They feed on grass, musci and tender shoots; during the winter they are compelled to eat branches and lichens. They are diurnal.

(b) Muntiacinae, known in Europe from the Lower Miocene to the Lower Pleistocene, in Asia from the Upper Miocene to the Recent, and in the East Indies in the Pleistocene. The Muntiacinae are today represented by two genera: *Muntiacus*, with about half a dozen species occurring in Nepal, India, Ceylon, Sumatra, Java, Borneo and adjacent islands, to south China and Taiwan; they live in areas with dense vegetation and on mountainous chains, from sea level to medium altitudes; and *Elaphodus*, with a single species, found in South China and northern Burma at altitudes of 900 to 2600 m; they always live near water.

(c) Cervinae, known from the Lower Pliocene to the Recent of Europe, Asia, North America and northern Africa. Four genera are known at the present time: 1. *Dama*, known in Asia since the Upper Pliocene to the Recent, in Europe from the Lower Pleistocene to the

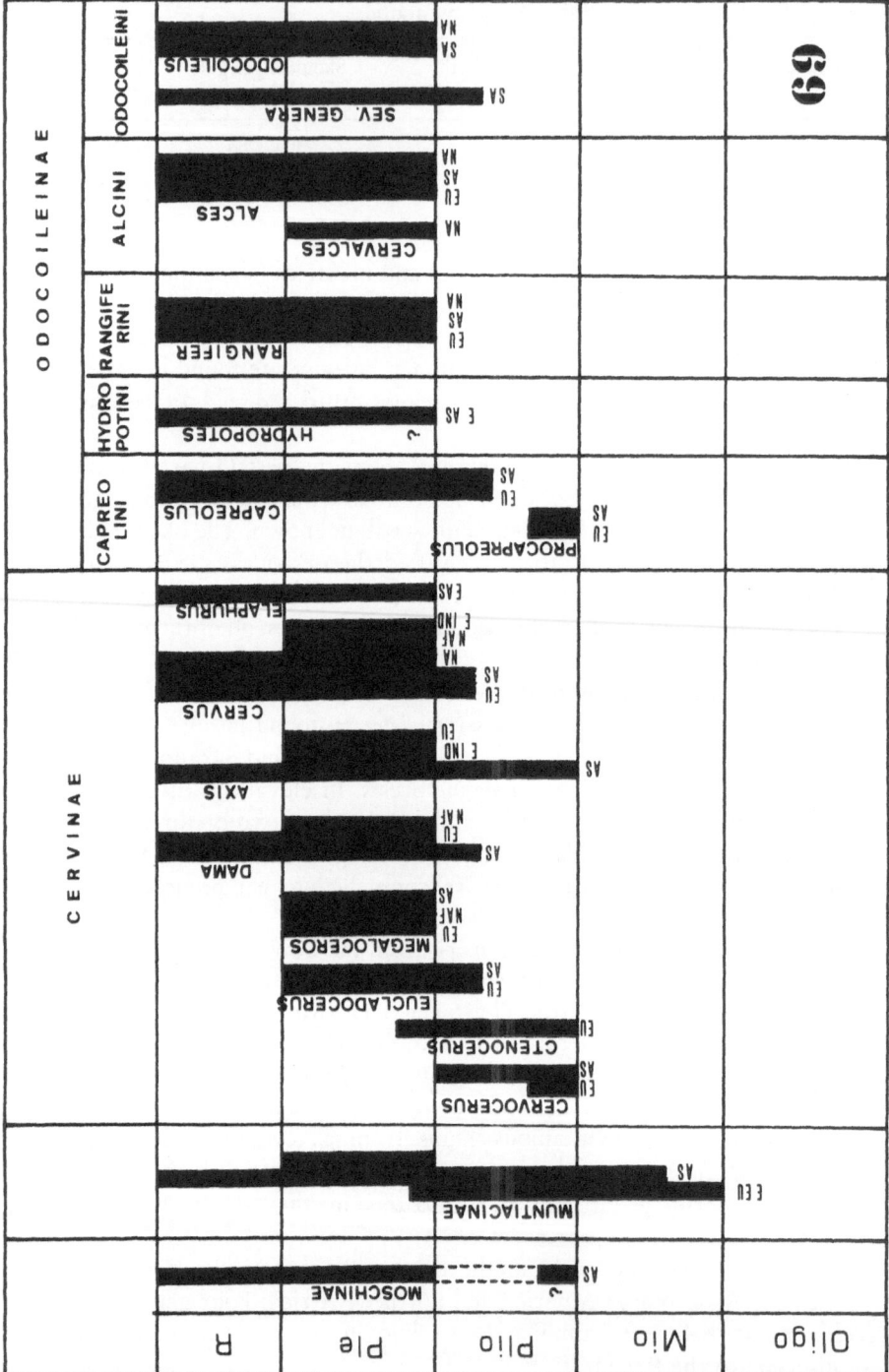

Fig. 69. Classification and time-range of the Cervidae (Artiodactyla, Ruminantia, Pecora, Cervoidea).

Recent, and in northern Africa during the Pleistocene. Two recent species are known, originally distributed through the Mediterranean region of southern Europe to Asia Minor; *Dama dama* was largely introduced and is now found feral all over western Europe, western Ukraine, Baltic countries, Great Britain; *Dama mesopotamica*, native from Persia and Iraq, is probably extinct, though a few individuals still exist in captivity; 2. *Axis*, known in Asia since the Lower Pliocene to the Recent, and in Europe and the East Indies in the Pleistocene; it now occurs in India and Ceylon, Calamian Islands and the Philippines, in Thailand and Indochina; 3. *Cervus*, known in Asia and Europe from the Upper Pliocene to the Recent, and in North America from the Lower Pleistocene to the Recent; it is also known from northern Africa and the East Indies from the Pleistocene. It now contains some 15 species, occurring in Europe, Asia, the Oriental Region, the Philippines and North America. The genus was divided into about 8 subgenera, of which two (*Cervus* and *Sika*) have Palaearctic and Nearctic species, and the other six (*Rusa, Rucervus, Thaocervus, Panolia, Przewalskium*, and *Hyelaphus*) occur exclusively in the Oriental Region (TATE, 1947: 337-346); 4. *Elaphurus*, known only from Asia, from the Lower Pleistocene to the Recent; the only species, *davidianus*, 'is reported to be now little more than a memory in its native land of China. It was obtained originally from the gardens of the Summer Palace at Pekin, by the French missionary naturalist, ARMAND DAVID. It has not been known wild within the recorded memory of man' (TATE, 1947). Presently it is known from some 400 individuals in zoos, descendants from the original herd preserved at the Imperial Park of Hunting in Peking (WALKER *et al.*, 1964).

(d) Odocoileinae, divided into 5 tribes: 1. Odocoileini, from the Upper Pliocene to the Recent of North America. The recent representatives belong to the genera *Blastocerus, Blastoceros, Hippocamelus, Mazama* and *Pudu*, in South America; *Odocoileus* is common to both Americas; 2. Alcini, known from Eurasia and North America from the Pleistocene to the Recent; now only with one genus, *Alces*, which occurs in Alaska, Canada, northwestern United States (along Rocky Mountains), Norway and Sweden, eastwards across Russia and Siberia to Manchuria and Mongolia; 3. Rangiferini, also from the Pleistocene and Recent of Eurasia and North America, with one genus, *Rangifer*, from the arctic regions; 4. Hydropotini, from the Pleistocene and Recent of Eastern Asia, with the single genus *Hydropotes*, found in China (islands and banks of the Yang Tse River) and Korea; 5. Capreolini, known in Eurasia since the Lower Pliocene to the Recent, with the single genus *Capreolus*, largely distributed in Eurasia (the extreme north and the Oriental Region excepted).

Table 12. Family Cervidae (Moschinae, Muntiacinae, Cervinae).

Subfamily	Genera per Regions			
	Palaearctic	Nearctic	Oriental	Neotropical
Moschinae	*Moschus*			
Muntiacinae			*Muntiacus* *Elaphodus*	
Cervinae	*Dama** *Cervus** *Elaphurus*	*Cervus**	*Cervus*	

Cervidae are parasitized by a specialized subfamily, the Cephene-myiinae, restricted to the Holarctic region. Examination of Tables 12-13 shows that cervids from the Oriental and Neotropical regions are exempt from the parasite, probably because cephenemyiines do not cross the ecological barriers represented by tropical forests, which separate these two regions from the Palaearctic and Nearctic regions. But not all the genera and species of Palaearctic Cervidae are parasitized. The characteristics of these groups seem to provide an explanation for this fact: (i) *Moschus* – this animal is below the limit of weight required from a host of oestrid (Table 14); (ii) *Dama* – only *D. dama* is reported as host; the other species (*mesopotamica*) is perhaps extinct;

Table 13. Family Cervidae (Odocoileinae)

Subfamily	Tribe	Genera per Regions			
		Palaearctic	Nearctic	Neotropical	Oriental
	Odocoileini		*Odocoileus**	*Odocoileus* *Blastocerus* *Blastoceros* *Hippocamelus* *Mazama* *Pudu*	
Odocoileinae	Alcini	*Alces**	*Alces**		
	Rangiferini	*Rangifer**	*Rangifer**		
	Hydropotini	*Hydropotes*			
	Capreolini	*Capreolus**			

Table 14. Family Cervidae.

Subfamily	Tribe	Genera	Body Length (mm)	Shoulder Height (mm)	Weight (kg)
Moschinae		*Moschus*	1000	510-610	9-11
Cervinae		*Dama**	1300-1600	1000	40-80
		Cervus*	1650-2500	1200-1500	100-260
		Elaphurus	–	1443	–
Odocoileinae	Odocoileini	*Odocoileus**	1500-2100	800-1100	48-145
	Alcini	*Alces**	2500-3000	1400-1900	up to 825
	Rangiferini	*Rangifer**	1200-2200	1100-1400	up to 318
	Hydropotini	*Hydropotes*	775-1000	450-550	9-11
	Capreolini	*Capreolus**	950-1350	650-775	15-30

(iii) *Cervus* – Only the subgenera *Cervus* and *Sika* have species reported as parasitized; in the first subgenus, only the species *elaphus* (Palaearctic) and *canadensis* (Nearctic) are parasitized; *macneilli* and *wallichi*, occurring in the Himalayas and Cashmere (TATE, 1947: 344) are Oriental and therefore out of reach; the subgenus *Sika*, in its turn, has one species, *Cervus (Sika) nippon*, with the following subspecies: *kopschii* (South China Sika), *taioanus* (Taiwan Sika), *mandarinus* (North China Sika), *hortulorum* (Ussuri Sika) and *nippon* (Japanese Sika); only the subspecies *hortulorum* is parasitized; the explanation of this fact may reside in the different size of these subspecies – TATE (1947: 341) declares that *hortulorum* 'is the largest of the Sika deer. It reaches 43 inches at the shoulder.' The other subspecies are very probably below the minimum weight required from an oestrid host. (iv) *Elaphurus*, as already seen, is known only from zoos, and of course is not listed as a host; (v) *Odocoileus* – only Nearctic species are parasitized; the Sonoran Desert and the tropical forests of southern Mexico seem to be impassable barriers for the Cephenemyiinae, preventing them entering South America; (vi) *Alces, Rangifer* and *Capreolus* are always parasitized; Table 14 shows that all of them have sufficient weight; *Capreolus* shows that probably the minimum weight required is around 20 kg; (vii) *Hydropotes* is not parasitized, as it is below the minimum limit of weight.

C. Bovoidea. This superfamily comprises two families: Antilocapridae, exclusively North American, known from the Middle Miocene to the Recent (Fig. 70) and Bovidae, a rich family which has occupied the Old World and North America, without however reaching South America. Only some representatives of Bovidae are parasitized.

Table 15. Bovidae Classification.

According to SIMPSON (1945)		According to ALLEN (1939)
Bovinae	Strepsicerotini Bovini Boselaphini	Tragelaphinae Bovinae –
Cephalopinae		Cephalopinae
Hippotraginae	Reduncini Hippotragini Alcelaphini	Reduncinae Oryginae Alcelaphinae
Antilopinae	Neotragini	Oreotraginae Neotraginae Madoquinae
	Antilopini	Aepycerotinae Antilopinae

The classification of the Bovidae is still the subject of disputes. ALLEN (1939) considers as subfamilies most of the tribes of SIMPSON (1945) (Table 15).

The classification in subfamilies and tribes adopted here for convenience follows VIRET (1961). According to that author, the Bovidae are divided into 5 subfamilies: (i) Bovinae. With 3 tribes: (a) Strepsicerotini (= Tragelaphini), known from the Upper Miocene to the Recent of Eurasia and Africa, with 3 surviving genera, now found in Africa; (b) Boselaphini, from the Lower Miocene to the Recent, known in Eurasia (especially in Asia) and in Africa, with 2 recent genera, both in the Oriental Region; (c) Bovini, from the Lower Pliocene to the Recent, known from Eurasia, Africa and North America, with 6 recent genera, distributed in Eurasia, North America and Africa (Fig. 70). (ii) Cephalopinae (Fig. 71), an exclusively African group, known from the Pleistocene and Recent; (iii) Hippotraginae (Fig. 72), with 4 tribes: (a) Reduncini, from the Lower Pliocene to the Recent, predominantly Asian; in Africa 6 genera survive; (b) Hippotragini, from the Lower Pliocene to the Recent, from Eurasia and Africa, with 3 living genera restricted to Africa; (c) Alcelaphini, known from the Upper Pliocene to the Recent, in Europe, southern and southwestern Asia and Africa; the 5 recent genera are African; (d) †Menelikiini, exclusively African, only known from the Lower Pleistocene; (iv) Antilopinae (Fig. 71), with 2 tribes: (a) Neotragini, including the

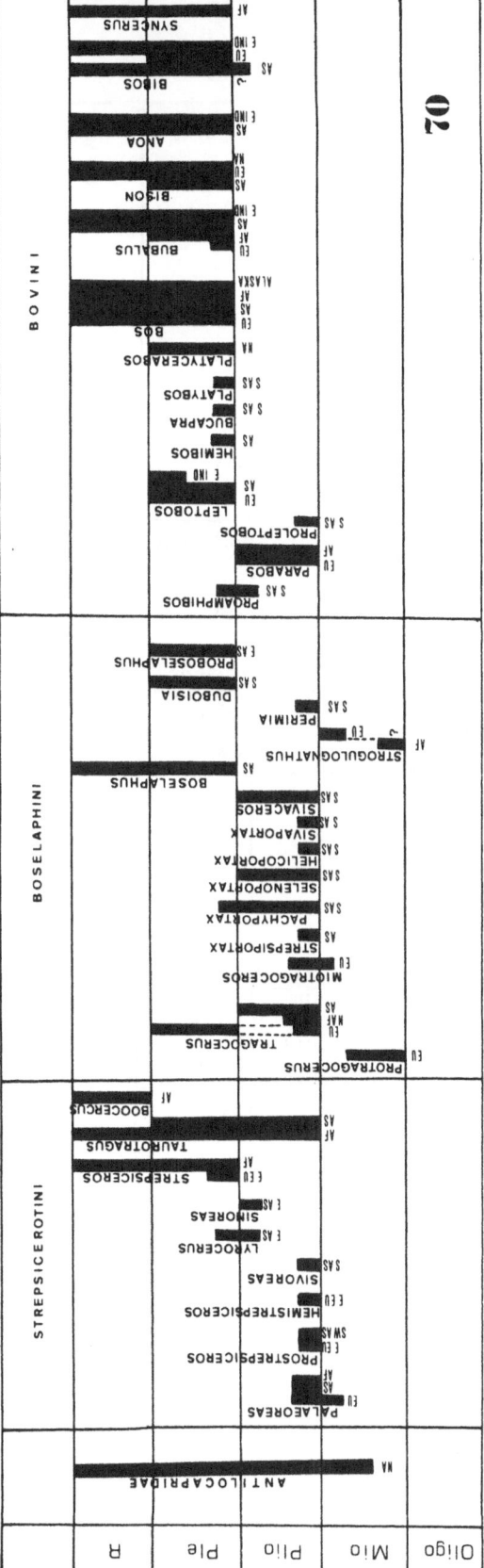

Fig. 70. Time-range of the Antilocapridae, and classification and time-range of the Bovinae (Bovidae; Artiodactyla, Ruminantia, Pecora, Bovoidea).

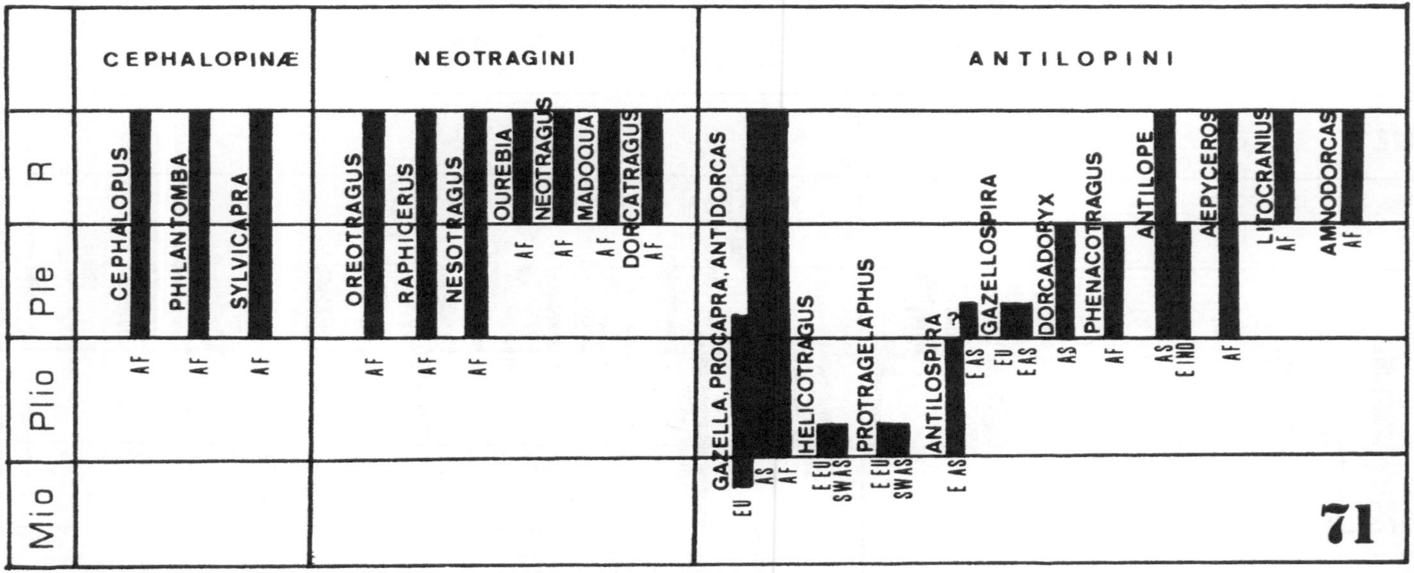

Fig. 71. Classification and time-range of the Cephalopinae and Antilopinae (Bovidae; Artiodactyla, Ruminantia, Pecora, Bovoidea).

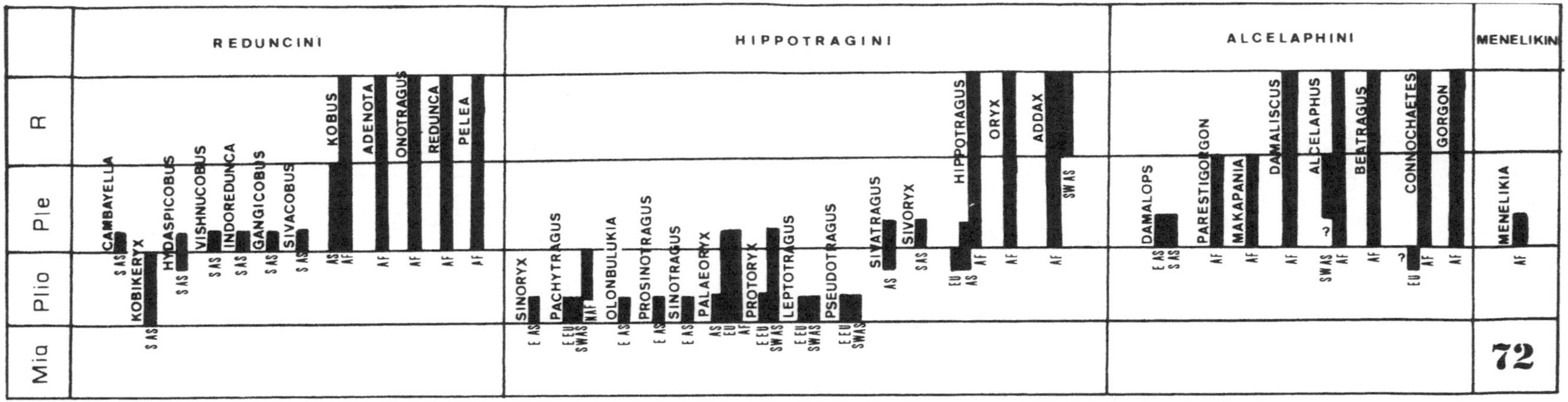

Fig. 72. Classification and time-range of the Hippotraginae (Bovidae; Artiodactyla, Ruminantia, Pecora, Bovoidea).

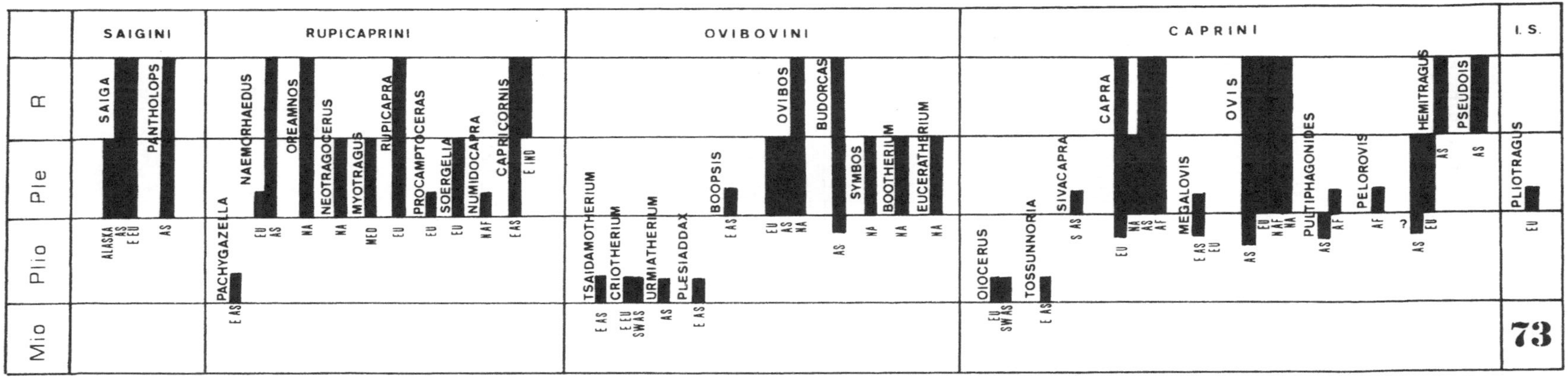

Fig. 73. Classification and time-range of the Caprinae (Bovidae, Artiodactyla, Ruminantia, Pecora, Bovoidea).

Table 16. Family Bovidae, Subfamily Bovinae.

Tribes	Genera per Regions			
	Ethiopian	Palaearctic	Nearctic	Oriental
Strepsicerotini	*Tragelaphus* *Boocercus* *Taurotragus*			
Boselaphini				*Boselaphus* *Tetracerus*
Bovini	*(Bos)** *(Bubalus)**	*Bos* *Bubalus* *Bison*	*Bos* *Bison*	*Bos* *Bubalus* *Anoa* *Bibos*
	Syncerus			

Table 17. Family Bovidae, Subfamily Cephalopinae.

Tribe	Genera per Regions			
	Ethiopian	Palaearctic	Nearctic	Oriental
Cephalopini	*(Cephalopus)*[1] *Sylvicapra*			

1. Including *Philantomba* and *Cephalopella*.

smallest antelopes known, the 'pocket' antelopes as called by VIRET (1961), known only from Africa, from the Pleistocene to the Recent, with 7 surviving genera; (b) Antilopini, known from the Upper Miocene to the Recent, with 7 surviving genera, distributed in Europe, Asia and Africa; (v) Caprinae (Fig. 73), with 4 tribes: (a) Saigini, from the Pleistocene and Recent, known from Alaska, Asia and Europe, with 2 surviving genera in the Palaearctic region; (b) Rupicaprini, from the Lower Pliocene to the Recent of Eurasia and North America and the Mediterranean region (including North Africa), with 4 recent genera, in the Holarctic and Oriental regions; (c) Ovibovini, from the Lower Pliocene to the Recent of Asia, Europe and North America, with 2 recent genera, in the Holarctic region; and (d) Caprini, from the Lower Pliocene to the Recent, known from Eurasia, Africa and North America, with 5 recent genera, in the Holarctic and Oriental regions.

Table 18. Family Bovidae, Subfamily Hippotraginae.

Tribes	Genera per Regions			
	Ethiopian	Palaearctic	Nearctic	Oriental
Reduncini	(Kobus)*[1] Redunca Pelea			
Hippotragini	Hippotragus* (Oryx)* Addax			
Alcelaphini	Damaliscus*[2] Alcelaphus* Connochaetes*[3]			

1. Including *Adenota* and *Onotragus*.
2. Including *Beatragus*.
3. Including *Gorgon*.

Table 19. Family Bovidae, Subfamily Antilopinae.

Tribes	Genera per regions			
	Ethiopian	Palaearctic	Oriental	Nearctic
Neotragini	Oreotragus Raphicerus Nesotragus Ourebia Neotragus Madoqua Dorcatragus			
Antilopini	Antidorcas* Aepyceros Litocranius Amnodorcas	Gazella Procapra*	Gazella Antilope	

Morphological (patterns of cerebral sulci) and quantitative (ence-phalization, neocorticalization) studies of the brain of several African antelopes by HAARMANN & OBOUSSIER (1972) and OBOUSSIER (1972, 1974; see bibliography in 1974) seem to indicate that ALLEN's classification of the Bovidae is closer to reality than that of SIMPSON.

Table 20. Family Bovidae, Subfamily Caprinae.

Tribes	Genera per Regions			
	Ethiopian	Palaearctic	Oriental	Nearctic
Saigini		Saiga Pantholops		
Rupicaprini		Nemorhaedus Capricornis Rupicapra	Nemorhaedus Capricornis	Oreamnos
Ovibovini			Budorcas	Ovibos
Caprini	(Capra) (Ovis)	Capra* Ovis* Pseudois Ammotragus	Hemitragus	Capra Ovis Pseudois

The Bovidae are parasitized by several groups of Oestrinae, and one genus of Bovidae (*Procapra*) by Cephenemyiinae. Examination of Tables 16-20 shows many interesting facts. (i) No bovid in the Oriental and Nearctic regions is parasitized; (ii) In the Palaearctic region only 3 (out of 15) genera are parasitized; two (*Capra* and *Ovis*) by Oestrinae (Oestrini and Rhinoestrini); the third (*Procapra*) by Cephenemyiinae; (iii) In the Ethiopian region the situation is paradoxical: basically we have only one group (Alcelaphini) heavily infested by several groups of Oestrinae (Kirkioestrini, Rhinoestrini, Oestrini and Gedoelstiini); then come the genera *Hippotragus* and *Antidorcas*, which are parasitized by several oestrines; all the other genera are either free from oestrids or only accidentally infested (*Kobus, Oryx, Cephalopus*, etc.), and in this case the larvae do not complete development.

The explanation of this limited parasitism of bovids in Africa is apparently very simple. It seems that the Oestrinae are of African origin, and that they invade the Palaearctic region in a limited measure. Therefore, there should be a higher level of infestation of Ethiopian groups of bovids than of late-coming invaders from the Palaearctic region.

According to WELLS (1957), there are two 'strata' of African antelopes: (a) an 'African' stratum, more ancient, with forms which must have evolved in Africa, including the Cephalopinae, Neotragini, and

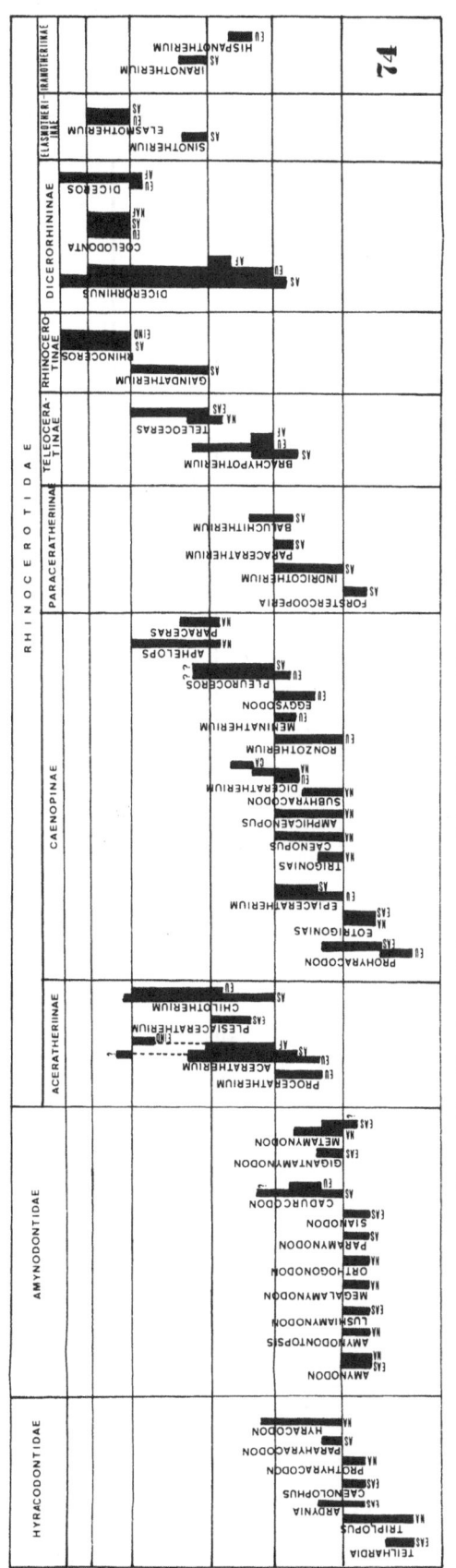

Fig. 74. Classification and time-range of the Rhinocerotoidea (Perissodactyla, Ceratomorpha).

195

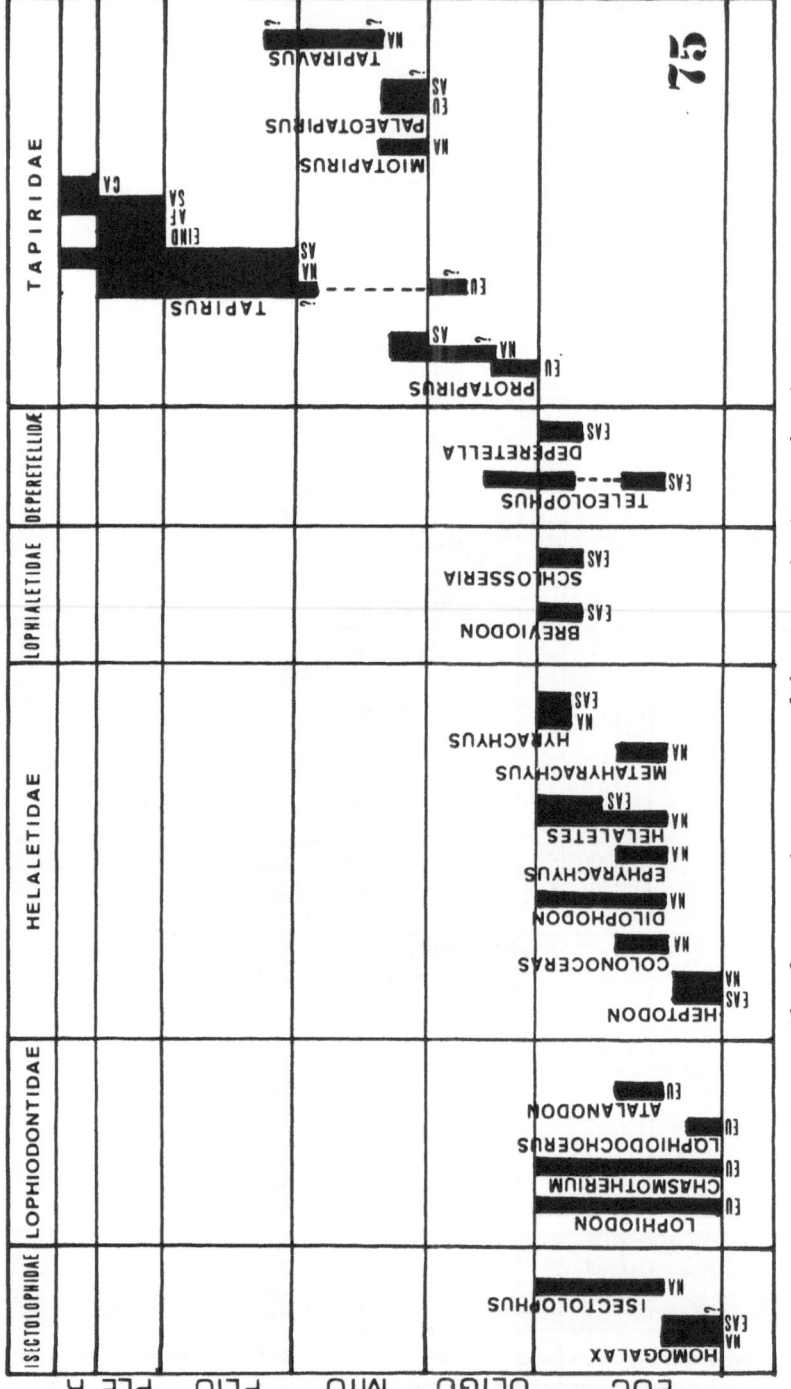

Fig. 75. Classification and time-range of the Tapiroidea (Perissodactyla, Ceratomorpha).

Alcelaphini; (b) a more recent stratum, 'Afro-Eurasiatic', with Antilopini, Strepsicerotini (Tragelaphini), Hippotragini and Reduncini.

From Tables 21-25 it can be immediately seen that the Cephalopinae and Neotragini are ruled out as hosts, as they are below the minimum limit of weight. It follows that African oestrids should be restricted to Alcelaphini and indeed this group is heavily parasitized.

On the other hand, Asiatic invaders of Africa would be parasitized only if they had been available to the flies for a long period of time. This is probably the case with *Antidorcas* and *Hippotragus*, although the fossil record does not corroborate this point. *Capra* and *Ovis*, being extremely common animals, had a strong probability of becoming infested. Further discussions about the Bovidae will be presented in Part D.

4. Perissodactyla

The Perissodactyla are arranged in 3 suborders: Ceratomorpha, Ancylopoda and Hippomorpha.

4.1. Suborder Ceratomorpha (Figs. 74-75)

This includes two superfamilies: (a) Rhinocerotoidea, with 3 families: †Hyracodontidae, from the Middle Eocene to the Lower Miocene of Asia and North America, †Amynodontidae, from the Upper Eocene to the Lower Miocene of Asia and North America, and Rhinocerotidae (Middle Eocene to Recent), known from Eurasia, North America and Africa with 4 surviving genera, two in the Oriental and 2 in the Ethiopian region. The Ethiopian genera are not parasitized, possibly because they not only reached Africa too recently, but also because they are not gregarious and thus not easily available to oestrids; the

Table 21. Family Bovidae, Subfamily Bovinae.

Genera	Body Length (mm)	Shoulder Height (mm)	Weight (kg)
Tragelaphus	1350	660-1092	112-270
Boocercus	–	1400	155-220
Taurotragus	1800-3400	1000-1800	900
Bubalus	2500-3000	1500-1800	726-816
Bos	–	900-1100	450-900
Syncerus	2100-3000	1100-1500	600-900

Table 22. Family Bovidae, Subfamily Cephalopinae.

Genera	Body Length (mm)	Shoulder Height (mm)	Weight (kg)
Cephalopus[1]	550–900	357–457	5–6.5
Sylvicapra	900–1100	570–670	14–17

1. Including Philantomba and Cephalopella.

Table 23. Family Bovidae, Subfamily Hippotraginae.

Genera	Body Length (mm)	Shoulder Height (mm)	Weight (kg)
Kobus[1]	1400–2100	760–1340	up to 272
Redunca	1100–1400	685–960	23–91
Pelea	1100–1200	710–787	20–23
Hippotragus*	1900–2100	1397–1650	204–280
Oryx	–	1000–2200	up to 210
Addax	2000	1000–1100	up to 120
Damaliscus*[2]	1300–1800	880–1200	114–136
Alcelaphus*	1500–2000	1200–1500	160–180
Connochaetes*[3]	1500–2000	1000–1300	230–275

1. Including Adenota and Onotragus.
2. Including Beatragus.
3. Including Gorgon.

Table 24. Family Bovidae, Subfamily Antilopinae.

Genera	Body Length (mm)	Shoulder Height (mm)	Weight (kg)
Oreotragus	770–920	450–600	11–16
Ourebia	1000	500–700	14–21
Raphicerus	700–850	500	7–14
Nesotragus	580–620	300	8–9
Neotragus	500	250–355	–
Madoqua	520–670	305–405	3–5
Dorcatragus	762–865	500–760	9–11
Aepyceros	1100–1500	775–1000	65–75
Amnodorcas	1170	762–890	27–34
Litocranius	1500	900–1000	43–50
Gazella	1000–1200	510–890	14–75
Antidorcas*	1200–1400	730–870	32–36
Procapra*	1000–1300	760	around 20

Table 25. Family Bovidae, Subfamily Caprinae.

Genera	Body Length (mm)	Shoulder Height (mm)	Weight (kg)
Saiga	1200-1700	750-800	36-69
Pantholops	1300-1400	790-810	40-50
Namorhaedus	900-1300	584-711	22.5-32
Capricornis	1400-1500	850-900	100-140
Rupicapra	900-1300	760-810	24-50
*Capra**	1200-1400	600-850	50-120
Pseudois	1300-1400	700-890	55-73
Ammotragus	1300-1900	915-1000	50-115
*Ovis**	1200-1800	350-480	75-200

Oriental forms are out of reach; (b) Tapiroidea, with 6 families (Fig. 75) known from the Lower Eocene to the Recent. Only 2 species in the same genus (*Tapirus*) survive, one in the Neotropical, the other in the Oriental region, and both are automatically exempt from oestrids.

4.2. Suborder Ancylopoda (Fig. 76)

With two families, known from the Lower Eocene to the Pleistocene, in Eurasia and North America.

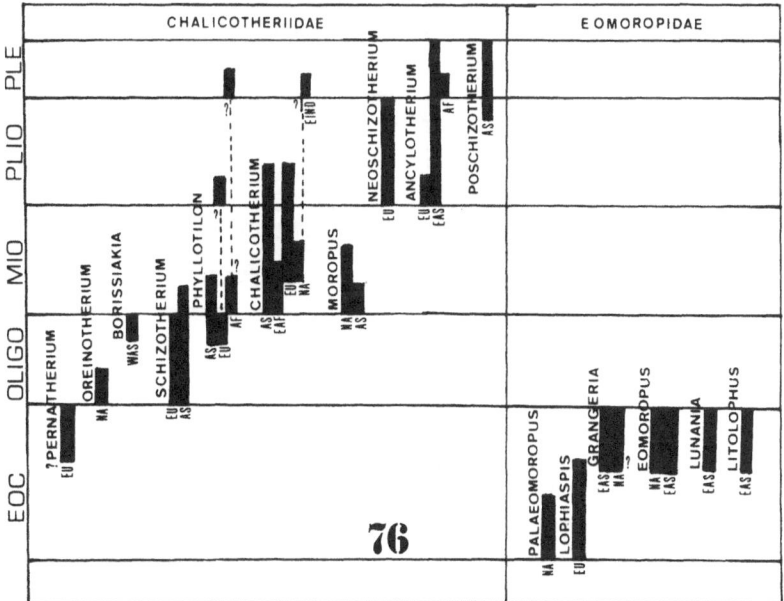

Fig. 76. Classification and time-range of the Ancylopoda (Perissodactyla).

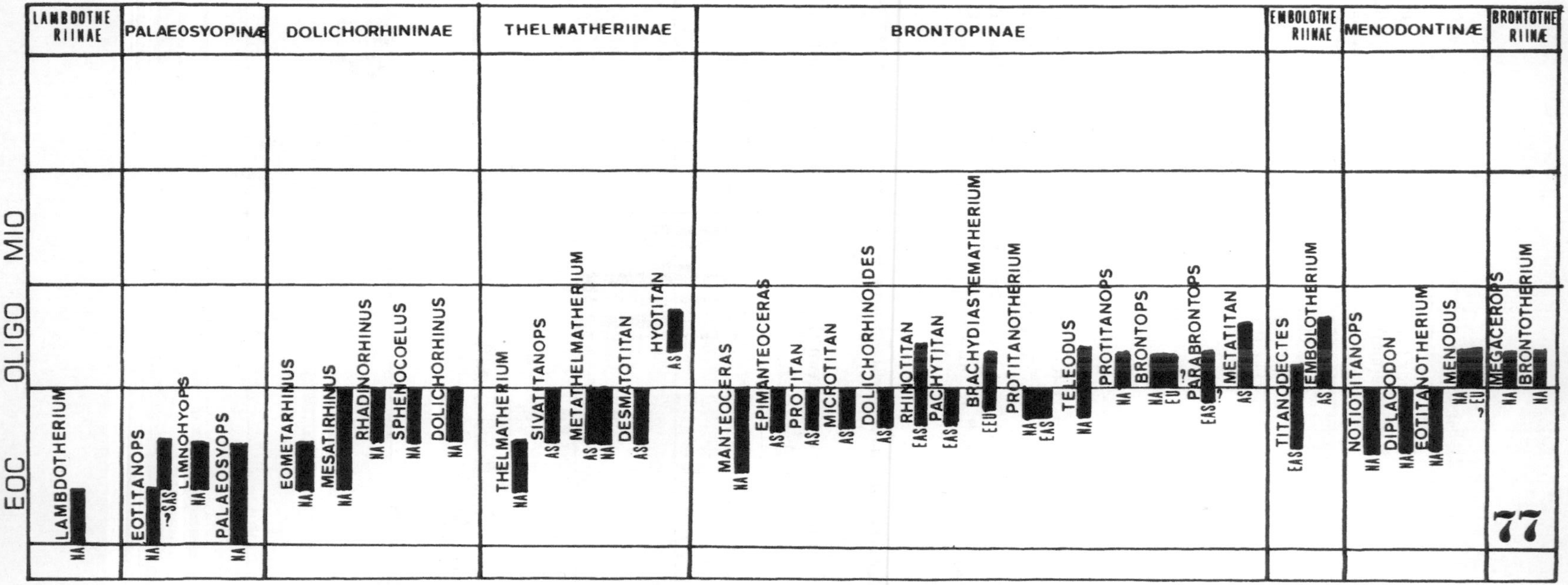

Fig. 77. Classification and time-range of the Brontotherioidea (Perissodactyla, Hippomorpha).

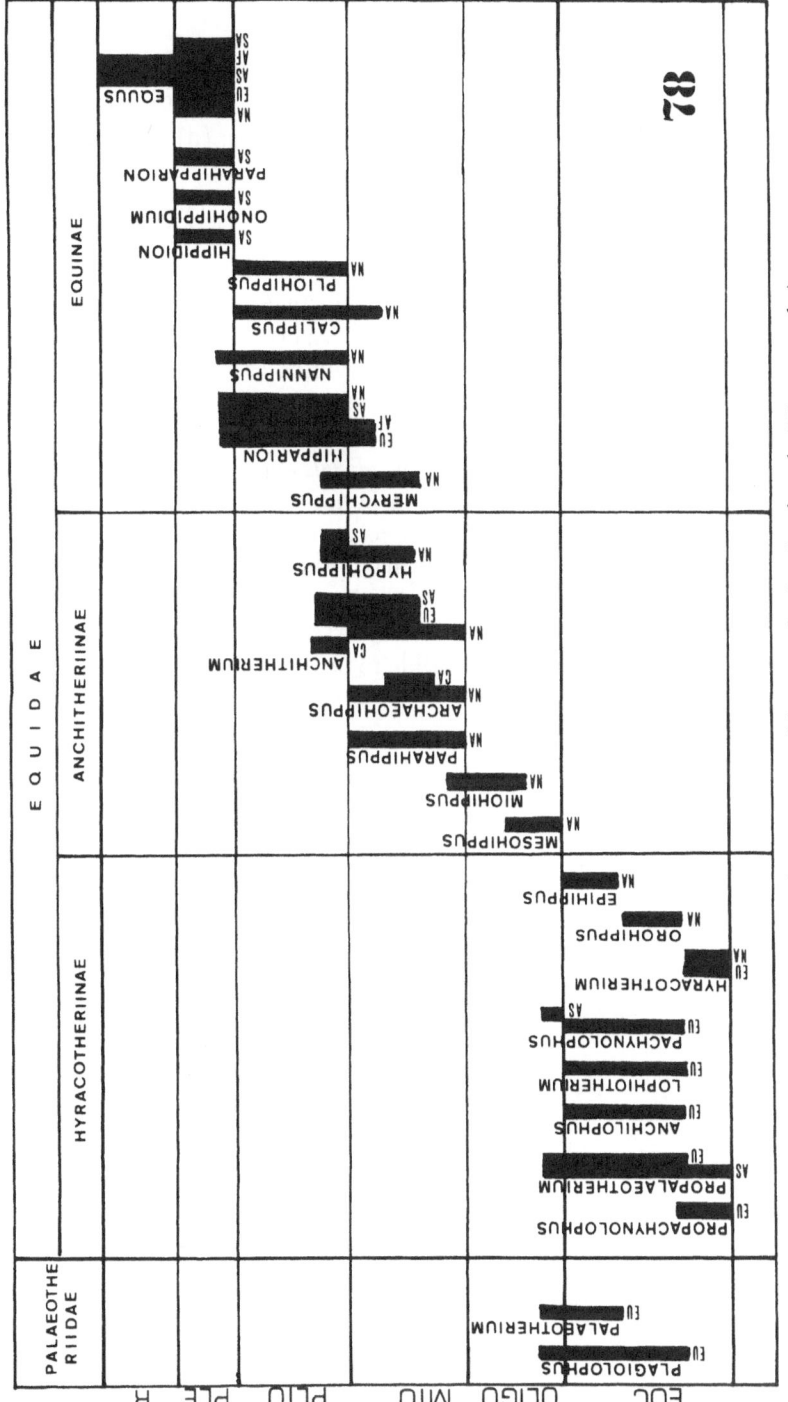

Fig. 78. Classification and time-range of the Equoidea (Perissodactyla, Hippomorpha).

With two superfamilies: (i) †Brontotherioidea (Fig. 77), with only one family which flourished during the Eocene and Lower and Middle Oligocene, in North America, Asia and Europe; (ii) Equoidea (Fig. 78), with two families: †Palaeotheriidae, represented only by two European genera from the Middle Eocene to the Lower Oligocene; and Equidae, known from Europe, Asia and North America since the Lower Eocene (†Hyracotheriinae). From the Oligocene to the Middle Miocene their history is confined to North America; two genera of †Anchitheriinae appear in Europe and Asia from the Middle Miocene to the Lower Pliocene; the Equinae, which are known from North America since the Middle Miocene appear in Europe, Africa and Asia in the Upper Miocene, represented by †*Hipparion*, which became extinct during the Lower Pleistocene; *Equus* appears in North America and the Old World, as well as in South America, since the Lower Pleistocene and becomes extinct, except in Asia and Africa.

Equus is parasitized by Oestrinae (Rhinoestrini). Probably these rhinoestrines became first adapted to *Hipparion* (in Africa since at least the Upper Miocene), and later transferred to *Equus* which is gregarious and therefore more susceptible to being parasitized.

III. REFERENCES

ALLEN, G. M., (1939): A checklist of African mammals. *Bull. Mus. comp. Zool. Harv.* 83: 1-763.

ANDERSON, S. & J. K. JONES, Jr., (1967): Recent mammals of the world. A synopsis of families, viii + 452 pp., 70 figs. N. York.

BENSLEY, B. A., (1903): On the evolution of the Australian Marsupialia: With remarks on the relationships of the marsupials in general. *Trans. Linn. Soc. Lond. (Zool.)* 9: 83-217.

BIGGERS, J. J. & E. D. DE LAMATER, (1965): Marsupial spermatozoa pairing in the epididymis of American forms. *Nature, Lond.* 208: 402-404, 3 figs.

BOOTH, A. H., (1954): The Dahomey Gap and the mammalian fauna of the West African forests. *Rev. Zool. Bot. afr.* 50(3-4): 305-314.

BOOTH, A. H., (1956): The distribution of primates in the Gold Coast. *Jl W. Afr. Sci. Ass.* 2: 122-133.

BUTLER, P. M. & Z. KIELAN-JAWOROWSKA, (1973): Is *Deltatheridium* a marsupial? *Nature, Lond.* 245: 105-106, 1 fig.

CLAY, T., (1970): The Amblycera (Phthiraptera: Insecta). *Bull. Br. Mus. nat. Hist. (Ent.)* 25(3): 75-98, 9 figs., 1 table, 5 pls.

CLEMENS, W. A., (1966): Fossil mammals of the type Lance Formation, Wyoming. Part II: Marsupialia. *Univ. Calif. Publs geol. Sci.* 62: 1-122.

CLEMENS, W. A., (1968): Origin and evolution of marsupials. *Evolution, Lancaster,* Pa. 22 (1): 1-18.

COOKE, H. B. S., (1968): Evolution of mammals on southern continents. II. The fossil mammal fauna of Africa. *Q. Rev. Biol.* 43: 234-264, 17 figs.

COOKE, H. B. S., (1972): III. The fossil mammal fauna of Africa, pp. 89-139, 17 figs., 2 tables. In: A. KEAST *et al.,* q. v.

COUTO, C. DE P., (1952a): Fossil mammals from the beginning of the Cenozoic in Brazil. Marsupialia: Polydolopidae and Borhyaenidae. *Am. Mus. Novit.* 1559: 1-27, 8 figs.

COUTO, C. DE P., (1952b): Idem. Marsupialia: Didelphidae. *Ibid.* 1567: 1-26.

COUTO, C. DE P., (1975): Marsupial dispersion and continental drift. *Anais Acad. bras. Cienc.* 46(1): 103-126, 9 figs. (1974).

COX, C. B., (1970): Migrating marsupials and drifting continents. *Nature, Lond.* 226: 767-770, 2 figs.

COX, C. B., (1973): Systematics and plate tectonics in the spread of marsupials, pp. 113-119. In: N. F. HUGHES, (ed.). Organisms and continents through time. Spec. Pap. Palaeont. 12 [= Syst. Ass. Publ. 9]: vi + 334 pp. London.

CRACRAFT, J., (1973): Continental drift, palaeoclimatology and the evolution and biogeography of birds. *J. Zool.* 169: 455-545, figs.

CRANWELL, L. M., (1959): Fossil pollen from Seymour Island, Antarctica. *Nature, Lond.* 184: 1782-1785, 2 figs.

CROIZAT, L., G. NELSON & D. E. ROSEN, (1974): Centers of origin and related concepts. *Syst. Zool.* 23(2): 265-287.

CRUSAFONT PAIRÓ, M., (1961): Super-famille: Giraffoidea Simpson, 1931, pp. 1022-1037, figs. 120-135. In: J. PIVETEAU, q. v.

DARLINGTON, P. J., Jr., (1963): Zoogeography: The geographical distribution of animals, xi + (2) + 675 pp., 80 figs., 21 tables. J. Wiley & Sons, N. York.

DUSÉN, P., (1908): Wiss. Ergebn. schwed. Südpolarexp. 3(3).

FOODEN, J., (1972): Breakup of Pangaea and isolation of relict mammals in Australia, South America and Madagascar. *Science, N.Y.* 175: 894-898, 1 fig., 2 tables.

GILL, E. D., (1957): The stratigraphical occurrence and palaeoecology of some Australian Tertiary marsupials. *Mem. nat. Hist. Mus. Victoria* 21: 135-203, 18 figs., 4 pls.

HAARMANN, K. & H. OBOUSSIER, (1972): Morphologische und quantitative Neocortexuntersuchungen bei Boviden, ein Beitrag zur Phylogenie dieser Familie. II. Formen geringen Körpergewichts (13 Kg–25 Kg) aus der Subfamilien Cephalopinae und Antilopinae. *Mitt. hamb. zool. Mus. Inst.* 68: 231-269.

HAYMAN, D. L., J. A. W. KIRSCH, P. G. MARTIN & P. F. WALLER, (1971): Chromosomal and serological studies of the Caenolestidae and their implication for marsupial evolution. *Nature, Lond.* 231: 194-195.

HARRISON, L., (1916): Bird parasites and bird phylogeny. *Ibis* (10) 4: 254-263.

HARRISON, L., (1922): On the mallophagan family Trimenoponidae, with the description of a new genus from an American marsupial. *Aust. Zool.* 2: 154-158.

HOFFSTETTER, R., (1970a): L'histoire biogéographique des marsupiaux et la dichotomie marsupiaux-placentaires. *C. r. hebd. Séanc. Acad. Sci., Paris* 271: 388-391.

HOFFSTETTER, R., (1970b): Radiation initiale des mammifères placentaires et biogéographie. *Ibid.* 270: 3027-3030.

HOPKINS, G. H. E., (1949): The host-associations of the lice of mammals. *Proc. zool. Soc. Lond.* 119(2): 387-604.

HUGHES, R. L., (1965): Comparative morphology of spermatozoa from five marsupial families. *Aust. J. Zool.* 13: 533-543, 1 fig., 1 pl.

IREDALE, T. & E. LE G. TROUGHTON, (1934): A check-list of the mammals recorded from Australia. *Mem. Aust. Mus.* 6: 1-122.

JARDINE, N. & D. MCKENZIE, (1972): Continental drift and the dispersal and evolution of organisms. *Nature, Lond.* 235: 20-24.

JONES, F. W., (1924): The mammals of South Australia. Part II (The bandicoots and the herbivorous marsupials). Government Printers, Adelaide.

KEAST, A., (1968a). Evolution of mammals on southern continents. I. Introduction: The southern continents as backgrounds for mammalian evolution. *Q. Rev. Biol.* 43(3): 225-233, 1 fig.

KEAST, A., (1968b): Idem. IV. Australian mammals: Zoogeography and evolution. *Ibid.* 43(4): 373-408, 13 figs.

KEAST, A., (1971): Continental drift and the evolution of the biota on southern continents. *Ibid.* 46(4): 335-378, 4 figs.

KEAST, A., F. C. ERK & B. GLASS, (1972): Evolution, mammals, and southern continents, 543 pp., illus. St. Univ. of N.Y. Press, Albany.

KIRSCH, J. A. W., (1968): Prodromus of the comparative serology of Marsupialia. *Nature, Lond.* 217: 418-420, 1 fig.

LILLEGRAVEN, J. A., (1969): Latest Cretaceous mammals of upper part of Edmonton Formation of Alberta, Canada, and review of marsupial-placental dichotomy in mammalian evolution. *Paleont. Contr. Univ. Kansas* 50 (= *Vertebrata* 12): 1-122.

MARTIN, P. G., (1970): The Darwin Rise hypothesis on the biogeographic dispersal of marsupials. *Nature, Lond.* 225: 197-198.

MATTHEW, W. D., (1915): Climate and evolution. *Ann. N.Y. Acad. Sci.* 24: 171-318.

MAXSON, L. R., V. M. SARICH & A. C. WILSON, (1975): Continental drift and the use of albumin as an evolutionary clock. *Nature, Lond.* 255: 397-399, 1 fig., 1 table.

McGOWRAN, B., (1973): Rifting and drift of Australia and the migration of mammals. *Science, N.Y.* 180: 759-761, 1 fig.

MOIR, R. J., M. SOMMERS & H. WARING, (1956): Studies in marsupial nutrition. I. Ruminant-like digestion in a herbivorous marsupial (*Setonix brachyurus* Quoy & Gaimard). *Aust. J. biol. Sci.* 9(2): 293-304, 3 figs.

MURRAY, M. D. & J. H. CALABY, (1971): Appendix II. The host relations of the Boopidae, pp. 81-82. In: S. VON KÉLER, A revision of the Australasian Boopidae (Insecta: Phthiraptera), with notes on the Trimenoponidae. *Aust. J. Zool.* (Suppl. Ser.) 6: 1-126, 135 figs.

MYKYTOWYCZ, R., (1964): Occurrence of bot-fly larvae *Tracheomyia macropi* Froggatt (Diptera: Oestridae) in wild red kangaroos, *Megaleia rufa* (Desmarest). *Proc. Linn. Soc. N.S.W.* 88[3(= n° 403)](1963): 307-312, 2 figs., 2 tables.

NELSON, G. J., (1969): The problem of historical biogeography. *Syst. Zool.* 18: 243-246.

OBOUSSIER, H., (1972): Morphologische und quantitative Neocortexuntersuchungen bei Boviden, ein Beitrag zur Phylogenie dieser Familie. III. Formen über 75 Kg Körpergewicht. *Mitt. hamb. zool. Mus. Inst.* 68: 271-292.

OBOUSSIER, H., (1974): Zur Kenntnis der Hippotraginae (Bovidae-Mammalia) unter Berücksichtigung von Körperbau, Hypophyse und Hirn. *Ibid.* 71: 203-233, 15 figs., 7 tables.

PILGRIM, G. E., (1941): The dispersal of the Artiodactyla. *Biol. Rev.* 16: 134-163.

PIVETEAU, J. (ed.), (1958): Traité de Paléontologie, 6(2) (Mammifères, évolution): 962 pp., 1040 figs., 1 pl. Masson, Paris.

PIVETEAU, J. (ed.), (1961): Idem, 6(1) (Mammifères, origine reptilienne, évolution): 1138 pp., 970 figs., 1 pl. Masson, Paris.

RAVEN, H. C. & W. K. GREGORY, (1946): Adaptive branching of the kangaroo family in relation to habitat. *Am. Mus. Novit.* 1309: 1-33.

RIDE, W. D. L., (1964): A review of the Australian fossil marsupials. *Proc. R. Soc. West. Aust.* 47: 97-131.

RICK, E. F., (1970): Lower Cretaceous fleas. *Nature, Lond.* 227: 746-747, 1 fig.

RODHAIN, J., (1926): Larves d'oestrides cavicoles chez *Okapia johnstoni* Scl. *Rev. Zool. Bot. afr.* 14: 137-139.

RODHAIN, J. & J. BEQUAERT, (1916): Matériaux pour une étude monographique des diptères parasites de l'Afrique. Deuxième partie. Révision des Oestrinae du Continent Africain. *Bull. scient. Fr. Belg.* 50(1-2): 53-165, 30 figs., pl. 2, figs. 1-4.

ROMER, A. S., (1967): Vertebrate palaeontology, ix + 468 pp., 443 figs., 4 tables. Univ. Chicago Press.

RUSSELL, L. S., (1952): Cretaceous mammals of Alberta. *Bull. natn. Mus. Can.* 126: 110-119, 2 pls.

SIMPSON, G. G., (1941): The affinities of the Borhyaenoidea. *Am. Mus. Novit.* 1118: 1-6.

SIMPSON, G. G., (1945): The principles of classification and a classification of mammals. *Bull. Am. Mus. nat. Hist.* 85: xvi + 350 pp.

SIMPSON, G. G., (1950): History of the fauna of Latin America. *Am. Scient.* 38: 361-389, 10 figs.

SIMPSON, G. G., (1967): The beginning of the age of mammals in South America. Part 2. *Bull. Am. Mus. nat. Hist.* 137: 1-259.

SLAUGHTER, B. H., (1968a): Early known marsupials. *Science, N.Y.* 162: 254-255.

SLAUGHTER, B. H., (1968b): *Holoclemensia* instead of *Clemensia*. *Ibid.* 162: 1306.

STIRTON, R. A., R. H. TELFORD & A. H. MILLER, (1961): Cenozoic stratigraphy and vertebrate paleontology of the Tirari Desert, South Australia. *Rec. S. Aust. Mus.* 14: 19-61, 4 figs., 1 table.

STIRTON, R. A., R. H. TELFORD & M. O. WOODBURNE, (1968): Australian Tertiary deposits containing terrestrial animals. *Univ. Calif. Publs geol. Sci.* 77: 1-30.

TATE, G. H. H., (1947): Mammals of Eastern Asia, xiv + 366 pp., 79 figs. McMillan, N.Y.

TATE, G. H. H., (1948): Results of the Archbold Expeditions n° 59. Studies on the anatomy and phylogeny of the Macropodidae (Marsupialia). *Bull. Am. Mus. nat. Hist.* 91(2): 233-352, 3 figs.

VIRET, J., (1961): Artiodactyla, pp. 887-1084, 166 figs. In: J. PIVETEAU, q. v.

WALKER, E. P., F. WARNICK, K. I. LANGE, H. E. UIBLE, S. E. HAMLET, M. A. DAVIS & P. F. WRIGHT, (1964): Mammals of the World, 1: xlviii + 644 pp., illus.; 2: viii + pp. 647-1500, illus. J. Hopkins Press, Baltimore.

WELLS, L. H., (1957): Speculations on the palaeogeographic distribution of antelopes. *S. Afr. J. Sci.* 53: 423-424.

WEMYSS, C. T., Jr., (1952): A preliminary study of marsupial relationships as indicated by precipitin tests. *Zoologica, N.Y.* 38: 173-181, 7 figs., 5 tables.

ZUMPT, F., (1965): Myiasis in man and animals in the Old World, xv + 267 pp., illus. London.

PART D. HYPOTHETICAL HISTORY OF OESTRIDAE EVOLUTION

Before producing a hypothetical history of Oestridae evolution, some consideration must be given to several problems of Oestridae biology, as well as some basic postulates.

1. From the available evidence, it seems justified to assume that oestrids were originally forest-dwellers and polivoltine (some recent African genera still are). Colonization of open (and drier and more changing) formations such as prairies, steppes, savannas and deserts requires special adaptations from the adult and pupa and is certainly an apoecous condition.

2. Forest-inhabiting oestrids must take a long time to adapt to open formations; once this adaptation is achieved, forests are impassable barriers for them (DOLLO's Rule). Species of open formations have, on the other hand, more chances to be widely spread by their hosts and by the normal mechanisms of dispersal.

3. An oestrid's host must be a terrestrial herbivorous mammal, with a certain size and weight, available for a long period of time. It is not known how oestrids recognize their hosts, or which chain of stimuli leads to larviposition. This would represent an extremely interesting field for ethologists. It is logical to presume that within the habitat of the fly a limited number of mammal species exist; from these only a small fraction presents the four basic requirements; again, these latter species must have populations with different sizes, and consequently different probabilities of becoming infested; then other stimuli, visual and olfactory, must intervene, until the female fly larviposits.

4. It may be assumed that female oestrids must make some 'mistakes', larvipositing in 'wrong' hosts, with the subsequent loss of many individuals. This is apparently true, since several aborted attempts at infestation are reported in the literature. Man, being a very common animal, with sufficient size and weight, is not infrequently infested; of course, larvae do not develop.

5. Within certain limits, these 'errors' by the female fly are extremely important for the preservation of the group – the only way of conquering new hosts and increasing the distribution area. A narrow specialization to a particular host would inevitably lead to extinction of the oestrid.

6. Adaptation to a new host must be a rare event, since (i) the adult

fly has to adjust to a new ecology; (ii) a great number of unsuccessful larvipositions must take place, until a pre-adapted larva succeeds in parasitizing a new host, adjusting to its biochemistry; (iii) the pupa also has to adjust to a probably different kind of soil and environment; (iv) the new host must be very common, available for a long period of time and must have an ecology similar to that of the former host.

7. These 'errors' in larviposition may lead to parasitism of secondary hosts entirely unrelated to the former one; therefore, one could not compare the phylogeny of the parasite with that of the hosts (FAHREN-HOLZ's Rule); in this case, similar ecologies would be more important than phylogeny.

8. A great number of extinctions must have occurred within the Oestridae; their history is now extremely fragmentary. Mammals have undoubtedly suffered great extinctions (their taxonomy suffers much from this and other factors). Consequently, much of the history discussed here is highly speculative.

I. ORIGIN AND DIVERSIFICATION OF THE FAMILY

1. Evidence obtained from the Tracheomyiinae

Extremely important for dating the appearance of oestrids is the presence of Tracheomyiinae in Australia. As already mentioned (Part C), two possibilities exist to explain their presence there: (i) oestrids colonized Australia from the Oriental Region by a long-range, island-hopping dispersal, a sweepstake route; (ii) oestrids in Australia are relics of an ancient, peri-Antarctic, Gondwanan distribution.

Which of these possibilities reflects the truth can best be decided by a consideration of the past history of Australia. RICH (1975: 77-78) has given a good summary of recent advances in our knowledge:

'During the Mesozoic and as late as mid-Eocene (ca. 50 m. y. B. P.), geological and paleobotanical data suggest that a southern dispersal route across East Antarctica and archipelagic West Antarctica was the most likely connection between Australia and other continents. Southeast Asia during this period was probably far to the north, per-haps as much as 30-50° of latitude. At least many of the islands making up West Ant-arctica, as well as parts of East Antarctica, supported substantial forests dominated by several kinds of gymnosperms, including *Araucaria* and *Podocarpus*, as well as a number of angiosperms including *Nothofagus*. Part of this route, possibly along the Victoria Land and Wilkes Land coast of East Antarctica and appropriate islands in West Ant-arctica, may have lain north of the Antarctic Circle, thus providing a route having some daylight hours throughout the entire year. By the end of the Miocene, Australia had severed its connections with East Antarctica and drifted north to within 10° of its present position, strengthening the dominance of the Indomalaysian dispersal route, which has been most influential during the latter part of the Cenozoic. Simultaneous with northward drift, climatic changes affecting the Australian continent brought about a shift from the early Cenozoic temperate flora dominated by proteads and *Nothofagus* to the mid and late Cenozoic (including the present) zoned vegetation com-posed of arid (in interior Australia), tropical (in New Guinea and Northern Australia), and cool temperate (in southeastern Australia and Tasmania) elements. Such marked changes in Australia's geographic position, its climate and floras, and its links with the rest of the world, undoubtedly had a tremendous effect on its Cenozoic vertebrate faunas.'

Therefore, if oestrids colonized Australia from the Oriental region, they could have entered this continent with Eutherians of large size and weight only after the upper Miocene, more probably during the Pliocene, and then must have undergone total extinction in the Oriental region, while several groups of potential hosts were flourish-ing there! Their hypothetical Eutherian hosts would also have become extinct in Australia, leaving no trace in the fossiliferous record (from the Miocene to the Recent). This possibility has to be entirely discarded, and only one explanation is left – the Tracheomyiinae represent a relict of a past south Gondwanan distribution.

Further indirect evidence for dating the appearance of oestrids can be obtained from the data afforded by India. It is justified to assume that this continent never had oestrids, and the explanation is that it became detached from the Supercontinent of Pangaea before these flies came on to the scene. HEIRTZLER et al. (1973) place the initial opening of the Indian Ocean at the Tithonian (or latest Jurassic, about 140 m. y. B. P.), India then moving in a generally westerly direction for about 80 million years after rifting.

Consequently, the date of origin of the Oestridae could be situated between the late Jurassic and early Cretaceous.

2. Diversification of the family

The phylogeny of this family (Part B) points to a very early dichotomy: (i) the Cephenemyiinae on one side, and (ii) the Pharyngobolinae-Oestrinae-Tracheomyiinae group on the other. This is probably the result of an early segregation of the family in the two main blocks of Pangaea – the Cephenemyiinae in Laurasia, the Pharyngobolinae-Oestrinae-Tracheomyiinae in Gondwana (after the separation of India).

If this is accepted, then there is no reason why oestrids could not have been in South America (where they are not represented at present), since fragmentation of West Gondwana began with the initial opening of the South Atlantic in the Valanginian (early Cretaceous, or 125 to 130 m. y. B. P.) (LARSON & LADD, 1973), and the final break between Brazil and Africa came, according to REYMENT (1969) only in the upper Lower Turonian (see also REYMENT & TAIT, 1972). As pointed out by CRACRAFT (1973: 468):

'This estimate accords reasonably well with seafloor spreading data. LE PICHON & HAYES (1971) estimate that spreading rates were three times faster in the far South Atlantic than at the equator. By extrapolating measured spreading rates they believe the South Atlantic began opening about 140 m. y. ago, and that after 80 m. y. (about early Santonian) the original mechanical constraints produced by the joined continental blocks were lost and that the previous differences in spreading rates were reduced. This would seem to correspond fairly well with the Turonian date of REYMENT which would represent a period of rift valley formation when the sea ingressed but there had not yet been a complete separation of the continental crusts. This separation would then have taken place along a plate boundary marked by a transform fault.'

It is possible, then, that the Pharyngobolinae and Oestrinae now known from Africa (and secondarily Palaearctic) were also present in South America from the Cretaceous onwards.

On the other hand, the Australian Tracheomyiinae were probably members of some peri-Antarctic group, also probably (but not necessarily) present in southern South America (an A-S group; see HENNIG, 1966), but most certainly present in Antarctica. Colonization of Australia from Africa is precluded by the early separation of Africa and East Antarctica. The relationship of Australia to the other parts of Gondwanaland were discussed in some detail by RICH (1975: 67-72), as follows:

'Fragmentation within Gondwanaland began as early as the Triassic, but actual severance in terrestrial continuity between these parts of the southern continents connected in the late Palaeozoic (approx. 225 m. y. B.P.) was a late Mesozoic and even early Tertiary phenomenon (approx. 70-130 m. y. B. P.). The first complete separation was apparently between Africa and East Antarctica in the late Jurassic or early Cretaceous (HEIRTZLER et al., 1968; FRANCHETEAU & SCLATER, 1969; DIETZ & SPROLL, 1970; SMITH & HALLAM, 1970; DIETZ & HOLDEN, 1970; DINGLE & KLINGER, 1971; HEIRTZLER & BURROUGHS, 1971; TARLING, 1971; SOWERBUTTS, 1972; JONES, 1972; DINGLE, 1973). Although not so well understood, the severance of any close alliance between India and Australia probably occurred in the early Cretaceous (HEIRTZLER et al., 1968; VEEVERS, 1971: TARLING, 1971; VEEVERS, JONES & TALENT, 1971: CRAWFORD, 1971), certainly not later than late Mesozoic (...). Thus the shortest dispersal route between Australia and the remaining continents during the late Mesozoic and much of the Paleogene was across Antarctica. Asia, at this time, was as much as 30-50° north of Australia (DIETZ & HOLDEN, 1970; TEDFORD, 1973; JARDINE & McKENZIE, 1972; RAVEN & AXELROD, 1972).

In the Mesozoic, East Antarctica and Australia seemingly formed a continuous land mass (JONES, 1971; LE PICHON & HEIRTZLER, 1968; SMITH & HALLAM, 1970; McELHINNY, 1970; HEIRTZLER et al., 1968; VEEVERS, 1971; McGOWRAN, 1971). Connection at this time to South America, as to New Zealand, appears to have been archipelagic. Similarly a West Antarctican archipelago may have lain between South America and New Zealand (FLEMING, 1962; TEDFORD, 1973; ELLIOT, 1972). Yet unresolved and critical to late Mesozoic and Cenozoic reconstructions, however, is the precise relationship of West Antarctica to South America and East Antarctica. Many workers suggest that during much of the Mesozoic West Antarctica and New Zealand were all part of one plate's compressive margin (ELLIOT, in press, 1972), and must have developed in close proximity to a continental sediment source, East Antarctica and Australia (LINDEN, 1969), with the southern Andean Cordillera and the Antarctic Peninsula, although not necessarily a linear chain (DALZIEL & ELLIOT, 1971, 1973), forming a link between these two continents. Severe disruption of this perhaps archipelagic cordillera may have begun prior to 82 m. y. B. P. (DALZIEL & ELLIOT, 1971, 1973), with transforms of transcurrent faulting rotating the Ellsworth Mountains (now situated between East Antarctica and the Antarctic Peninsula) (DALZIEL & ELLIOT, 1971, 1973; ELLIOT, 1972; SCHOPF, 1969). During the late Mesozoic and early Cenozoic, South America and West Antarctica may have approached one another more closely than at present (DALZIEL & ELLIOT, 1973). Some workers, however, suggest that East and West Antarctica were independent entities during the Mesozoic, becoming closely apposed only in the Cenozoic (BECK, 1972). Others further suggest that the post-Jurassic tectonic histories of southern South America and West Antarctica have been decidedly independent and that deformation of the Scotia Arc occurred prior to the late Mesozoic (KATZ, 1973). ELLIOT (pers. comm., 1974) has pointed out that this does not necessarily mean that insular linkage was lacking between the two areas during the Mesozoic and part of the Cenozoic, however. More general agreement attends the positioning of New Zealand decidedly closer to Marie Byrd Land in West Antarctica until about 80 million years ago, when it apparently began its northward drift with the foundering of the Campbell Plateau and opening of the Tasman Sea (GRIFFITHS, 1971; GRIFFITHS & VARNE, 1972). The opinion of most geologists, however, despite the differing details of their reconstructions, is that any dispersal route involving West Antarctica from mid-Mesozoic to the Recent would probably have been archipelagic (ELLIOT, 1972; JARDINE & MCKENZIE, 1972; DALZIEL, pers. comm., 1973).

Only in the Tertiary, by the early to mid-Eocene (50-55 m. y. B. P.) and certainly by the late Oligocene (KENNETH et al., 1972), did Australia sever its connection with East Antarctica (LE PICHON & HEIRTZLER, 1968; JONES, 1971; VEEVERS, 1971; SMITH & HALLAM, 1970; DAVIES & SMITH, 1971; MCGOWRAN, 1971; VOGT & CONNOLLY, 1971; GRIFFITHS & VARNE, 1972; WEISSEL & HAYES, 1972) and begin its northward journey, which is still in progress. From a position of 60-70° south latitude (measured from Canberra) in the Cretaceous, Australia moved to within 10° of its present position by the Miocene (WELLMAN, MCELHINNY & MCDOUGALL, 1969; VEEVERS, JONES & TALENT, 1971; RAVEN & AXELROD, 1972). New Zealand continued separating from both Australia and West Antarctica, increasing the isolation that presently characterizes it. Thus, during the late Mesozoic and early Tertiary, the separation of Australia and New Zealand from their present Asian and most Indonesian neighbours was great with close apposition occurring only after the Oligocene (see HAMILTON, 1972a-b; HAILE & MCELHINNY, 1972; STAUFFER & CORBETT, 1972; contra RIDD, 1971, and AUDLEY-CHARLES et al., 1972).'

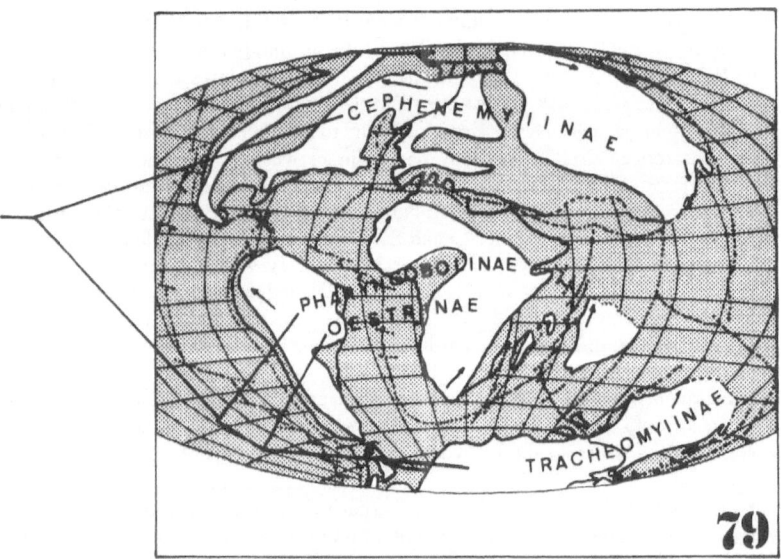

Fig. 79. Continental arrangement during Late Cretaceous (65 m. y. B.P.) (after RICH, 1975: 70, fig. 4), showing relationships and probable geographical history of the Oestridae subfamilies.

The relationships of the oestrid subfamilies and their probable geographical history in relation to the fragments of the Supercontinent of Pangaea, based on the hypothesis suggested above, are represented in Fig. 79.

3. The Mesozoic hosts of Oestridae

If it is accepted that oestrids can trace their origin back to the latest Jurassic or early Cretaceous, the next problem posed is: which mammals could serve as their hosts?

Unfortunately, this question cannot be answered, as Palaeontology has nothing to offer us in this regard. However, if we examine Fig. 80, we can see that during this period only Docodonta, Triconodonta, Multituberculata, Symmetrodonta and Pantotheria, and Insectivora, are reported, none of these groups serving as hosts.

The Docodonta and Triconodonta are very little known; they comprise in all 4 families with 13 genera, these represented mainly in the northern hemisphere (with the exception of one genus which is South African). These orders were probably carnivorous; thus, PIVETEAU (1961: 569) says about the Triconodonta: 'A un tel régime [carnivore] correspondaient alors de nombreuses proies. On trouve dans les gisements jurassiques, associés aux Triconodontes, des restes de petits Multituberculés herbivores, de Symmétrodontes ou de Pantothériens, omnivores ou insectivores, et des Reptiles de petite taille. On a pensé également qu'ils pouvaient se nourrir d'oeufs de Dinosauriens.'

The Symmetrodonta, known from very few genera in North America, Europe and Manchuria, had 'presumably predaceous habits (of a modest sort)' (ROMER, 1967: 21).

The Pantotheria, known from North America, Europe and East Africa, are also eliminated as potential hosts, as their feeding habits were 'comparable aux Opossums et nombreux Insectivores' (PIVETEAU, 1961: 582).

The Multituberculata are known from the northern hemisphere (especially from North America). They 'appear to have been the first herbivorous mammals, with skull and tooth specializations somewhat analogous to those seen among the later rodents, for the treatment of vegetable food; (...) they appear in the late Jurassic and, unlike contemporary orders, survived beyond the end of the Mesozoic to the true Eocene before being wiped out, presumably by the competition of advanced placentals – notably the rodents, which usurped the ecological niche long occupied by the Multituberculata' (ROMER, 1967: 200). If indeed the multituberculates were so similar to rodents, they would

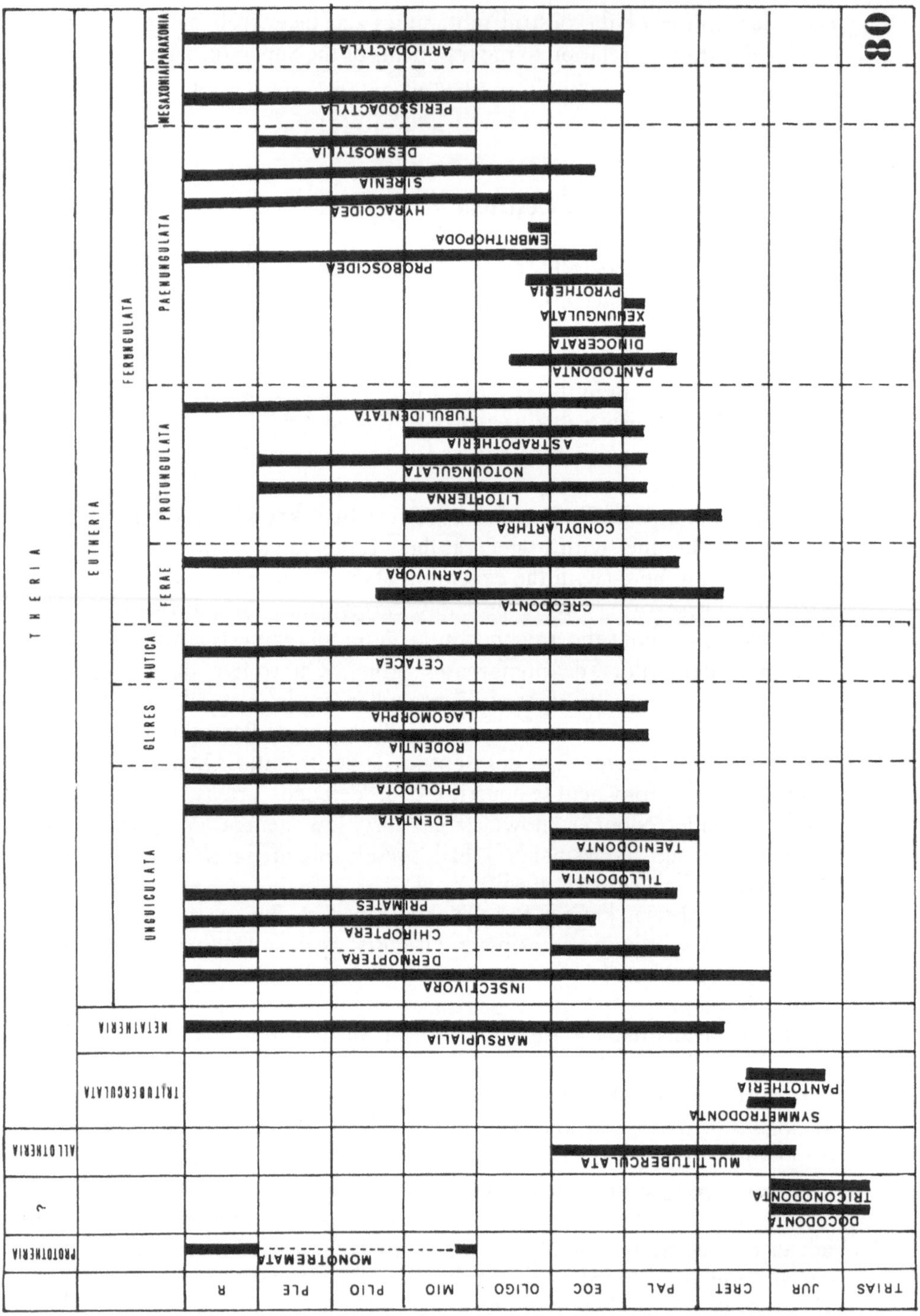

Fig. 80. Classification and time-range of the mammalian orders.

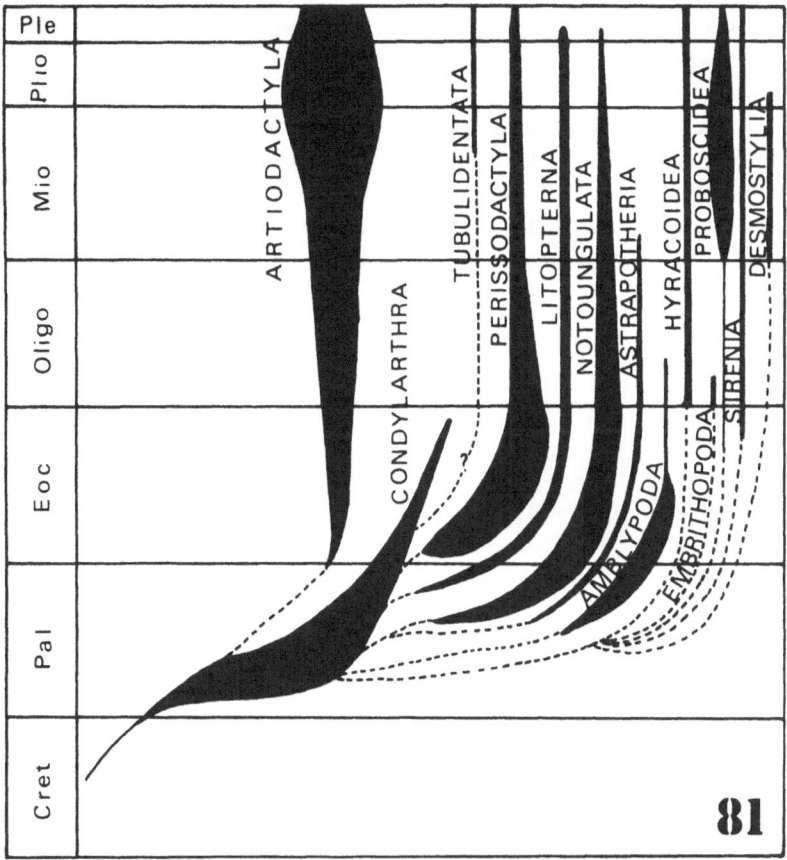

Fig. 81. Relationships of the 'ungulate' orders (after ROMER, 1967).

have to be ruled out as oestrids' hosts (see Part C).

Representatives of the Cohorts Unguiculata, Glires and Mutica are also ineligible as hosts. So we are only left with the 'ungulate' orders of the Cohort Ferungulata, which appear, however, much too late in the palaeontological record. Simpsonians consider that the 'ungulates' evolved from the Condylarthra. Thus, ROMER (1967: 241) declares: 'the Condylarthra appear to represent a truly basic stock from which many, at least, of the further ungulate orders may well have been derived. The boundaries of this group are none too well defined [some genera are included in Creodonta and Insectivora by ROMER, 1967, whereas PIVETEAU, in his 'Traité' places them in Condylarthra]. The Condylarthra appear to be an assemblage of forms transitional from ancestral insectivores to true ungulates.' The Simpsonian 'phylogeny' of the 'ungulates' is represented in Fig. 81. Of course this cannot be accepted as a true phylogeny, in the Hennigian sense.

By the Eocene we already have extremely diversified and specialized orders of 'ungulates', more or less segregated on the different continents.

It must be accepted, therefore, that oestrids parasitized some unknown ancestral group(s) of the several orders of ungulates. With the breakup of Pangaea and the subsequent isolation of the continents, these ancestral forms became restricted to those continents, giving rise to the ungulate orders characteristic of each of them.

II. THE PROBLEM OF THE PRESENT ABSENCE OF OESTRIDAE IN SOUTH AMERICA

If it is admitted that oestrids existed in South America since the Cretaceous, this leads to two problems: (i) which mammals could be parasitized, and (ii) what is the explanation of the complete disappearance of oestrids from that continent.

As already seen in Part C, marsupials could not be their hosts. However, among the Eutheria, at least seven orders of 'ungulates' were already present in South America by the Paleocene or Eocene; they were well diversified and six were apparently endemic: Condylarthra, Xenungulata, Notoungulata, Astrapotheria, Litopterna and Trigonostylopoidea (Figs. 82-84). Some of these groups, as will be seen below,

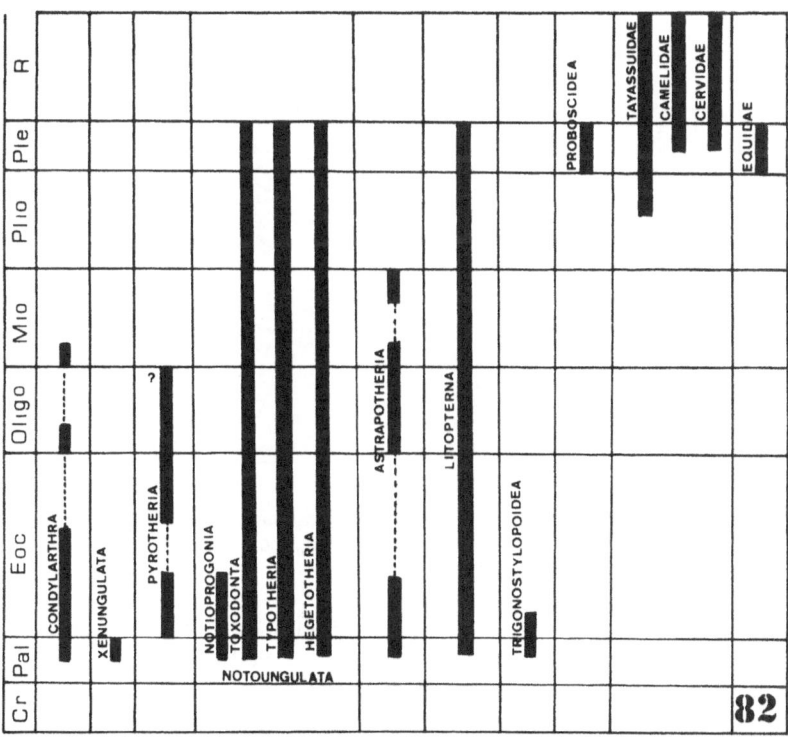

Fig. 82. Time-range of South American 'ungulates'.

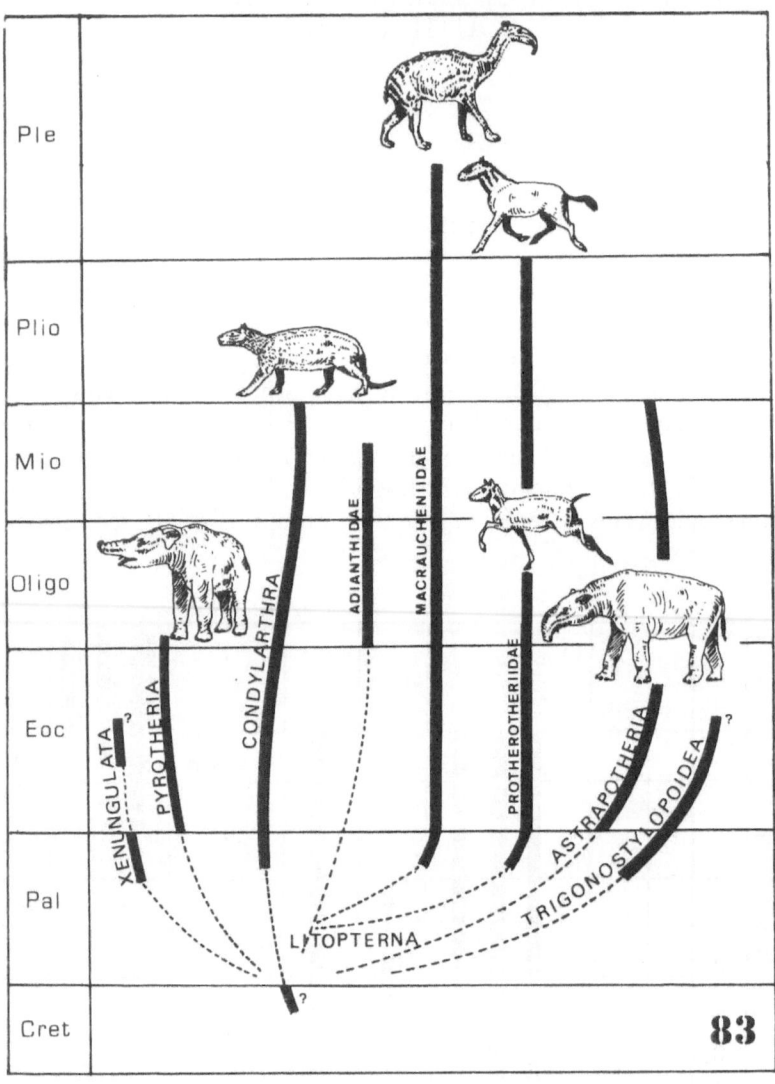

Fig. 83. South American endemic 'ungulates'; Time-ranges and 'phylogeny' (adapted from **Patterson & Pascual**, 1968).

could perfectly well harbour oestrids in their nasopharyngeal passages.

1. The Condylarthra are considered as a specifically North American group; a single family, Didolodontidae, which is not very rich in species and is considered as a descendant from a North American stock, is known from South America.

218

2. The Xenungulata are 'known essentially from one genus, the Riochican (upper Paleocene) *Carodnia*. This was a large mammal with partially bilophodont cheek teeth, strong and sharp canines, rather chisel-like incisors, relatively slender limbs and pentadactyl feet. Members of the order no doubt played a role in the early faunas that was at least roughly comparable to those of pantodonts and uintatheres in the northern hemisphere' (PATTERSON & PASCUAL, 1968: 430; 1972: 273).

3. '*Pyrotherium*, of the Oligocene, is a member of a small, short-lived South American group, the Pyrotheria. This beast not only had grown at that early time in the Tertiary to the size of an elephant but anatomically had paralleled the proboscideans to a remarkable degree. (...). The many similarities in body, skull, and dentition all suggest that *Pyrotherium* was really related to the proboscideans, then developing to the east in Africa. But these resemblances are surely a case of exceedingly close parallelism, for the group apparently had arisen in South America. Fragmentary remains of forms apparently related to *Pyrotherium* have been discovered in Eocene rocks of that continent. But beyond this fact we have no clue as to the origin of the pyrotheres, although obviously they come from some early ungulate stock' (ROMER, 1967: 246). However, PATTERSON & PASCUAL (1968: 430; 1972: 273) are not so enthusiastic: 'Given this combination of characters it is not surprising that some earlier students viewed the pyrotheres as related to the Proboscidea. Improved knowledge of pyrothere anatomy effectively disposed of the notion. The degree of convergence between the two orders is not very impressive. Nothing really convergent to a proboscidean has evolved anywhere; (...) reexamination (of the skull) reveals enough in the way of notoungulate characters to suggest the possibility that pyrotheres may constitute another suborder of that remarkable group.'

4. The Litopterna (Fig. 83) 'underwent a minor radiation into three families, each somewhat diversified within itself. The least known of these is the Adianthidae, a group of small, evidently delicately-built creatures, within which three lineages can dimly be discerned. The Prototheriidae, medium sized forms, were equid-like in build and foot structure, much less so in dentition; an enlarged sharp pair of incisors probably served as defensive weapons. The feet of the three-toed forms, such as the Santacruzian (mid-Miocene) *Diadiaphorus*, were strikingly (if only superficially) similar to those of anchitherine horses, while the contemporary *Thoatherium* was monodactyl, going beyond equine horses in degree of reduction of the lateral digits. The Macraucheniidae, alone in the order, achieved large size, terminal forms equalling the larger true camels. They were the only South American

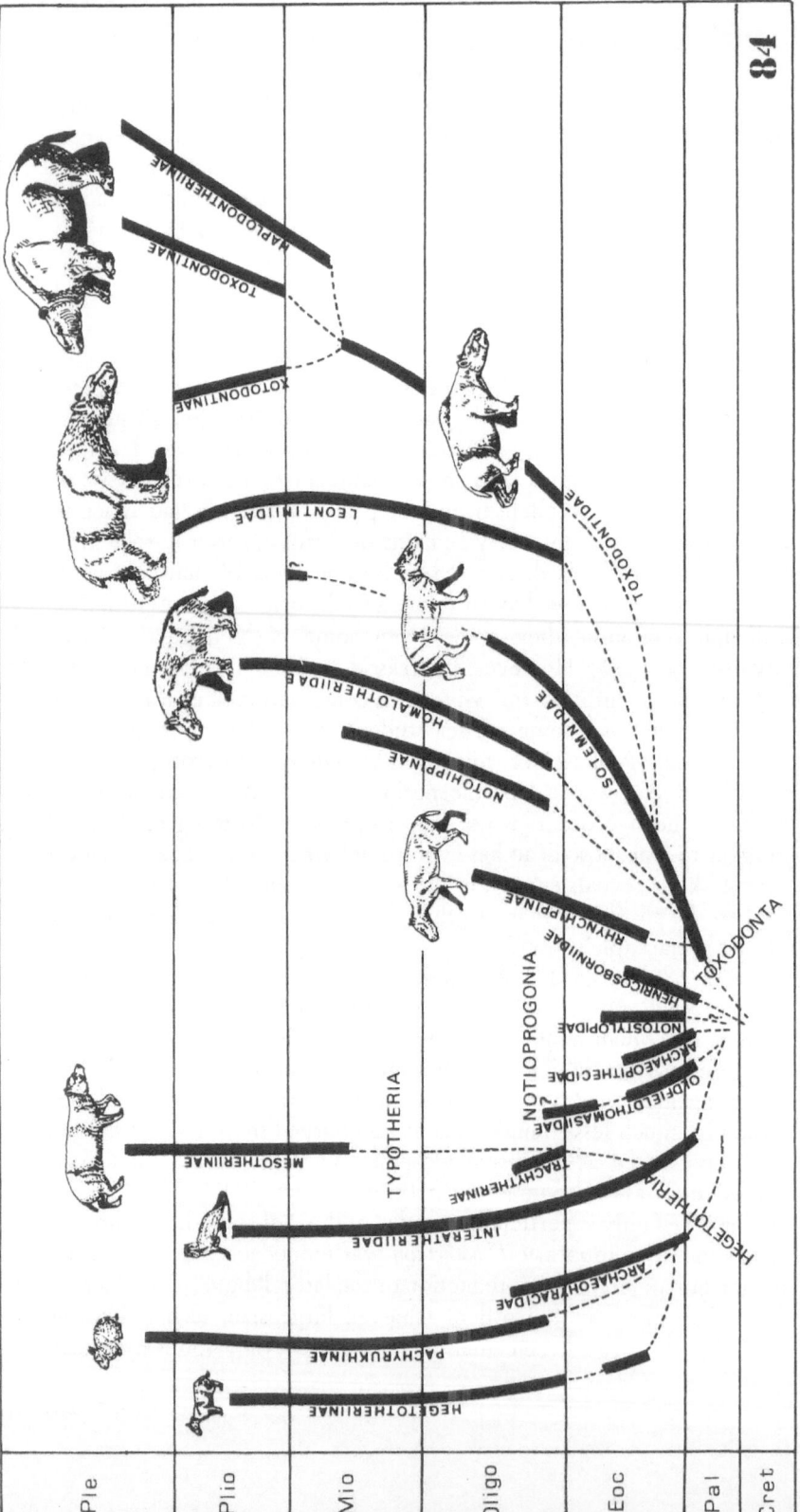

Fig. 84. Time-range and 'phylogeny' of South American Notoungulata (adapted from Patterson & Pascual, 1968).

84

ungulates known to have evolved a long, graceful neck, which accounts in large part for their rather camel-like appearance. The chief peculiarity displayed by them was a progressive reduction of the nasal bones accompanied by unmistakable evidence of the evolution of a short proboscis. Prototherids and macrauchenids seem to have thrived in the evolving pampas areas, whereas adianthids, if any survived beyond the Miocene, have not as yet been found there' (PATTERSON & PASCUAL, 1972: 270).

5. The Astrapotheria 'were extraordinary looking animals with undershot jaws, large and sharp tusks, no upper incisors, a domed head and an inflated nasal region. From the Deseadan (lower Oligocene) on they come in two sizes, large and very large, the two apparently representing different lineages. The lower tusks were probably used for rooting; along their front edges they frequently display grooves worn into them by dirt-covered plants that had been held against them and the lower incisors by the upper lip, and then pulled out by jerks of the head. No astrapothere is known to have survived beyond Friasian (upper Miocene) time. There is some degree of convergence between astrapotheres and amynodont rhinoceroses in the skull and dentition' (PATTERSON & PASCUAL, 1972: 271).

6. The Trygonostylopoidea, 'a poorly known group, formerly included in the Astrapotheria, was recently raised to ordinal rank by SIMPSON (1967). Resemblances to astrapotheres, such as the presence of moderately enlarged upper and lower tusks followed by diastemata, appear to be more than offset by differences in skull structure, especially in the basicranium. Nothing is known of the postcranial skeleton' (PATTERSON & PASCUAL, 1972: 271).

7. The Notoungulata (Fig. 84). 'The ancestral stock of the enormously successful order Notoungulata underwent an early adaptational dichotomy into an ungulate series, the suborder Toxodonta (and possibly Pyrotheria also), and a rodent- and rabbit-like series, comprising the suborders Typotheria and Hegetotheria. One family, Notostylopidae, of the ancestral suborder Notioprogonia, also evolved in a rodent-like direction. Evolutionary trends among the rodent- and rabbit-like forms included enlargement of the anterior incisors, development of diastemata between these and the cheek teeth, acquisition by some of a sciurid-like zygomasseteric structure, retention of a rabbit-like size generally (the approximately pig-sized mesotherids were the only fairly large forms), and attainment, in the pachyrukhine hegetotherids, of a hopping gait. The feet of the Santacruzian (mid-Miocene) interatherids and hegetotherids are of a rather generalized, four-toes type, but those of *Miocochilius*, an interatherid from the La Venta fauna of Colombia, are highly specialized, the pes resembling

that of tayassuid artiodactyls to a rather striking degree. Typotheres and hegetotheres were well adapted to their niches by Deseadan (lower Oligocene) time and the groups then in existence do not appear to have been affected by the arrival of the rodents.'

'On the ungulate, or toxodont, side a number of families descended from the generalized, highly variable, earlier Tertiary Isotemnidae. The Notohippidae show vague but far from precise resemblances to horses in the skull and the cropping incisor teeth. Homalodotheres converged chalicotheres, agriochoerids and the pantodont *Titanoides* in their possession of secondarily acquired claws in place of hoofs. These families and the lumbering leontiniids disappeared from the scene by or shortly after the close of the Miocene. It was otherwise with the toxodontids. Three subfamilies that arose from the ancestral Nesodontinae flourished during the Pliocene in the new environment afforded by the evolving pampas area, and one of them, Toxodontinae, survived until the end of the Pleistocene. Alone among South American ungulates, the Toxodontidae include forms with small dermal horns resembling those of certain rhinoceroses. Members of the family also have a sharp, moderately projecting pair of lower incisors, again as in rhinoceroses with small horns' (PATTERSON & PASCUAL, 1972: 274).

It can only be imagined to what extent oestrids could have differentiated in South America. The Pharyngobolinae could very well have been parasites of Pyrotheria; the Oestrinae must have had a very rich fauna of hosts, specializing in the Litopterna and toxodont Notoungulata, etc.

The present absence of oestrids in South America may be reasonably explained by two factors: (i) they were eliminated when their primitive hosts disappeared – 'these ungulate groups reached the climax of their development in the Oligocene and Miocene; with the Pliocene and Pleistocene came the advent of placental carnivores, to which they probably fell an easy prey, and of higher ungulate types, with which they could not successfully compete for pasturage. By the end of the Pleistocene this once extremely numerous assemblage had vanished entirely' (ROMER, 1967: 255); (ii) if they had survived until the Pliocene, South American oestrids were not able to infest the latecomers from North America (Camelidae, Tayassuidae, Cervidae, Proboscidea and Equidae; the latter two also became extinct during the Pleistocene), due to the relatively insignificant time which these mammals remained in contact with the autochtonous ungulates.

III. THE PROBLEM OF THE PRESENCE OF OESTRIDAE IN AUSTRALIA

It is very difficult to explain how oestrids could have survived in Australia, if it is admitted that they are relicts of a past, south Gondwanan distribution.

From the considerations made about marsupials in Part C, we can extract the following data: 1. Mammal-inhabiting fleas are reported from the Lower Cretaceous of Australia; 2. Serological data indicate that the separation of the Australian and New World faunas of marsupials took place at some 73 m. y. B. P. (Upper Cretaceous, Campanian-Maestrichtian transition); 3. Australia became detached from East Antarctica only at the mid-Eocene, some 43 m. y. B. P.; 4. Marsupials were already differentiated in Australia by the Miocene, some 26 m. y. B. P.; 5. The Dasyuroidea and Perameloidea were and are ruled out as potential hosts, and among the Phalangeroidea only the Diprotodontidae and Macropodidae have developed herbivorous forms, which must have appeared during the Miocene (their fossil records begins however in the Pliocene).

Therefore, oestrids had no herbivorous marsupial available in Australia until the Miocene (or Pliocene), so far as is known. If it is true that herbivorous forms did not appear until that late date, it must then be admitted that herbivorous Eutheria, probably some 'ungulate' related to South American and/or Antarctic forms, has entered Australia, surviving (as a group, naturally) until the Miocene, when it became extinct leaving no trace. From the beginning of the Tertiary to the Lower Eocene Australia could have received 'ungulates' from southern South America and/or Antarctica. The morphology of *Tracheomyia* indicates that this group is very closely related to the Oestrinae, a group which I accept has existed since the Lower Cretaceous in Africa and South America.

IV. EVOLUTION IN AFRICA

1. The early hosts of African Oestridae

Africa and South America became completely separated during the Turonian (Upper Cretaceous, 90 m. y. B. P.). The first fossils of mammals appear in Africa only during the Eocene (COOKE, 1972: 96). From what is known, it appears that an African group of orders had been developing in that continent for a longer time than is suggested by the fossil record. These orders, Sirenia, Proboscidea, Hyracoidea, Tubulidentata and Embrithopoda, had so many primitive characters in common that they were united by SIMPSON (1945) in the Superorder Paenungulata. All other groups of Artiodactyla and Perissodactyla (see Fig. 85) are accepted as immigrants from Eurasia, having entered Africa at different times when land connections were established with various parts of Eurasia. COOKE (1972: 130) has summarized the palaeogeographic and palaeontological history of Africa as follows:

'The African continent has long been regarded as a "refuge" for the survival of archaic forms of life, but in recent years evidence has been accumulating which serves to emphasize the essentially indigenous nature of the living and extinct mammalian faunas. (...) Through most of geological time Arabia has been basically part of the African continent, the Red Sea rift developed as a terrestrial trough in the Oligocene, was invaded from the Mediterranean in the Miocene, and connected through the Indian Ocean in the Pliocene. The pre-Mediterranean Tethys Sea cut Arabo-Africa off from Eurasia during much of the Mesozoic and Tertiary. Temporary land or island connections occurred during the Paleocene elevation of the Arabo-African block, during the late Oligocene orogenesis in the Atlas and the Alps, and again near the end of the Miocene. At those times there was opportunity for limited faunal interchange between Arabo-Africa and Eurasia. From a Paleocene ferungulate stock, already possessing early antracotheres and hyaenodonts, the proboscideans, hyracoids and sirenians developed as new elements in Africa. (...). In the late Oligocene there was an export of African stock to Eurasia in exchange for importation of some perissodactyls, fissiped carnivores, and perhaps basic suid, tragulid, and palaeomerycid-bovid elements. Lesser interchanges took place in the later Miocene, contributing African forms to Eurasia and admitting hipparionids to Africa. During the Pliocene great diversification took place among the African Proboscidea, Bovidae and Suidae in particular, but the Ethiopian region was effectively isolated from Eurasia until the end of the period. However, Arabia became firmly linked to Asia and probably furnished much African material to that continent.'

Oestrids, isolated in Africa after the complete separation of this portion of West Gondwana, became restricted to the ancestors of the Paenungulata. The Proboscidea certainly and probably the Embritho-

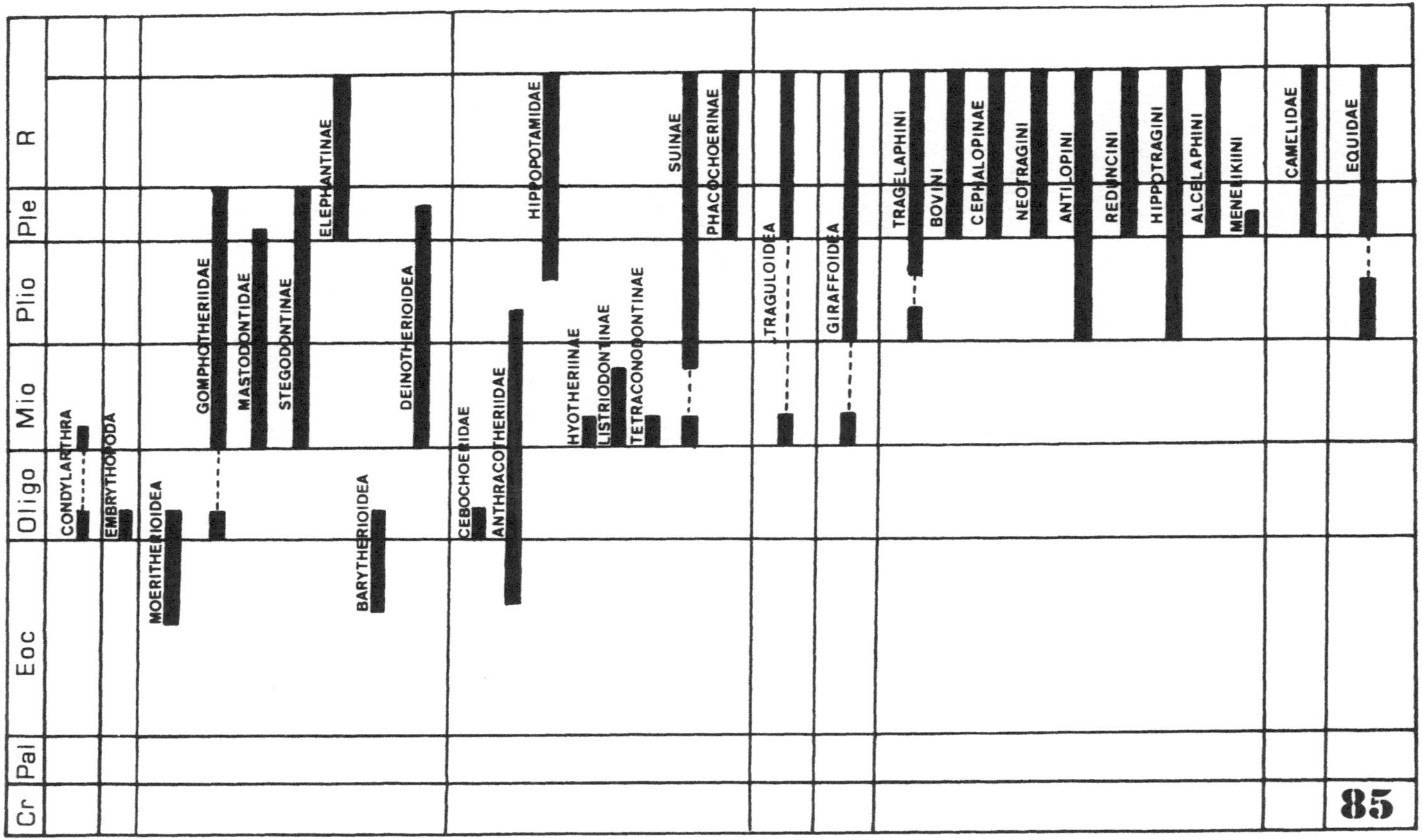

Fig. 85. Time-range of African 'ungulates'.

poda and some Hyracoidea were their hosts.

1. The Proboscidea, which have remained in Africa and the Oriental region, after a spectacular radiation, are parasitized by a specialized subfamily, the Pharyngobolinae. These, however, are restricted to the equatorial forest belt in Africa and nothing can be speculated about their history.

2. The Embrithopoda are known from a single genus, *Arsinoitherium*, 'an animal outwardly resembling an African rhinoceros but with a pair of large bony horns whose bases extended from behind the nostrils to the top of the braincase. Its affinities are somewhat uncertain but it is most probably related fairly distantly to the hyracoids. It is certainly not a perissodactyl but perhaps occupied an ecological place similar to that of the Asiatic rhinoceros' (COOKE, 1972: 100). But Y. SHAWI MOUSTAFA (1955), studying the skeleton of *Arsinoitherium*, reached an opposite conclusion; he says that this animal could move only very slowly on land and that its true habitat must be the marshes and swamps, where it led a life comparable to the hippopotamus or the sauropod dinosaurs. This genus, which is the only representative of the Order Embrithopoda, is only known from the Lower Oligocene of the Fayum (Egypt).

3. The Hyracoidea include two families, Geniohyidae and Hyracidae (Procaviidae). The Geniohyidae are represented by the genera *Bunohyrax, Geniohys, Megalohyrax (Mixohyrax)* and *Titanohyrax*, all from the Lower Oligocene of North Africa. The Hyracidae are known from two genera from the Lower Oligocene (*Pachyhyrax* and *Sagatherium*, both from North Africa), two genera from the Lower Miocene (*Pliohyrax* and *Meroehyrax*, from West and East Africa, respectively), and one genus, *Hyrax (Procavia, Prohyrax)*, which has existed in Africa from the Lower Miocene to the Recent, and in the Recent of southwestern Asia. According to COOKE (1972). 'The hyracoids are varied in size and character, the largest (*Titanohyrax*) being considerably bigger than a domestic pig and about as large as the contemporary *Moeritherium*, while the smallest (*Saghatherium*) is much the same size as the living *Procavia* or a large domestic cat. The cleavage of the hyracoids into two quite distinct families reflects a long period of evolution and adaptation to different ecological niches and in some forms the suine structures of the skull, jaw musculature, and teeth may indicate a habit somewhat like that of river-hog.' DECHASEAUX (1958: 331) also says: 'Les Hyracoïdes constituent parmi les Mammifères, un groupe tout à fait singulier. Leurs traits généraux sont donnés par la juxtaposition de caractères primitifs, de caractères spéciaux, absolumment inconnus chez les autres Mammifères, et de caractères évoquant des groupes variés: Rongeurs, Condylarthres, Notongulés, Embrithopodes, Pro-

boscidiens, Périssodactyles et Artiodactyles.'

I believe that the suid-like Geniohyid (or their ancestors) were the original hosts of the Oestrinae, and that these secondarily infested the true Suina, which came to Africa from Eurasia during the Oligocene. It is interesting to note that the most primitive Rhinoestrini are parasites of Suina (pigs and hippos; see Part B).

2. Diversification of the African Oestrinae

Five groups of genera exist within this subfamily, considered as tribes (see Part B): Kirkioestrini, Rhinoestrini, Oestrini, Gedoelstiini and Cephalopinini.

The Oestrinae, passing to Suina during the Oligocene, probably then underwent their first differentiation. A later adaptation to the African savannahs, probably during the Miocene, permitted infestation of several other groups of Artiodactyla and Perissodactyla, which had arrived from Eurasia (see Fig. 85). As the available potential hosts in the savannahs were relatively few, although common, much competition and exclusion must have taken place among the different oestrine stocks. On the other hand, as seen in Part C, among the Bovidae it is accepted that the Alcelaphini antelopes had been evolving in Africa for a longer period of time than is shown by the fossil record, and these should be heavily infested by Oestrinae. This is corroborated by the fact that Kirkioestrini, Oestrini and Gedoelstiini parasitize these antelopes. The Hippotragini are considered as invaders from Eurasia where they became extinct, being now represented only in Africa. Many groups of Artiodactyla and Perissodactyla were and are common to both Africa and Eurasia (notably equids). Therefore, oestrids adapted to open formations could pass from Africa to Eurasia, as hosts would be found all along their route.

Based on these considerations and on the hypothetical phylogeny of the Oestrinae presented in Part C, I consider that the Oestrinae include two main stocks: (i) a group which is predominantly African, represented by the Kirkioestrini, Rhinoestrini and Oestrini, with most of their history taking place in that continent; (ii) another group, which, although originally African, passed very early to Eurasia, radiating there and only much later recolonizing Africa – the Gedoelstiini and Cephalopinini

Let us now consider the hypothetical histories of these two groups:
1. The Kirkioestrini are now represented by a single genus, with only two species distributed over the northern and southern savannahs of Africa; their larvae parasitize Alcelaphini and Hippotragini. I believe

that this group is a relict; its past history can no longer be discovered. Once adapted to the open formations of Africa they became specialized at first in the endemic African Alcelaphini, passing secondarily to the invading Hippotragini. 2. The Rhinoestrini are the only tribe with a fairly complete and well documented history. They include 3 subtribes: (i) Suinoestrina with a single species, *Suinoestrus nivarleti*, limited to the equatorial forest belt of Africa; the larvae are parasitic on *Potamochoerus* (Suidae), an ancestral biological character preserved by this group; (ii) Rhinoestrina with the genus *Rhinoestrus*, and two species-groups:

a. the *hippopotami*-group, with the following species – *hippopotami*, also parasitic on Suina, and several species adapted to hosts of open formations: *antidorcitis* (in *Antidorcas marsupialis*, Antilopini and now restricted to South Africa and Namibia), *steyni* (in zebras, also restricted to South Africa and Namibia), and *latifrons* (in Equidae, occurring in European USSR, Kazakhstan, Uzbekistan, Mongolia and China); this group illustrates very well the process described above – from a Suina-based stock, the *hippopotami*-group became adapted to open formations, parasitizing Antilopini and Equidae; with the latter family they were able to colonize Eurasia; retractions of the hosts and ecologies led to their now more or less restricted distributions;

b. the *purpureus*-group; this group also probably has its beginnings in Suina (*Rh. phacochoeri* may belong here; see Part B), later colonizing open formations and also infesting Antilopini (*vanzyli*, restricted to South Africa and Namibia) and Equidae (*usbekistanicus* and *purpureus*); as with the first species-group, the *purpureus*-group was able to colonize Eurasia with Equidae hosts – *usbekistanicus* is still widely distributed over the Ethiopian and Palaearctic regions, while *purpureus* is now exclusively Palaearctic. It is interesting to note that two species in the genus *Rhinoestrus, vanzyli* and *antidorcitis*, parasitize the same host, the Antilopini *Antidorcas marsupialis*, in the same area (South Africa and Namibia); however, they must have different niches inside the host, as the entirely different larval morphology of the two species seems to indicate – *vanzyli* has the ventral armature of the segments with quadrangular scales forming 2-4 uninterrupted rows on the anterior part of each segment; the dorsal armature is absent, except on the pseudocephalon; on the other hand, in *antidorcitis*, the ventral armature consists of the usual spines, the dorsal armature is present, the lateroventral bulges of the segments present small groups of spines, etc. Therefore, it may be assumed that these two species do not compete for the same host and this is probably a case of 'character displacement' (BROWN & WILSON, 1956). (iii) Gruniniina, with a single species, *Gruninia tshernyshevi*, now isolated in Central Asia (Tadzhikistan and

Kazakhstan), the larvae parasitizing Caprinae; this group probably originated from an African ancestor which colonized Eurasia; the early steps of its history are now lost and nothing further can be said about this group.

3. The Oestrini (genera *Loewioestrus, Oestroides* and *Oestrus*) also appear to be an African group which first became established in the Alcelaphini, later passing to the invading Hippotragini; probably with these they gained access to Eurasia where they were able to remain (*Oestrus caucasicus* and *Oestrus ovis*), infesting several Caprinae.

The second group of Oestrinae probably arose from an African ancestor which very early was able to colonize Eurasia. There, one branch (Gedoelstiini) became specialized in the Hippotragini; the other (Cephalopinini) in the Camelidae, which arrived from North America during the Pliocene (or possibly earlier). Later, the Hippotragini (and their parasites) colonized Africa, becoming extinct in Eurasia; on their arrival in Africa, the Hippotragini were secondarily parasitized by Kirkioestrini and Oestrini, which were primarily parasites of Alcelaphini; the Gedoelstiini were almost eliminated probably by competition, now surviving with only two species. The Cephalopinini, being restricted to Camelidae, also receded with their hosts and are now distributed in the Palaearctic region only, in the area occupied by camels and dromedaries.

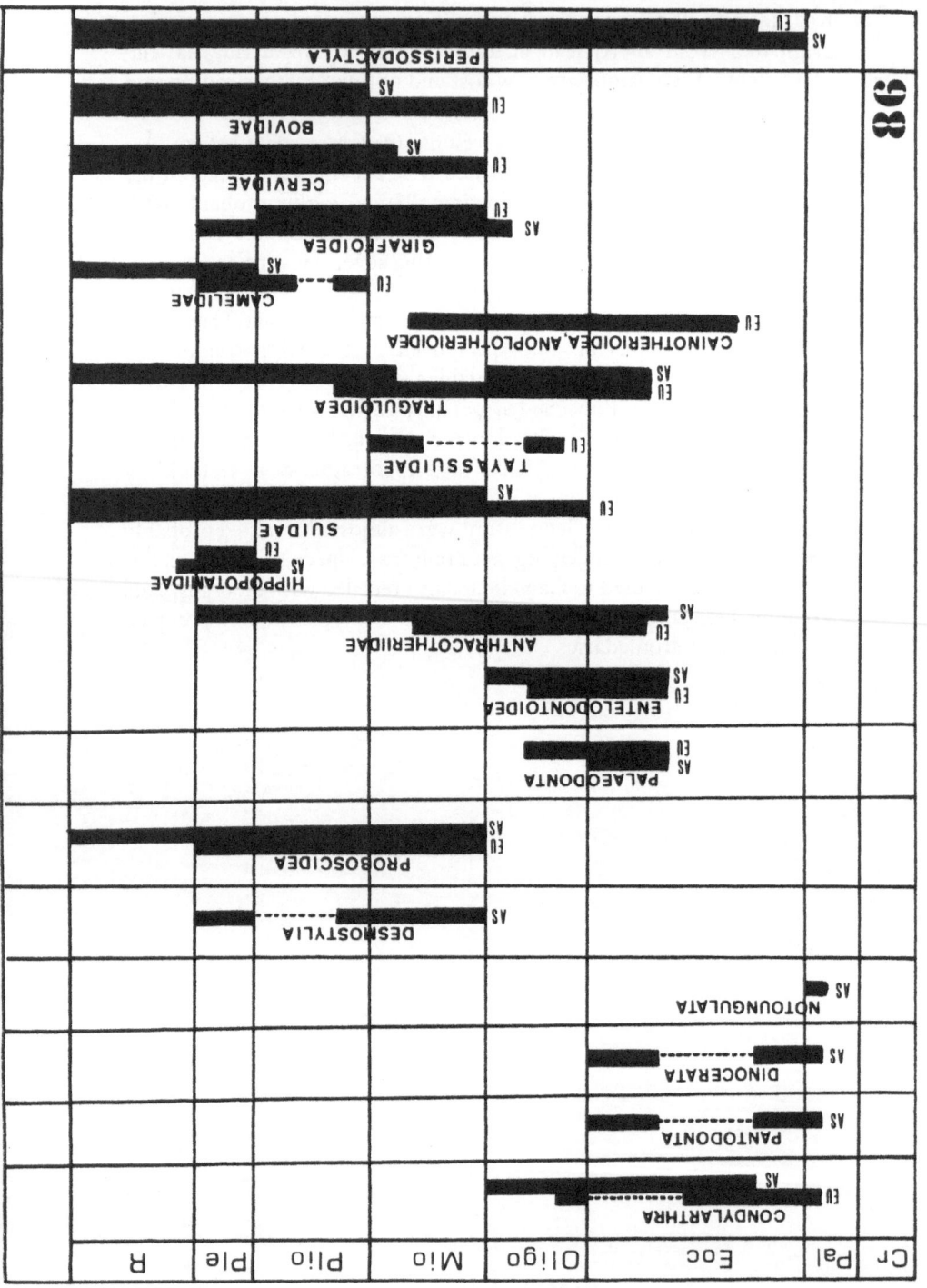

Fig. 86. Time-range of Eurasian 'ungulates'.

V. EVOLUTION IN EURASIA AND NORTH AMERICA

Several groups of mammals in the Northern Hemisphere could have served as hosts for the Cephenemyiinae (Figs. 86-87). This family has probably had a very complex history, with the production of many different branches. What we now have, however, is only the very tip of one of these branches, which is exclusively restricted to certain groups of Cervidae (and, secondarily, to one genus of Bovidae). The past history of the Cephenemyiinae remains a mystery and can only be traced down to the Pliocene.

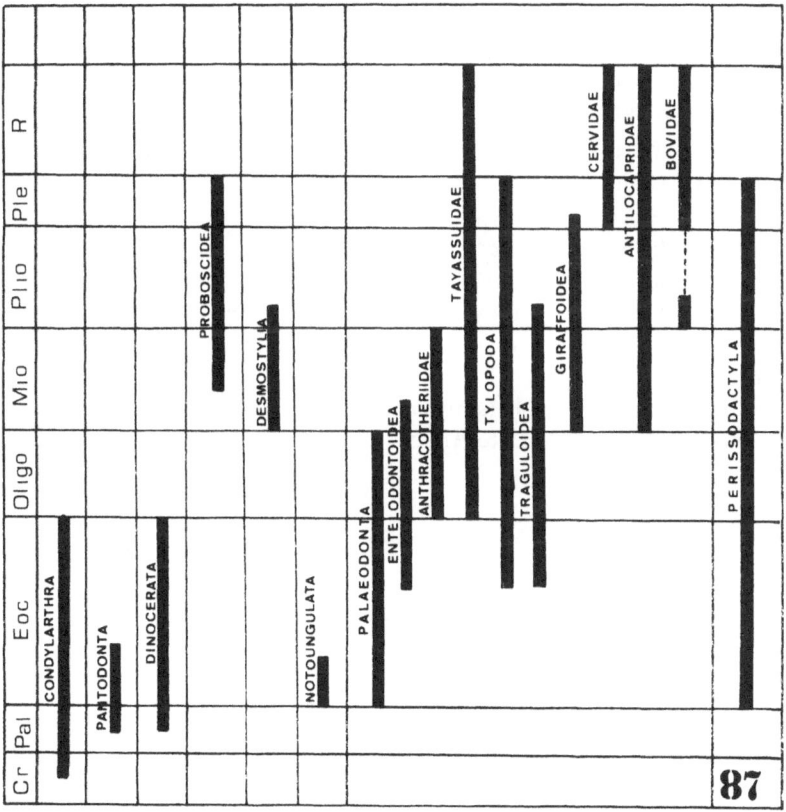

Fig. 87. Time-range of North American 'ungulates'.

The relationships between Cephenemyiinae and Cervidae are summarized in Fig. 54. *Procephenemyia* and *Cephenemyia* specialize in Odocoileinae (*Capreolus* and *Odocoileus* and *Rangifer*, respectively); *Acrocomyia* and *Pharyngomyia picta* in Cervinae (*Cervus*); the other branch of the Pharyngomyiini, '*Pharyngomyia*' *dzerenae*, became specialized in Antilopini (*Procapra gutturosa*).

The history of *Cephenemyia* is of particular interest. The earlier branch is represented by *phobifera* which occupies the eastern United States, parasitizing *Odocoileus* (*Odocoileus*) *virginianus* Zimmermann; the remaining species are divided into two groups – one occupying the western United States, with *pratii* (southwestern USA) and *apicata* (northwestern USA), and primarily parasitizing *Odocoileus* (*Eucervus*) *hemionus* (Rafinesque); the other group with two branches – one with *trompe* (circumpolar, parasitic on *Rangifer*) and *ulrichi* (Palaearctic, on *Alces*), the second with *jellisoni* (western USA, on *Odocoileus*).

What we have therefore is a very recent history, with many instances of complementary distributions between the Palaearctic and Nearctic regions. The 'division' of *Odocoileus* among the several species of *Cephenemyia* is significant – in the eastern USA only one species occurs, *virginianus*, and only one species also of *Cephenemyia* – *phobifera*; in the western USA there are two species of *Odocoileus* – *virginianus* and *hemionus*, and three species of *Cephenemyia*: *pratti* (southwest), *apicata* (northwest) and *jellisoni* (the entire western USA). Therefore, there is probably no competition among these species and their number is probably the maximum permissible by the hosts.

The American Cephenemyiinae have probably not adapted to the Sonoran Desert, or, if they became adapted, they could not pass the forests of Central America and consequently have not invaded South America.

The Palaearctic Cephenemyiinae became restricted to that region, as cervids did not penetrate Africa.

VI. REFERENCES

AUDLEY-CHARLES, M. G., D. J. CARTER & J. S. MILSOM, (1972): Tectonic development of eastern Indonesia in relation to Gondwanaland dispersal. *Nature, Lond.* 239(90): 35-39.

BECK, M. E., (1972): Palaeomagnetism and magnetic polarity zones in the Jurassic Dufek Intrusion, Pensacola Mountains, Antarctica. *Geophys. J. R. astr. Soc.* 28: 49-63.

BROWN, W. L. & E. O. WILSON, (1956): Character displacement. *Syst. Zool.* 5: 49-64.

COOKE, H. B. S., (1972): III. The fossil mammal fauna of Africa, pp. 89-139, figs. 1-17, 2 tables. In: KEAST *et al.*, q.v.

CRACRAFT, J., (1973): Continental drift, paleoclimatology, and the evolution and biogeography of birds. *J. Zool.* 169: 455-545, 20 figs., 4 tables.

CRAWFORD, A. R., (1971): Gondwanaland and the growth of India. *J. geol. Soc. India* 12(3): 205-221.

DALZIEL, I. W. D. & D. H. ELLIOT, (1971): Evolution of the Scotia Arc. *Nature. Lond.* 233: 246-252.

DALZIEL, I. W. D. & D. H. ELLIOT, (1973): The Scotia Arc and Antarctic margin. In: A. E. M. NAIRN & F. G. STEHLI, (eds.), The ocean basins and margins, 1: 171-246.

DAVIES, H. L. & I. E. SMITH, (1971): Geology of Eastern Papua. *Bull. geol. Soc. Am.* 82: 3299-3312.

DECHASEAU, C., (1958): Hyracoidea, pp. 319-332, 18 figs. In: J. PIVETEAU, q.v.

DIETZ, R. S. & J. C. HOLDEN, (1970): Reconstruction of Pangaea: Breakup and dispersion of continents, Permian to present. *J. geophys. Res.* 75(26): 4939-4956, figs.

DIETZ, R. S. & W. P. SPROLL, (1970): Fit between Africa and Antarctica: A continental drift reconstruction. *Science, N.Y.* 167(3925): 1612-1614.

DINGLE, R. V., (1973): Mesozoic palaeogeography of the southern Cape, South Africa. *Palaeogeogr., Palaeoclim., Palaeoecol.* 13(3): 203-214.

DINGLE, R. V. & H. C. KLINGER, (1971): Significance of upper Jurassic sediments in the Knysna Outlier (Cape Province) for timing the breakup of Gondwanaland. *Nature, Lond.* 232: 37-38.

ELLIOT, D. H., (1972): Aspects of Antarctic geology and drift reconstructions. *Contr. Inst. Polar Stud.*, Ohio State Univ., Columbus.

FLEMING, C. A., (1962): New Zealand biogeography, a palaeontologist's approach. *Tuatara* 10(2): 53-108.

FRANCHETEAU, J. & J. G. SCLATER, (1969): Paleomagnetism of the southern continents and plate tectonics. *Earth planet. Sci. Lett.* 6: 93-106.

GRIFFITHS, J. R., (1971): Reconstructions of the southwest Pacific margin of Gondwanaland. *Nature, Lond.* 234: 203-207.

GRIFFITHS, J. R. & R. VARNE, (1972): Evolution of the Tasman Sea, Macquarie Ridge and Alpine Fault. *Ibid.* 235: 83-86.

HAILE, N. S. & M. W. McELHINNY, (1972): The potential value of paleomagnetic studies in restraining romantic speculation about the geological history of Southeast Asia. [Geol. Soc. Malaysia] *Reg. Conf. Geol. SE Asia, Abstr., Annex to Newsletter* 34: 16.

HAMILTON, W., (1972a): Plate tectonics of Southeast Asia and Indonesia. *Ibid.*: 19.

HAMILTON, W., (1972b): Tectonics of the Indonesian region, 13 pp. (U.S. Dept. Int., Project Rep., Indonesian Invest. (IR), IND-20).

HEIRTZLER, J. R. & R. H. BURROUGHS, (1971): Madagascar's paleoposition: New data from the Mozambique Channel. *Science, N.Y.* 174: 488-490.

HEIRTZLER, J. R., J. V. VEEVERS, H. M. BOLLI, A. N. CARTER, P. J. COOK, V. A. KRASHE-NINNIKOV, B. K. McKNIGHT, F. PROTO-DECIMA, G. W. RENZ, P. T. ROBINSON, K. ROCKER, Jr. & P. A. THAYER, (1968): Marine magnetic anomalies, geomagnetic field reversals and motions of the ocean floor and continents. *J. geophys. Res.* 73(6): 2119-2136.

HEIRTZLER, J. R., et al., (1973): Age of the floor of the eastern Indian Ocean. *Science, N.Y.* 180: 952-954.

HENNIG, W., (1966): The Diptera fauna of New Zealand as a problem in systematics and zoogeography [Transl. from German by P. WYGODZINSKY]. *Pacif. Insects Mon.* 9: 1-81, 27 figs., 7 tables.

JARDINE, N. & D. McKENZIE, (1972): Continental drift and the dispersal and evolution of organisms. *Nature, Lond.* 235: 20-24.

JONES, J. G., (1971): Australia's Cenozoic drift. *Ibid.* 230: 237-239.

JONES, J. G., (1972): Significance of upper Jurassic sediments in the Knysna Outlier (Cape Province). *Ibid.* 235: 59-60.

KATZ, H. R., (1973): Contrasts in tectonic evolution of orogenic belts in the south-east Pacific. *Jl R. Soc. N.Z.* 3(3): 333-362.

KEAST, A., F. C. ERK & B. GLASS, eds., (1972): Evolution, mammals, and southern continents, 543 pp., illus. Univ. of N.Y. Press, Albany.

KENNETH, J. P., et al., (1972): Australian-Antarctic continental drift, paleocirculation changes and Oligocene deep-sea erosion. *Nature, Lond.* 239: 51-66.

LARSON, R. L. & J. W. LADD, (1973): Evidence for the opening of the South Atlantic in the Early Cretaceous. *Ibid.* 246: 209-212.

LE PICHON, X. & D. E. HAYES, (1971): Marginal offsets, fracture zones, and the early opening of the South Atlantic. *J. geophys. Res.* 76: 6283-6293.

LE PICHON, X. & J. R. HEIRTZLER, (1968): Magnetic anomalies in the Indian Ocean and sea-floor spreading. *Ibid.* 73: 2101-2117.

McELHINNY, M. W., (1970): Formation of the Indian Ocean. *Nature, Lond.* 228: 977-979.

McGOWRAN, B., (1971): [Australia-Antarctica separation and the Eocene transgression in southern Australia. Dept. Mines S. Aust., G.S. n° 4655. Unpubl.].

MOUSTAFA, Y. S., (1955): An interpretation of *Arsinoitherium. Bull. Inst. Égypte* 36(1).

PATTERSON, B. & R. PASCUAL, (1968): The fossil mammal fauna of South America. *Q. Rev. Biol.* 43: 409-451, figs.

PATTERSON, B. & R. PASCUAL, (1972): Idem, pp. 247-309, 13 figs., 12 tables. In: KEAST et al., q.v.

PIVETEAU, J., (ed.), (1958): Traité de Paléontologie, 6(2): 962 pp., 1040 figs., 1 pl. Masson, Paris.

PIVETEAU, J., (ed.), (1961): Idem, 6(1): 1138 pp., 970 figs., 1 pl. Masson, Paris.

RAVEN, P. H. & D. I. AXELROD, (1972): Plate tectonics and Australasian paleobiogeography. *Science, N.Y.* 176(4042): 1379-1386.

REYMENT, R. A., (1969): Ammonite biostratigraphy, continental drift and oscillatory transgressions. *Nature, Lond.* 224: 137-140.

REYMENT. R. A. & E. A. TAIT, (1972): Biostratigraphical dating of the early history of the South Atlantic Ocean. *Phil. Trans. R. Soc.* 264B: 55-95.

RICH, P. V., (1975): Antarctic dispersal routes, wandering continents, and the origin of Australia's non-passeriform avifauna. *Mem. natn. Mus. Vict.* 26: 63-125, 10 figs., 3 tables.

RIDD, M. F., (1971): South-east Asia as a part of Gondwanaland. *Nature, Lond.* 234: 531-533.

ROMER, A. S., (1967): Vertebrate Paleontology, ix + 468 pp., 443 figs., 4 tables. Univ. Chicago Press.

SCHOPF, J. M., (1969): Ellsworth Mountains: Position in West Antarctica due to sea floor spreading. *Science, N.Y.* 164: 63-66.

SMITH, A. G. & A. HALLAM, (1970): The fit of southern continents. *Nature, Lond.* 225: 139-144.

SOWERBUTTS, W. T. C., (1972): Rifting in eastern Africa and the fragmentation of Gondwanaland. *Ibid.* 235: 435-437.

STAUFFER, P. H. & D. J. GOBBETT, (1972): Southeast Asia a part of Gondwanaland? *Ibid.* 240: 139.

TARLING, D. H., (1971): Gondwanaland, palaeomagnetism and continental drift. *Ibid.* 229: 17-21, 71.

TEDFORD, R. H., (1973): Marsupials and the new biogeography. In: C. A. Ross, (ed.), Paleogeographic provinces and provinciality (S.E.P.M. Symposium, Denver, Colo., 1972).

VEEVERS, J. J., (1971): Phanerozoic history of Western Australia related to continental drift. *J. geol. Soc. Aust.* 18(2): 87-96.

VEEVERS, J. J., J. G. JONES & J. A. TALENT, (1971): Indo-Australian stratigraphy and the configurations and dispersal of Gondwanaland. *Nature, Lond.* 229: 383-388.

VOGT, P. R. & J. R. CONOLLY, (1971): Tasmanian guyots, the age of the Tasman Basin, and motion between the Australian plate and the mantle. *Bull. geol. Soc. Am.* 82(9): 2577-2584.

WEISSEL, J. K. & D. E. HAYES, (1972): Magnetic anomalies in the southeast Indian Ocean. In: D. E. HAYES, (ed.), Antarctic Oceanology. II. The Australian-New Zealand sector. [Amer. Geophys. Union, Antarctic Res. Ser. Mon. 19]: 165-196.

WELLMAN, P., M. W. McELHINNY & I. McDOUGALL, (1969): On the polar-wander path for Australia during the Cenozoic. *Geophys. J. R. astr. Soc.* 18: 371-395.

ACKNOWLEDGMENTS

During the elaboration of this work I was greatly helped by the colleagues of the Museu de Zoologia da Universidade de São Paulo, Drs. J. H. GUIMARÃES, L. R. GUIMARÃES, N. BERNARDI and R. L. ARAÚJO.

The late Dr. KAROL LENKO was extremely kind in providing much interesting information and bibliography for the section on the popular knowledge about Oestroidea.

Dr. LLOYD V. KNUTSON (U.S. National Museum, Washington, D.C.) was most helpful and patient in correcting some parts of the manuscript.

Mrs. JUVENTINA DOS SANTOS prepared the coloured drawings.

ABSTRACT

The family Oestridae (Oestroidea Cavicolae) is divided into 4 sub-families: (i) Tracheomyiinae, new (only genus, *Tracheomyia* Townsend; Australian; larvae parasitic on the kangaroo); (ii) Pharyngobolinae, new (only genus, *Pharyngobolus* Brauer; African; larvae parasitic on the African elephant); (iii) Cephenemyiinae Townsend (4 genera; Holarctic; larvae parasitic on Cervidae and Bovidae); and (iv) Oestrinae (several genera; African and Palaearctic; larvae parasitic on some groups of Artiodactyla and Perissodactyla).

The Cephenemyiinae are further subdivided into: (i) Cephenemyiini, with 2 subtribes – Acrocomyiina, new (with the only genus *Acrocomyia*, new, type-species *Oestrus auribarbis* Meigen) and Cephenemyiina (with two genera: *Procephenemyia*, new, type-species, *Oestrus stimulator* Clark, and *Cephenemyia* Latreille); (ii) Pharyngomyiini Townsend (only genus, *Pharyngomyia* Schiner).

The Oestrinae are subdivided into: (i) Kirkioestrini, new (only genus, *Kirkioestrus* Rodhain & Bequaert; African); (ii) Rhinoestrini, new (with three subtribes – Suinoestrina, with the only genus *Suinoestrus*, new, type-species, *Rhinoestrus nivarleti* Rodhain & Bequaert; Gruniniina, with the genus *Gruninia*, new, type-species, *Rhinoestrus tshernyshevi* Grunin; and Rhinoestrina, with *Rhinoestrus* Brauer); (iii) Oestrini Leach (with 2 subtribes – Oestrina, with the genera *Oestroides* Gedoelst and *Oestrus* Linnaeus; and Loewioestrina, new, with the only genus *Loewioestrus* Townsend); (iv) Gedoelstiini, new (only genus, *Gedoelstia* Rodhain & Bequaert, African); and (v) Cephalopinini, new (only genus, *Cephalopina* Strand; Palaearctic).

A tentative phylogeny is proposed. A bibliography with about 1,000 references is included.

Only 25 (out of 998) genera of recent mammals have been reported as being parasitized by Oestridae. These 25 genera belong to only 4 (out of 18) mammalian orders. These orders (Marsupialia, Proboscidea, Artiodactyla and Perissodactyla) make up a total of 170 genera; only 25 (14.7%) are parasitized. A study of the characteristics of these four orders showed that to act as a host of Oestridae a mammal must present the following conditions: 1. Be a terrestrial herbivore; 2. Have a certain length and weight (which are assumed to be, as a minimum, 1 meter in body length and around 20-25 kg in weight); 3. Be in the area (geographical and ecological) of the parasite for a certain lapse of time, so an adaptation of the parasite to the host can be obtained; this period of time is calculated to be, at least, from the Upper Miocene to the Upper Pliocene. 4. Gregariousness is also required for mammals inhabiting open formations; large herds available to the flies make infestation more liable to occur.

These four principles, applied to the world fauna of mammals, satisfactorily explain the scarce number of hosts available to oestrids.

It is further postulated that oestrids appeared in the Supercontinent of Pangaea during the Upper Jurassic or Early Cretaceous. Fragmentation of the Supercontinent into Laurasia and Gondwana brought about the first dichotomy of the family – the Cephenemyiinae in Laurasia, the Pharyngobolinae-Oestrinae-Tracheomyiinae group in Gondwana (after the separation of India). The Pharyngobolinae and Oestrinae are assumed to have colonized West Gondwana (South America and Africa) and the Tracheomyiinae Antarctica and Australia. South American oestrids were eliminated after the Pleistocene, when their hosts, the endemic 'ungulate' orders Litopterna, Pyrotheria, Noto-ungulata, etc., became extinct, thus explaining their total absence in that continent during the Recent. The Australian Tracheomyiinae probably reached that continent with an unknown placental, passing on to macropodids during the Miocene. The African Pharyngobolinae specialized in Proboscidea and are now restricted to the Congo forest. Oestrinae probably began infesting geniohyid Hyracoidea, passing to Suina during the Oligocene, and then, becoming adapted to open formations, probably during the Miocene, became specialized in various groups of Artiodactyla and in Equidae (Perissodactyla), also colonizing Eurasia. The Cephenemyiinae are discussed in relation to the history of Cervidae; their early history remains unknown. The Oriental region lacks Oestridae because (i) India became detached from Pangaea before the appearance of the family; (ii) the Oriental region is protected by ecological barriers against the invasion of Palaearctic Cephenemyiinae.

INDEX